Laser Surface Engineering for Tribology

Laser Surface Engineering for Tribology

Editors

Xiulin Ji
Yong Sun

MDPI

Basel • Beijing • Wuhan • Barcelona • Belgrade • Novi Sad • Cluj • Manchester

Editors
Xiulin Ji
Department of Mechanical Engineering, Shantou University
Shantou
China

Yong Sun
School of Engineering and Sustainable Development, De Montfort University
Leicester
UK

Editorial Office
MDPI
St. Alban-Anlage 66
4052 Basel, Switzerland

This is a reprint of articles from the Special Issue published online in the open access journal *Lubricants* (ISSN 2075-4442) (available at: https://www.mdpi.com/journal/lubricants/special_issues/M85KH3V382).

For citation purposes, cite each article independently as indicated on the article page online and as indicated below:

Lastname, A.A.; Lastname, B.B. Article Title. *Journal Name* **Year**, *Volume Number*, Page Range.

ISBN 978-3-7258-0953-0 (Hbk)
ISBN 978-3-7258-0954-7 (PDF)
doi.org/10.3390/books978-3-7258-0954-7

Contents

About the Editors

Xiulin Ji

Xiulin Ji is a Professor of Surface Engineering in the Department of Mechanical Engineering at the School of Engineering, Shantou University, China. With extensive experience in materials engineering education and research, he earned his PhD degree in Materials Science from Southeast University, China, in 2008, and subsequently joined the School of Mechanical & Electrical Engineering at Hohai University. Since 2021, he has been actively contributing to Shantou University. Dr. Ji has also served as a visiting scholar at prestigious institutions, including Southern Methodist University, Oklahoma State University, and the University of North Texas in the USA. His primary research areas encompass surface engineering, tribology, corrosion, and tribocorrosion. Dr. Ji holds editorial roles on the boards of journals such as *Nonferrous Metals Science and Engineering, Materials Protection, Surface Science and Technology*, among others. His scholarly contributions extend to the publications of over 80 journal and conference papers.

Yong Sun

Yong Sun is a Professor of Surface Engineering at the School of Engineering and Sustainable Development in the Faculty of Computing, Engineering and Media, De Montfort University, Leicester, U.K. Prof. Sun has many years of experience in materials engineering education and research. Prof. Sun has worked at several institutions, including the University of Birmingham in the U.K., the Nanyang Technological University in Singapore, and De Montfort University. His major research areas include surface engineering, tribology, corrosion, and tribocorrosion. Prof. Sun has published more than 200 journal and conference papers.

Preface

Laser surface engineering has been an evolving area of research, particularly in the context of tribology, which focuses on the study of the friction, wear, and lubrication of interacting surfaces in relative motion. Laser surface engineering techniques, such as laser surface melting, alloying, and cladding, have been employed to modify the surface properties of materials, thus enhancing their resistance to wear and friction. Furthermore, researchers have explored the use of lasers to induce nanostructuring on surfaces, aiming to improve hardness, reduce friction, and enhance wear resistance at the nanoscale level. On the other hand, laser surface engineering has been applied to create functionally graded materials, where the composition and properties vary gradually across the material's surface, optimizing the performance under specific tribological conditions. In addition, surface texturing through the use of lasers has gained attention for its potential to control friction and improve lubrication by creating micro-dimples or patterns on surfaces, and thus affecting the contact and sliding behavior.

Despite these advancements, there may still be gaps in the knowledge and challenges in the field of laser surface engineering for tribology. A deeper understanding of the underlying mechanisms governing the interactions between laser-modified surfaces and tribological conditions is necessary, including the effects of temperature, microstructure, and material composition. At the same time, the lack of standardized testing methods and characterization techniques for laser-modified surfaces poses challenges in comparing the results across different studies and establishing reliable performance metrics. Moreover, the long-term durability and stability of laser-engineered surfaces under varying operating conditions require further investigation to ensure practical applicability.

This reprint contains 12 articles and 1 review regarding a wide array of topics in the field of laser surface engineering for tribology, including preparing methods (additive manufacturing, preheating, and supersonic assistant laser deposition), post-machining, and surface texturing. Meanwhile, this reprint pays more attention to the laser surface-modified chemical composition and microstructure. The wear resistance of laser-clad coatings was significantly improved by alloying, ceramic addition, or phase transformation in high-entropy alloy, cast iron, steel, and Ti-based, Cu-based, Al-based, and Ni-based alloys. Finally, the review presented some effective surface modification strategies for the laser cladding of metal-based alloys, ceramic-reinforced composites, amorphous, and high-entropy alloys on aluminum alloys.

Laser surface engineering for tribology is a sustainable approach that minimizes resource consumption and waste generation. For future research, emphasis could be placed on application-specific optimization, tailoring laser surface engineering approaches for specific applications, and considering the diverse requirements of different industries, such as automotive, aerospace, and manufacturing. Meanwhile, it is important to develop comprehensive multiscale models that integrate the effects of laser processing at the micro and nano levels to predict tribological performance accurately. In addition, it is advantageous to utilize advanced characterization techniques, such as in situ microscopy and spectroscopy, in order to better understand the dynamic changes in laser-modified surfaces during tribological testing. These future research directions can contribute to the continued evolution of laser surface engineering for tribological applications, addressing existing challenges and expanding the practical implementation of these techniques.

We aim for this reprint to provide readers with a comprehensive understanding of the latest state-of-the-art developments in these rapidly evolving research areas, along with an introduction to some of the most recent techniques developed within this field. We seize this opportunity to express our gratitude to all the contributors who have actively participated in this reprint. Simultaneously, we extend our thanks to the reviewers for their dedicated time and efforts in enhancing the quality of the submitted papers.

Xiulin Ji and Yong Sun
Editors

lubricants

Editorial

Special Issue on Laser Surface Engineering for Tribology

Xiulin Ji [1,*] and Yong Sun [2]

1 Department of Mechanical Engineering, Shantou University, Shantou 515063, China
2 School of Engineering and Sustainable Development, Faculty of Computing, Engineering and Media, De Montfort University, Leicester LE1 9BH, UK; ysun01@dmu.ac.uk
* Correspondence: xiulinji@gmail.com

1. Introduction

The realm of laser surface engineering has seen continual development, particularly within the realm of tribology, a field dedicated to the examination of friction, wear, and lubrication between interacting surfaces in relative motion. Various laser surface engineering techniques, such as laser surface melting [1,2], alloying [3,4], and cladding [5,6], have been employed to alter material surface properties, enhancing their resistance to wear and friction [7]. Additionally, researchers have delved into the application of lasers to induce nanostructuring on surfaces, aiming to elevate hardness, diminish friction, and augment wear resistance at the nanoscale [8–10]. Concurrently, laser surface engineering has found utility in crafting functionally graded materials [11,12], wherein composition and properties gradually shift across the material's surface, optimizing performance in specific tribological conditions. Furthermore, laser-induced surface texturing [13,14] has garnered attention for its potential to regulate friction and enhance lubrication by creating microdimples or patterns on surfaces, thereby influencing contact and sliding behavior. Hence, laser-induced manufacturing holds vast potential applications in the realm of tribology. The combination of laser surface engineering and tribology promises to yield outstanding insights and contributions that illuminate the academic and engineering field.

2. An Overview of the Published Articles

The Special Issue entitled "Laser Surface Engineering for Tribology" includes thirteen papers, featuring a review and twelve research articles. These contributions delve into cutting-edge topics, including innovative preparation methods such as additive manufacturing, preheating, and supersonic-assisted laser deposition. The coverage extends to post-machining techniques and surface texturing, providing a comprehensive exploration of advancements in the field.

Firstly, Zhao et al. (Contribution 1) present a comprehensive review focusing on the laser cladding of aluminum alloys. They believe that laser cladding technology is a highly effective strategy for enhancing the surface hardness and wear resistance of aluminum alloys. This review delves into six distinct types of coatings, namely, Al-based, Ni-based, Fe-based, ceramic-based, amorphous glass, and high-entropy-alloy coatings. The focus is particularly on aspects such as microstructure, hardness, wear resistance, and corrosion resistance. The findings of this work contribute significantly to our understanding of coating design and manufacturing perspectives for aluminum alloys, offering valuable insights for future research and development in this domain.

Reinforcement coating is a common method used to improve tribological properties through laser surface engineering. Ji et al. (Contribution 2) explore laser-clad AlCr2FeCoNiNbx (x = 0, 0.5, 1.0, 1.5, 2.0) high-entropy-alloy (HEA) coatings on Q345 carbon steel, focusing on the impact of Nb incorporation on reciprocating sliding wear resistance. The microstructure evolves from a single Face-Centered Cubic (FCC) solid solution (x = 0) to hypoeutectic and hypereutectic states ($x \geq 1.0$), with increasing Laves

Citation: Ji, X.; Sun, Y. Special Issue on Laser Surface Engineering for Tribology. *Lubricants* **2024**, *12*, 98. https://doi.org/10.3390/lubricants12030098

Received: 20 February 2024
Accepted: 13 March 2024
Published: 16 March 2024

phase and decreasing FCC. Despite a decline in corrosion resistance, wear resistance improves, and friction decreases with Nb addition. Nb0.5 exhibits remarkably low wear rates, outperforming Nb0 by 3.6 to 7.2 times under different loads. This work demonstrates that introducing Laves-phase-forming elements for hardness modulation could be an effective strategy to enhance the wear resistance of HEA coatings. Rajaei et al. (Contribution 3) discuss the potential of Fe_3Al coatings as an eco-friendly alternative to brake disc coatings. A 500 µm Cr–Mo steel buffer layer is applied to the coating on a gray cast-iron disc to enhance coating quality and prevent hot crack formation. While pin-on-disc tests reveal a coefficient of friction comparable to that of an uncoated disc, the Fe_3Al-coated disc exhibits a significant reduction in particulate matter (PM) emissions. This reduction holds particular importance in light of impending international standards, underscoring the environmental advantages associated with Fe_3Al coatings. The experimental results would be more convincing if standard friction equipment for automobile brake discs was used. Huang et al. (Contribution 4) suggest a laser-clad Ti_5Si_3/Ti_3Al composite coating to enhance the wear resistance of titanium alloys. A double-layer presetting method is employed for the laser cladding using a Ti-63 wt.% Al mixed powder layer and a Si powder layer. The coating predominantly consists of a Ti_5Si_3 primary phase and a Ti_5Si_3/Ti_3Al eutectic structure. The coating's microhardness is approximately 668 HV, more than 3 times higher than the matrix's. Finally, the coating significantly improves the wear resistance of TA2, and the mass wear rate of the coating is less than 1/5 of that of TA2. Liu et al. (Contribution 5) discuss the benefits of in situ NbC-reinforced laser cladding Ni45 coatings, emphasizing their high bond strengths, low dilution rates, small heat-affected zones, and excellent wear resistance. Unlike previous methods using costly pure niobium powder, the authors successfully synthesize NbC in situ in Ni45 powder using affordable FeNb65 and Cr_3C_2. The laser-clad coating is mainly composed of NbC and $Cr_{23}C_6$ phases and exhibits a uniform microstructure and outstanding wear resistance. The coating with 25 wt.% FeNb65 and Cr_3C_2 demonstrates the highest microhardness at 776.3 HV0.2 and exceptional wear resistance, surpassing Cr12MoV steel and the Ni45 coating by about 60 and 24 times, respectively. It provides an attractive strengthening strategy for Ni-based alloys. Zhang et al. (Contribution 6) notice that the CoCrFeNi high-entropy alloy (HEA) is known for its impressive mechanical properties but lacks sufficient wear resistance, particularly under high loads. Their study addresses the limited research on adjusting the deformation mechanism to enhance wear resistance by employing boron doping. The introduction of boron successfully regulates the deformation mechanism, resulting in a remarkable 35-fold improvement in wear resistance for the CoCrFeNi HEA. The authors observe the subsurface microstructure and find a reduction in shear bands and the formation of nanostructured mixed layers, contributing significantly to the enhanced wear resistance. So, this boron doping strategy provides an effective way to improve wear resistance in high-entropy alloys (HEAs).

The post-treatment stage has a strong influence on the mechanical and tribological properties of laser-clad coatings. Zhang et al. (Contribution 7) investigate the post-heat treatment of a laser-clad T15 coating on 42CrMo steel. The microstructural uniformity is enhanced using the author-suggested post-heat treatment in the temperature range of 1100 to 1240 °C. Quenching at 1100 °C results in a lower wear rate, while tempering increases it. During high-temperature quenching, carbides at grain boundaries decompose and integrate into the matrix, while tempering precipitates carbides within the grain. The authors show that heat treatment significantly increases the content of martensite and alloy carbide. The cladding layer's microhardness reaches 910 HV after quenching and 750 HV after tempering. The wear resistance is enhanced by quenching, although no significant change in the friction coefficient is observed.

Machining is another kind of post-treatment for laser additively manufactured parts. Ozaner et al. (Contribution 8) describe the machinability of Inconel 718 (IN718), a widely used Ni-based superalloy in the aerospace, nuclear, and chemical industries, specifically focusing on parts produced through laser metal deposition (LMD). As the machinability of LMDed IN718 has received little attention, this work comprehensively evaluates its machinability in both dry and minimum-quantity lubrication (MQL) cutting environments, considering LMD process variables. The results reveal challenges in machining closer to the substrate due to hardness variations. MQL significantly improves machining, and laser power is identified as a crucial parameter. The study provides practical guidelines for optimizing the machining process and emphasizes the potential economic benefits of MQL in terms of tool longevity without substantial energy cost increase, supporting the successful adoption of LMD in the additive–subtractive hybrid manufacturing of metallic parts.

In recent years, combination with other technologies has been a hot spot in the development of laser processing technology. Utilizing supersonic laser deposition (SLD) technology, Zhang et al. (Contribution 9) explore the fabrication of promising diamond/copper composite coatings for wear-resistance applications. The wettability of Ti-coated diamond is enhanced through optimal parameters in a salt bath. The diamond's surface features nano-spherical titanium carbides, fostering a favorable interface bond with a copper matrix. A protective transition layer acts as a buffer, preventing diamond breakage within the coating. SLD prevents graphitization due to its low processing temperature. The collaborative action of laser and diamond metallization significantly enhances tribological properties, resulting in a microhardness of approximately 173 $HV_{0.1}$, surpassing that of cold-sprayed copper. At a laser power of 1000 W, the diamond/copper composite coating demonstrates low friction (0.44) and a minimal wear rate (11.85 $\mu m^3 \cdot N^{-1} \cdot mm^{-1}$), underscoring SLD technology's substantial potential for creating wear-resistant composite coatings.

Laser power, scanning speed, powder feeding rate, and preheating temperature are important laser processing parameters. Liu et al. (Contribution 10) notice that laser cladding faces challenges on copper due to its high thermal conductivity and reflectivity. The authors explore the impact of preheating temperatures (100, 200, 300, and 400 °C) on laser-clad Ni-based coatings on copper substrates. The microstructures and properties are analyzed, revealing a uniform distribution of Ni-based alloy powder elements, creating good metallurgical bonding. The microstructure consists of cellular, dendrite, and plane crystals, with the main reinforced phases being γ (Fe, Ni), $Cr_{0.09}Fe_{0.7}Ni_{0.21}$, WC, and Ni_3B. The coating preheated at 200 °C exhibits the highest hardness (~942 $HV_{0.5}$) and superior wear resistance and corrosion resistance due to reinforced phases and fine crystals.

Recently, laser-assisted surface texturing has attracted the attention of tribology researchers. Phun et al. (Contribution 11) address the challenge of tool wear due to friction in metal-forming processes and propose a solution through the surface modification of SKH51 tool steel. Using a nanosecond pulse laser with an average power of 25 W, a hexagonal array of micro-dimples is created on the tool surface, facilitating lubricant retention and hydraulic pressure during contact. This work investigates the impact of dimple density and sliding speed on the coefficient of friction through pin-on-disc tests. At a 35% dimple density, the laser-textured surface exhibits a 12.6% lower friction coefficient (0.087) than the untextured surface at a sliding speed of 15 cm s^{-1}. Furthermore, the laser-textured surface shows minimal wear compared to the untextured sample. These findings provide valuable guidelines for tool and die surface treatment, enhancing friction and wear reduction in metal-forming processes. Zhang et al. (Contribution 12) employ laser processing technology to create micro-textures on 42CrMo steel, enhancing the wear resistance in high-load conditions, and particularly addressing tooth plate wear in oil drilling wellhead machinery. Finite element analysis guides the selection of optimal texture shapes and parameters. Three types of textures (micro-dimples, micro-grooves, and reticular grooves) are applied, and dry friction experiments reveal improved wear resistance, with micro-dimples outperforming the other shapes. Specifically, micro-dimples with certain dimensions reduce wear by over 80% in ring-block dry friction. Laser hardening enhances surface hardness, and micro-dimples store abrasive particles, mitigating furrow formation and reducing abrasive wear on tooth plates.

Finally, to resist the tribocorrosion of mild steel in a NaCl-containing solution, Sun et al. (Contribution 13) propose a γ'-Fe_4N nitride layer, which is prepared through plasma nitriding with about 5 μm thickness. At a cathodic potential of −700 mV, the γ'-Fe_4N layer experiences 37% less total material removal (TMR) compared to untreated mild steel, demonstrating resistance to mechanical wear. At an open circuit potential, the TMR is 34% lower, while at an anodic potential of −200 mV, the γ'-Fe_4N layer remarkably reduces the TMR by 87%. This improvement is attributed to the layer's high hardness, corrosion resistance, and ability to resist both mechanical wear and corrosion, reducing wear–corrosion synergism.

3. Conclusions

This Special Issue compiles significant contributions addressing challenges in laser surface engineering for tribology, suggesting a future research focus on application-specific optimization. This involves tailoring laser surface engineering approaches to meet the diverse requirements of various industries, including automotive, aerospace, and manufacturing. Additionally, the utilization of advanced characterization techniques like in situ microscopy and spectroscopy is advantageous for a deeper understanding of dynamic changes in laser-modified surfaces during tribological testing. These proposed research directions aim to further advance laser surface engineering for tribological applications, tackling current challenges and enhancing the practical implementation of these techniques.

Author Contributions: X.J. and Y.S. jointly developed the concept and co-wrote this editorial article. All authors have read and agreed to the published version of the manuscript.

Funding: This work is supported by the National Natural Science Foundation of China (51875169) and the Natural Science Foundation of Guangdong Province, China (2024A1515010125).

Acknowledgments: We express our heartfelt gratitude to the authors and peer reviewers for their invaluable contributions to the Special Issue entitled "Laser Surface Engineering for Tribology". Additionally, we extend our thanks to the entire staff and all individuals who played a role in bringing this Special Issue to fruition.

Conflicts of Interest: The authors declare no conflict of interest.

List of Contributions:

1. Zhao, P.; Shi, Z.; Wang, X.; Li, Y.; Cao, Z.; Zhao, M.; Liang, J. A Review of the Laser Cladding of Metal-Based Alloys, Ceramic-Reinforced Composites, Amorphous Alloys, and High-Entropy Alloys on Aluminum Alloys. *Lubricants* **2023**, *11*, 482. https://doi.org/10.3390/lubricants11110482.
2. Ji, X.; Guan, K.; Bao, Y.; Mao, Z.; Wang, F.; Dai, H. Effect of Nb Addition on the Corrosion and Wear Resistance of Laser Clad AlCr2FeCoNi High-Entropy Alloy Coatings. *Lubricants* **2024**, *12*, 5. https://doi.org/10.3390/lubricants12010005.
3. Rajaei, H.; Amirabdollahian, S.; Menapace, C.; Straffelini, G.; Gialanella, S. Microstructure and Wear Resistance of Fe3Al Coating on Grey Cast Iron Prepared via Direct Energy Deposition. *Lubricants* **2023**, *11*, 477. https://doi.org/10.3390/lubricants11110477.
4. Huang, K.; Huang, W. Microstructure and Wear Resistance of Ti5Si3/Ti3Al Composite Coatings Prepared by Laser Cladding on TA2 Titanium Alloy. *Lubricants* **2023**, *11*, 213. https://doi.org/10.3390/lubricants11050213.
5. Liu, Y.; Wang, K.; Fu, H.; Zong, B.; Zhang, J. Wear Resistance of In Situ NbC-Reinforced Laser Cladding Ni45 Coatings. *Lubricants* **2023**, *11*, 316. https://doi.org/10.3390/lubricants11080316.
6. Zhang, H.; Miao, J.; Wang, C.; Li, T.; Zou, L.; Lu, Y. Significant Improvement in Wear Resistance of CoCrFeNi High-Entropy Alloy via Boron Doping. *Lubricants* **2023**, *11*, 386. https://doi.org/10.3390/lubricants11090386.
7. Zhang, Y.; Ma, Y.; Duan, M.; Wang, G.; Li, Z. The Improvement of the Wear Resistance of T15 Laser Clad Coating by the Uniformity of Microstructure. *Lubricants* **2022**, *10*, 271. https://doi.org/10.3390/lubricants10100271.
8. Ozaner, O.C.; Kapil, A.; Sato, Y.; Hayashi, Y.; Ikeda, K.; Suga, T.; Tsukamoto, M.; Karabulut, S.; Bilgin, M.; Sharma, A. Dry and Minimum Quantity Lubrication Machining of Additively

Manufactured IN718 Produced via Laser Metal Deposition. *Lubricants* **2023**, *11*, 523. https://doi.org/10.3390/lubricants11120523.
9. Zhang, Q.; Chen, Y.; Li, B.; Wang, C.; Wu, L.; Yao, J. Tribological Behavior of Ti-Coated Diamond/Copper Composite Coating Fabricated via Supersonic Laser Deposition. *Lubricants* **2023**, *11*, 216. https://doi.org/10.3390/lubricants11050216.
10. Liu, Y.; Jin, H.; Xu, T.; Xu, Z.; Du, F.; Yu, M.; Gao, Y.; Zhang, D. Effect of a Substrate's Preheating Temperature on the Microstructure and Properties of Ni-Based Alloy Coatings. *Lubricants* **2024**, *12*, 21. https://doi.org/10.3390/lubricants12010021.
11. Phun, C.; Daodon, W.; Septham, K.; Kumkhuntod, P.; Zhu, H.; Saetang, V. Laser-Fabricated Micro-Dimples for Improving Frictional Property of SKH51 Tool Steel Surfaces. *Lubricants* **2023**, *11*, 456. https://doi.org/10.3390/lubricants11110456.
12. Zhang, H.; Pei, X.; Jiang, X. Anti-Wear Property of Laser Textured 42CrMo Steel Surface. *Lubricants* **2023**, *11*, 353. https://doi.org/10.3390/lubricants11080353.
13. Sun, Y.; Bailey, R. Tribocorrosion Behavior of γ'-Fe4N Nitride Layer Formed on Mild Steel by Plasma Nitriding in Chloride-Containing Solution. *Lubricants* **2023**, *11*, 281. https://doi.org/10.3390/lubricants11070281.

References

1. Masood Arif Bukhari, S.; Husnain, N.; Arsalan Siddiqui, F.; Tuoqeer Anwar, M.; Abbas Khosa, A.; Imran, M.; Hassan Qureshi, T.; Ahmad, R. Effect of laser surface remelting on Microstructure, mechanical properties and tribological properties of metals and alloys: A review. *Opt. Laser Technol.* **2023**, *165*, 109588. [CrossRef]
2. Karimi, J.; Antonov, M.; Kollo, L.; Prashanth, K.G. Role of laser remelting and heat treatment in mechanical and tribological properties of selective laser melted Ti6Al4V alloy. *J. Alloys Compd.* **2022**, *897*, 163207. [CrossRef]
3. Staia, M.H.; Cruz, M.; Dahotre, N.B. Microstructural and tribological characterization of an A-356 aluminum alloy superficially modified by laser alloying. *Thin Solid. Films* **2000**, *377–378*, 665–674. [CrossRef]
4. Peng, L. Preparation and Tribological Properties of NiCrBSiC Reinforced Laser Alloying Layer. *Tribol. Trans.* **2013**, *56*, 697–702. [CrossRef]
5. Yang, Y.; Ren, Y.; Tian, Y.; Li, K.; Bai, L.; Huang, Q.; Shan, Q.; Tian, Y.; Wu, H. Microstructure and tribological behaviors of FeCoCrNiMoSix high-entropy alloy coatings prepared by laser cladding. *Surf. Coat. Technol.* **2022**, *432*, 128009. [CrossRef]
6. Li, J.; Cui, X.; Guan, Y.; Jin, G.; Zheng, W.; Su, W.; Wan, S.; Shi, Z. Effects of Cr content on microstructure and tribological properties of laser cladding Ti-based coatings. *Tribol. Int.* **2023**, *187*, 108744. [CrossRef]
7. Lisiecki, A. Tribology and Surface Engineering. *Coatings* **2019**, *9*, 663. [CrossRef]
8. Bonse, J.; Kirner, S.V.; Koter, R.; Pentzien, S.; Spaltmann, D.; Krüger, J. Femtosecond laser-induced periodic surface structures on titanium nitride coatings for tribological applications. *Appl. Surf. Sci.* **2017**, *418*, 572–579. [CrossRef]
9. Wang, Z.; Zhao, Q.; Wang, C. Reduction of Friction of Metals Using Laser-Induced Periodic Surface Nanostructures. *Micromachines* **2015**, *6*, 1606–1616. [CrossRef]
10. Veiko, V.P.; Odintsova, G.V.; Gazizova, M.Y.; Karlagina, Y.Y.; Manokhin, S.S.; Yatsuk, R.M.; Vasilkov, S.D.; Kolobov, Y.R. The influence of laser micro- and nanostructuring on the wear resistance of Grade-2 titanium surface. *Laser Phys.* **2018**, *28*, 086002. [CrossRef]
11. Lu, G.; Shi, X.; Liu, X.; Zhou, H.; Chen, Y.; Yang, Z.; Huang, Y. Tribological performance of functionally gradient structure of graphene nanoplatelets reinforced Ni3Al metal matrix composites prepared by laser melting deposition. *Wear* **2019**, *428–429*, 417–429. [CrossRef]
12. Ostolaza, M.; Zabala, A.; Arrizubieta, J.I.; Llavori, I.; Otegi, N.; Lamikiz, A. High-temperature tribological performance of functionally graded Stellite 6/WC metal matrix composite coatings manufactured by laser-directed energy deposition. *Friction* **2024**, *12*, 522–538. [CrossRef]
13. Bonse, J.; Kirner, S.V.; Griepentrog, M.; Spaltmann, D.; Krüger, J. Femtosecond Laser Texturing of Surfaces for Tribological Applications. *Materials* **2018**, *11*, 801. [CrossRef] [PubMed]
14. Kümmel, D.; Hamann-Schroer, M.; Hetzner, H.; Schneider, J. Tribological behavior of nanosecond-laser surface textured Ti6Al4V. *Wear* **2019**, *422–423*, 261–268. [CrossRef]

lubricants

Review

A Review of the Laser Cladding of Metal-Based Alloys, Ceramic-Reinforced Composites, Amorphous Alloys, and High-Entropy Alloys on Aluminum Alloys

Pengfei Zhao [1,2], Zimu Shi [3], Xingfu Wang [3], Yanzhou Li [4], Zhanyi Cao [1], Modi Zhao [3,*] and Juhua Liang [3,*]

1 Key Laboratory of Automobile Materials, Ministry of Education, College of Materials Science and Engineering, Jilin University, Changchun 130022, China; zhaopf16@mails.jlu.edu.cn (P.Z.); caozy@jlu.edu.cn (Z.C.)
2 College of Information Engineering, Jilin Vocational College of Industry and Technology, Jilin 132012, China
3 Key Laboratory of Materials Physics, Institute of Solid State Physics, Chinese Academy of Sciences, Hefei 230031, China; shizimu@issp.ac.cn (Z.S.); wangxingfu@issp.ac.cn (X.W.)
4 School of Mechanical and Vehicle Engineering, West Anhui University, Luan 237010, China; liyanzhou9336@163.com
* Correspondence: mdzhao@issp.ac.cn (M.Z.); liangjuhua0721@126.com (J.L.); Tel.: +86-15044211299 (J.L.)

Abstract: As one of the lightest structural metals, the application breadth of aluminum alloys is, to some extent, constrained by their relatively low wear resistance and hardness. However, laser cladding technology, with its low dilution rate, compact structure, excellent coating-to-substrate bonding, and environmental advantages, can significantly enhance the surface hardness and wear resistance of aluminum alloys, thus proving to be an effective surface modification strategy. This review focuses on the topic of surface laser cladding materials for aluminum alloys, detailing the application background, process, microstructure, hardness, wear resistance, and corrosion resistance of six types of coatings, namely Al-based, Ni-based, Fe-based, ceramic-based, amorphous glass, and high-entropy alloys. Each coating type's characteristics are summarized, providing theoretical references for designing and selecting laser cladding coatings for aluminum alloy surfaces. Furthermore, a prediction and outlook for the future development of laser cladding on the surface of aluminum alloys is also presented.

Keywords: aluminum alloy; laser cladding; material system; research status

Citation: Zhao, P.; Shi, Z.; Wang, X.; Li, Y.; Cao, Z.; Zhao, M.; Liang, J. A Review of the Laser Cladding of Metal-Based Alloys, Ceramic-Reinforced Composites, Amorphous Alloys, and High-Entropy Alloys on Aluminum Alloys. *Lubricants* **2023**, *11*, 482. https://doi.org/10.3390/lubricants11110482

Received: 4 October 2023
Revised: 28 October 2023
Accepted: 2 November 2023
Published: 8 November 2023

1. Introduction

Aluminum and its alloys are regarded as one of the most pivotal metal materials of the 21st century [1,2]. Its content in the earth's crust is as high as 7.73%, ranking third, following only oxygen and silicon, which makes it the most abundant of the metallic elements. The tensile strength of pure aluminum (with an aluminum content not less than 99.60%) is 70–130 MPa, its elongation rate is 6–43%, and its hardness is 35 HB [2–9]. An aluminum alloy is made by adding some other elements to pure aluminum. Thus, its performance outperforms pure aluminum. Aluminum alloys possess high specific stiffness, high specific strength, excellent corrosion resistance, and easy processability [10–12]. It is not only widely used in aerospace, rail transit, shipbuilding, automotive, and military fields, but its development trend is towards larger sizes, integration, complex shapes, and high precision, making it the first choice for lightweight design [12–22].

However, the surface of aluminum alloys is soft, lacking high wear and heat resistance [23–25]. This restricts its application in mechanical parts requiring high surface hardness, wear resistance, and heat resistance [26–28]. Meanwhile, with the rapid development of the equipment manufacturing industry, many critical components of mechanical equipment have failed and been scrapped in harsh and complex environments, causing tremendous economic losses and a high equipment failure rate. There is an urgent need to explore efficient, green, emerging manufacturing technologies to enhance the lifespan of crucial new parts and remanufacture and repair outdated equipment.

Surface modification technology can be used to improve the surface performance of these parts. Currently, relatively mature surface modification technologies include thermal spraying, physical vapor deposition, electroplating, and laser cladding [29–32]. Among these, the most developed technologies are thermal spraying, physical vapor deposition, and electroplating. Even though they can effectively improve the surface performance of aluminum alloys, they all have specific issues. For instance, thermal spraying can form a thick coating, but inclusions and pores may appear, leading to poor bonding. The coating produced via physical vapor deposition is thin, and the high-temperature working environment may cause organizational changes at the bottom of the coating, resulting in the peeling of the coating [28]. The coating produced via electroplating is not dense enough, and the electroplating solution causes significant environmental pollution. Laser cladding technology has emerged with the development of laser and surface modification technologies [33–38]. By utilizing a laser beam to melt the cladding material and the surface of the base body and allowing it to solidify to form a cladding layer that metallurgically bonds with the base body, it can be used for the surface strengthening of new parts and the remanufacturing and repair of failed parts [39–42].

Laser surface modification technology occupies an advanced position in surface engineering. Its research began in the 1960s, but it was not until the early 1970s that it was practically applied [37,38,43]. The technique is widely acknowledged for its efficiency, pollution-free nature, and low material consumption. At its core, it uses lasers as a heat source to treat the surface of the metallic materials, optimizing their microstructure, phase structure, and chemical composition to enhance surface properties. Unlike traditional whole-body heat treatments, the laser only affects the surface, preserving the overall material properties. Its key features include a high heating speed, high power density, minimal heat-affected zones, adaptability to complex parts, energy reduction, cost reduction, and suitability for automation. The main processes include laser phase transformation hardening, melting hardening, laser shock hardening, laser alloying, and laser cladding.

Laser phase transformation hardening involves rapidly scanning the workpiece with a high-energy laser beam, bringing the surface to a phase change point below the melting point at, a heating rate 10^4~10^6 °C/s [44,45]. The cooler workpiece substrate then undergoes rapid self-quenching at a cooling rate of 10^4~10^8 °C/s, creating the surface phase transformation hardening. Laser melting hardening uses a laser beam of extremely high power density to instantly heat a localized metal surface region above its melting point [46–48]. Then, through the cooling effect of the cold metal substrate, the thin melted metal surface rapidly solidifies, resulting in a modified surface layer that may or may not have the same composition as the substrate but possesses different properties. This technique applies to both ferrous and non-ferrous metals and not only corrects surface defects like pores, cracks, and impurities but also enhances surface performance.

Furthermore, it can be employed for remelting plasma-sprayed coatings that are wear-resistant, heat-resistant, and corrosion-resistant, and for remelting laser-clad coatings, making it highly versatile. Laser shock hardening irradiates the metal workpiece surface with a short-pulse laser of ultra-high power density, causing the surface material to rapidly vaporize and generate shockwaves with pressures up to 10^4 Pa [49–51]. This leads to a significant toughening deformation on the metal surface, resulting in hardening. The microscopic structure in the laser shock zone showcases a complex network of tangled dislocations, akin to the substructures seen in materials post-explosion or rapid planar shock, significantly enhancing the material's surface strength, yield strength, and fatigue life.

Laser alloying melts the substrate metal surface using a high-energy laser while introducing alloying elements [52,53]. This forms a concentrated, uniformly mixed alloy layer on the substrate surface, giving it resistance to wear, corrosion, and high-temperature oxidation. Due to the swift nature of laser alloying, solute elements primarily achieve uniformity through liquid-state convection. This means that the composition of the laser alloy can greatly differ from conventional alloy compositions. Moreover, the process can replace expensive whole alloy substrates with cheaper base materials, reducing manufacturing costs and substantially improving production efficiency. Methods like preset material, synchronous powder feeding, and laser gas alloying are employed during laser alloying.

Laser cladding, a branch of laser surface modification, was first researched in the 1970s, emerging with the development of high-power lasers. Here, cladding material is added to the workpiece surface. When exposed to a high-energy laser, both the cladding material and a thin layer of the substrate metal melt quickly, then rapidly solidify into a cladded layer, resulting in a modified or coated layer with the desired properties. The power density distribution for laser cladding ranges between 10^4 and 10 W/cm^2, positioning it between laser hardening and laser alloying in terms of application. Laser cladding, compared with traditional surface modification technologies, has many advantages: it applies to a wide range of material systems; enables rapid heating and swift cooling, leading to the refinement of the cladding layer composition; has a low dilution rate, and bonds well with the substrate; has a minimal heat-affected zone on the base material; offers high energy utilization rate; and is an environmentally friendly process [54,55]. Therefore, the laser cladding of aluminum alloys has broad development prospects. However, it is not without its challenges. The equipment and operational costs associated with laser cladding can be high. The technique also demands precise control over laser parameters; mismanagement can induce defects like cracking, porosity, or unwanted phase formations.

Additionally, there are potential risks of substrate damage from excessive laser energy, and the process requires skilled operators for optimal outcomes. Laser cladding has found extensive applications in sectors like oil and gas, aerospace, and automotives, enhancing wear and corrosion resistance despite its limitations. It is also invaluable for restoring or repairing complex or costly components, demonstrating its significance in contemporary material engineering.

Laser cladding technology integrates laser technology, computer-assisted manufacturing, and automated control technology. It is worth noting that aluminum alloys have a strong reflective effect on laser beams, with their reflectivity mainly inversely proportional to the wavelength of the laser [38,56]. Aluminum's interaction with electromagnetic radiation, particularly in the visible and infrared regions, is dictated by its dielectric function and interband transitions. Aluminum's high reflectivity arises from its conduction-electron solid response, governed by its free electrons. Spectrally, aluminum exhibits increased reflectivity for longer wavelengths, which is clearly evidenced by its response to different lasers. For instance, for a CO_2 laser with a wavelength of 10.6 μm, the laser reflectivity is as high as 90%. For a Nd: YAG laser with a wavelength of 1.06 μm, the absorption rate is seven times that of a CO_2 laser. Radiation photo-absorbent materials can enhance absorptivity to address this challenge, especially in more extended wavelength lasers. The application of laser cladding on aluminum alloys is closely tied to developing a high-quality, high-power laser.

Laser cladding technology uses the laser beam's energy to simultaneously melt the coating powder and the surface of the substrate, forming a well-bonded coating when the laser beam moves away and the molten pool solidifies quickly. Therefore, laser cladding technology has the advantages of high processing efficiency, high-quality coating preparation, and environmental friendliness.

Two types can be distinguished depending on the method of inputting cladding materials: the pre-placed powder method and the synchronous powder feeding method, as depicted in Figure 1 [38]. The pre-placed powder method is simple to operate. However, due to its lower-quality cladding layer, it is not suitable for industrial production. The synchronous powder feeding method, including synchronous side feeding and coaxial feeding, provides excellent-quality cladding layers and is conducive to automated control.

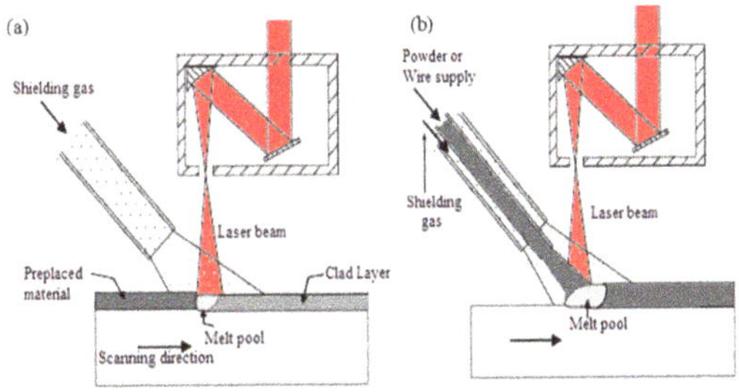

Figure 1. Schematic diagram of laser cladding ((**a**) is the pre-placed powder method; (**b**) is the synchronous powder feeding method) [38].

As shown in Figure 2, the geometric shape of the laser cladding layer typically includes the height (H), width (D), depth (h), wetting angle (θ), and dilution rate (y) of the cladding layer. The dilution rate y can be mathematically represented by the formula $y = h/(h + H)$. Selecting appropriate cladding materials and sensible cladding process parameters is vital to acquiring a well-formed, high-performance, high-quality cladding layer. The choice of cladding material will determine the performance of the cladding layer. In contrast, the cladding process parameters will directly influence the layer's geometric shape, structure, and performance. Standard parameters for the laser cladding process include laser power, scanning speed, spot diameter, overlap rate, degree of dilution, and the quantity of powder supplied, which is an influential factor in determining the thickness of the coating.

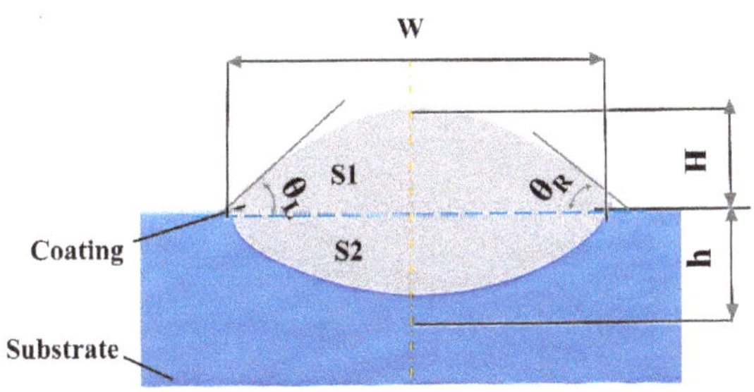

Figure 2. Cross-section diagram of the cladding layer.

Currently, the primary material systems for the laser cladding of aluminum alloys include metal-based alloys, ceramic-reinforced composites, and emerging materials such as amorphous and high-entropy alloys. These cladding materials typically come in powder, wire, or sheet form, of which the powder form is the most widely used in industrial applications. This article mainly compared aluminum alloy laser cladding based on the cladding material system. It details the process of using powder cladding materials under different experimental conditions and their forming and organizational performance. It identifies current issues with applying other component powders in the laser cladding of aluminum alloy surfaces. Finally, we summarize and forecast the future development of the laser cladding of aluminum alloys.

2. Metal Based Alloys

In the field of laser cladding technology, research on metal-based coatings for aluminum alloy surfaces predominantly focuses on three primary categories: Al-based, Fe-based, and Ni-based coatings.

2.1. Al-Based Alloys

Due to its low heat input and high automation, laser cladding technology has been widely applied in metal repair, especially in restoring Al-based alloys. Many scholars have reported on this.

Meinert et al. [57] first laser-clad the 4000 series aluminum alloy onto the surface of 6061 and 7075 aluminum alloys. They found that the tensile strength of the repaired specimens was 61% and 64% of the original 6061-T6 and 7075-T651 samples, respectively. Cottam et al. [58] studied the laser cladding of a 7075 aluminum alloy powder on the surface of a 7075 aluminum alloy. The hardness of the cladding layer they prepared was about 80% of the matrix. Heat treatment can bolster the hardness of the coating, yet it concurrently amplifies the residual stress within the coating. Figure 3 is pivotal in illustrating the sample's residual stress distribution, a key factor influencing the material's performance and durability. To be precise, Figure 3a–c map out the stress distribution in the substrate, the as-clad coating, and the heat-treated coating, respectively. A nuanced evaluation reveals that the substrate exhibits the most average residual stress. Importantly, this figure underscores the transformative effect of the heat treatment: the residual stress in the coating treatment is approximately double that of the as-clad coating. Such insights emphasize the critical nature of understanding and controlling residual stresses in various processing conditions.

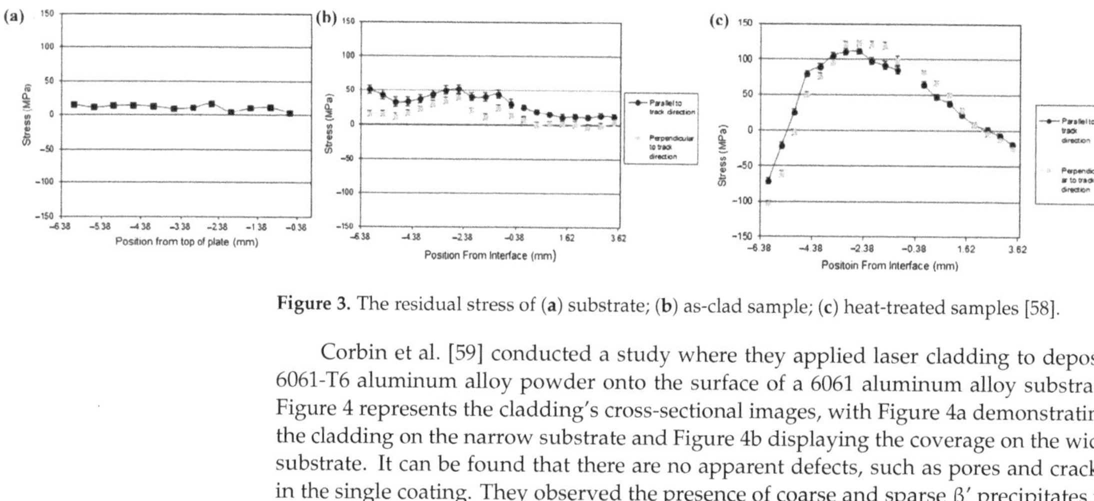

Figure 3. The residual stress of (**a**) substrate; (**b**) as-clad sample; (**c**) heat-treated samples [58].

Corbin et al. [59] conducted a study where they applied laser cladding to deposit 6061-T6 aluminum alloy powder onto the surface of a 6061 aluminum alloy substrate. Figure 4 represents the cladding's cross-sectional images, with Figure 4a demonstrating the cladding on the narrow substrate and Figure 4b displaying the coverage on the wide substrate. It can be found that there are no apparent defects, such as pores and cracks, in the single coating. They observed the presence of coarse and sparse β' precipitates in the cladding layer, resulting in a lower hardness than the substrate. In the heat-affected zone, fine and abundant β'' precipitates were maintained, which resulted in hardness levels similar to the substrate in that region. These studies indicate that in aluminum alloy repair technology, the performance of the cladding layer is generally lower than that of the aluminum matrix, and issues such as abrupt changes in hydrogen solubility, alloy element ablation, poor powder fluidity, and ease of oxidation exist. Therefore, some researchers have conducted in-depth studies on aluminum alloy repair technology for aspects such as powder feeding devices, process optimization, and post-welding treatment to improve the quality of the cladding layer.

Yang et al. [60] used a high-power diode laser to repair ZL205A aluminum alloy casting defects, which then went through a solution treatment in water at 535 °C for 11 h, followed by artificial aging at 175 °C for 5 h; the hardness of the coating was increased to 134 HV. In contrast to Corbin's research, significant pores were found in the cladding layer. Figure 5 depicts the cross-sectional morphology of the coating, with Figure 5a showing the macroscopic interface and Figure 5b highlighting the pore features.

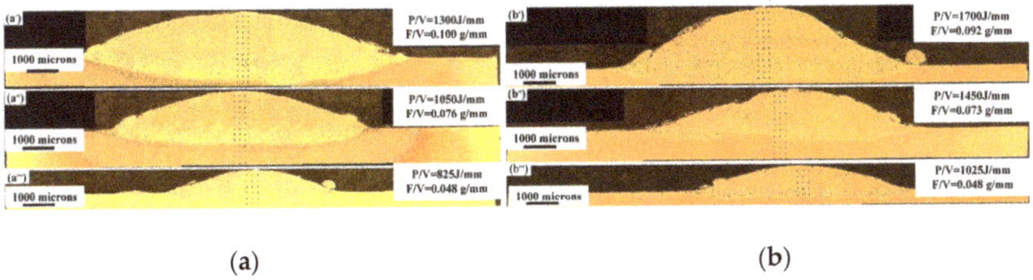

Figure 4. The cross-sectional images of the claddings produced on (**a**) narrow substrate and (**b**) wide substrate [59].

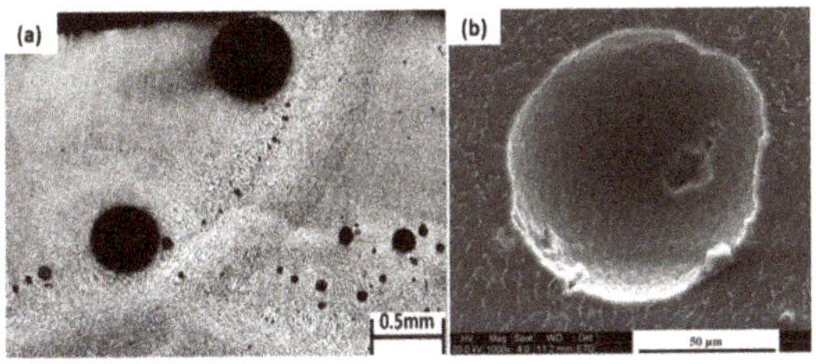

Figure 5. Coating morphology with (**a**) illustrating the cross-sectional view and (**b**) showing the pore SEM imaging [60].

Regarding process optimization, Song et al. [61] have conducted research to enhance the damage tolerance of aircraft skin. Their approach centers on exploring the benefits of laser cladding aluminum alloy structures for improved crack resistance. They developed a three-dimensional finite element simulation model, incorporating a double ellipsoid heat source and the sequentially coupled thermal–mechanical analysis method. Figure 6 depicts the residual stress distribution in cladded samples under varying laser powers. It was discovered that with increased laser power, the residual compressive stress in the models was enhanced due to an increase in heat energy input, which broadened the thermal influence range and intensified the temperature gradient in the heat-affected zone. This amplified the stress response, particularly in areas where the coating meets the substrate. Their findings suggest that the laser cladding samples reach peak crack resistance at a laser power of 1400 W and a laser scanning velocity of 10 mm/s. Remarkably, they discovered that a sample with a cladding coating length of 50 mm exhibited a fatigue life that was 2.93 times longer than an untreated sample. Moreover, their research points to the superior crack resistance performance of a linear pattern compared to other patterns, with the cladding angle of the linear design at "0" resulting in the highest performance. This research underscores the enhanced fatigue life of samples treated with laser cladding, a benefit attributed to the induced residual compressive stress from the laser cladding process.

Additionally, they found that modifying the laser parameters, such as increasing the laser power or reducing the laser scanning velocity, could further extend the fatigue life of laser-treated samples. The study concludes that various factors, including laser parameters, laser cladding patterns, and angles, can influence the fatigue life of the samples. Some researchers also prepared aluminum-based coatings on the surface of aluminum alloys to improve their surface performance.

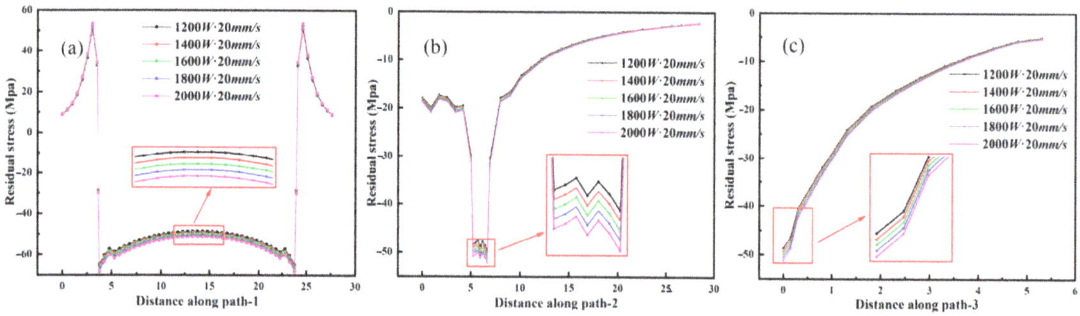

Figure 6. The distribution of residual stress under various laser power levels ((**a**) shows Path-1, (**b**) illustrates Path-2, and (**c**) represents Path-3 [61]).

Dubourg et al. [62] studied the mechanical properties of Al-Cu alloy coatings laser-alloyed onto the surface of pure aluminum. The results showed that when the Cu content was 40%, the hardness of the alloy was the highest, reaching 250 ± 10 $HV_{0.2}$.

Table 1 presents the key findings for Al-based alloys. In summary, the use of Al-based alloys to address defects in aluminum alloy components has improved considerably. Leveraging Al-based materials can result in cost reductions. Nonetheless, challenges persist when using Al-based alloys as modification coatings. The degree of hardness enhancement achieved through such coatings is often limited, and while heat treatment might intensify surface hardness, it concurrently amplifies the residual stress within the coating. Moreover, even though the integration of ceramic materials can significantly augment hardness, potential issues related to coating cracking must be considered.

Table 1. Key findings on Al-based alloy coatings.

Authors	Coating Material	Key Findings
Meinert et al. [57]	4000 series Al	Repaired specimens had tensile strengths of 61% and 64% of their original counterparts.
Cottam et al. [58]	7075 Al	The cladding layer's hardness was 80% of the matrix. Heat treatment improved the hardness but also increased residual stress.
Corbin et al. [59]	6061-T6 Al	Coatings produced were without apparent defects. The cladding layer had coarser β′ precipitates with reduced hardness.
Dubourg et al. [62]	Al-Cu alloy	At 40% Cu content, alloy hardness peaked at 250 ± 10 $HV_{0.2}$.

2.2. *Fe-Based Alloys*

Fe-based alloy powders, with excellent wear resistance, low cost, and wide availability, have been used by some researchers for laser cladding onto the surface of aluminum alloys to improve surface wear resistance and anti-deformation.

Jeyaprakash et al. [63] used laser cladding technology to prepare a FeCrMoVC (H13 steel) cladding layer on the surface of an AA6061 aluminum alloy. The coating structure contained large carbides and martensite, the carbides formed by vanadium and molybdenum significantly improve the hardness of the coating. Compared with the base material, the wear resistance of the layer increased by nine times, and the surface roughness of the coating after wear was reduced by three times. Ye et al. [64] prepared a Fe-Al intermetallic compound cladding layer on the surface of ZL114A aluminum alloy. The microhardness of the cladding layer was as high as 614 HV, which is 5 to 6 times higher than the base material. Figure 7 shows the microstructure of the coating transition zone. Figure 7a illustrates the interface between the melt pool and the matrix, highlighting evident grain refinement in the transition zone due to the rapid heating and cooling of the aluminum alloy surface. Figure 7b reveals the formation of dendritic grains along a specific growth direction at the bottom of the melt pool. In Figure 7c, these dendritic grains transform into equiaxial grains at the top of the melt pool. Cracks were found in the bonding zone, as shown in Figure 7d.

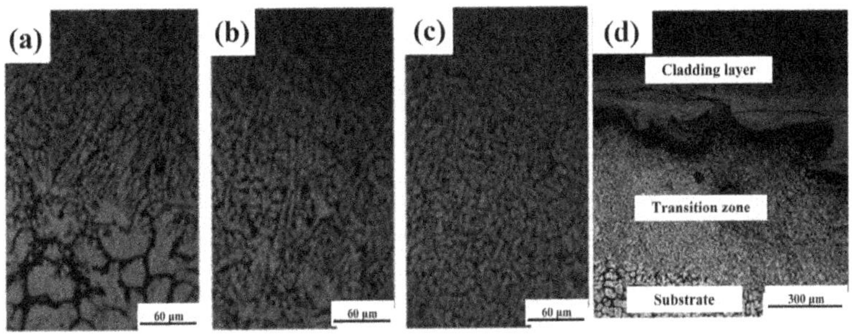

Figure 7. SEM images of the coating, (**a**) is the bottom of the transition zone, (**b**) is the middle of the transition zone, (**c**) is the top of the transition zone, and (**d**) is the macroscopic morphology of the bonding zone [64].

Tomida et al. [65] used a 2.2 kW continuous-wave CO_2 laser to prepare an xWt.%Al/Fe alloy coating on the surface of an A5052 aluminum alloy, including four Al-Fe mixed coatings with Al contents ranging from 10 wt.% to 40 wt.%. The results showed that the microstructure of the cladding layer consisted of α-Al + $FeAl_3$, $FeAl_3$, and Fe_2Al_5, and the hardness reached 600–1000 HV. The alloy layers with fine needle-like $FeAl_3$ and Fe_2Al_5 composite structures also have high hardness at a temperature of 673 K, 300 HV, and 800 HV, respectively. The wear resistance of such a coating is about 4–5 times higher than that of the aluminum alloy matrix. Mei et al. [66] found that the Fe-based coating prepared on the surface of an Al Si alloy contained austenite, Cr_7C_3, and $Cr_{23}C_6$; α-Al, $NiAl_3$, Fe_2Al_5, and $FeAl_2$ were present at the junction; and α-Al and Si were present in the heat-affected zone. Due to the generation of brittle intermetallic compounds, the coating was prone to cracking. Carroll et al. [67] performed the laser cladding of Fe powder prefabricated on the surface of a 319 aluminum alloy. The Al-Fe-Si intermetallic compound formed increased the hardness of the coating, but there was a significant increase in brittleness.

Table 2 presents the key findings for Fe-based alloys. In summary, using Fe-based alloy powder in laser cladding offers advantages like notable improvements in the mechanical properties of aluminum alloys and enhanced wear resistance at a low cost, but it also poses challenges. Specifically, Fe-based powder coatings on aluminum alloys can form various Fe-Al phase intermetallic compounds. This results in the frequent generation of defects such as pores and cracks in the bonding region, in addition to the challenges of pore formation in the cladding layer, limited enhancement in corrosion resistance, and the propensity for brittle intermetallic compounds. In response to these issues, future research may need to optimize cladding processes and alloy components further to improve the performance and stability of the coating.

2.3. Nickel-Based Alloys

Nickel-based alloys are known for their high hardness, wear resistance, corrosion resistance, and excellent resistance to high-temperature oxidation. They also have good wetting properties with aluminum substrates. Several advancements have been made in the laser cladding of aluminum alloys with nickel-based alloys.

Table 2. Key findings on Fe-based alloy coatings.

Authors	Coating Material	Key Findings
Jeyaprakash et al. [63]	FeCrMoVC	Coating has enhanced wear resistance; reduced surface roughness after wear.
Ye et al. [64]	Fe-Al intermetallic compound	Cladding layer has a hardness 5–6× that of the base material; cracks noted in bonding zone.
Tomida et al. [65]	Wt.%Al/Fe alloy (Al content: 10–40 wt.%)	The hardness of 600–1000 HV; wear resistance 4–5× greater than base matrix.
Mei et al. [66]	Fe-based coating	Contains multiple intermetallic compounds; prone to cracking due to brittle intermetallic compounds.
Carroll et al. [67]	Fe powder	Increased hardness with the Al-Fe-Si compound; significant increase in brittleness.

Wu et al. [68] applied a self-fluxing nickel-based powder coating to an $AlSi_7Mg$ aluminum alloy surface via laser cladding. They found that different microstructures were formed in various regions of the coating. The top of the cladding layer consisted of a substantial amount of block- and network-shaped intermetallic compounds; the middle part had $(Ni, Cr, Fe)_xC_y$ intermetallic compounds. At the same time, columnar α-Al dendrites with a distinct growth direction were prevalent at the bottom. The highest hardness, reaching 8200 MPa, was recorded in the top and middle parts of the coating. Adjusting the scanning speed could control the formation of pores and cracks in the coating.

He et al. [69] created a TiB_2-reinforced nickel-based composite coating on a 7005 aluminum alloy. The coating contained NiAl, Ni_3Al, Al_3Ni_2, TiB_2, TiB, TiC, CrB, and $Cr_{23}C6$ phases. Regarding the wear properties, a disk-type dry friction wear test was performed. Corundum, with a hardness of 9 on the Mons' scale, was the friction coupling, measuring Φ40 mm × 10 mm. The wear samples were fashioned into 4.5 mm diameter and 10 mm height cylindrical shapes. The test was executed at a rotating speed of 100 $r \cdot min^{-1}$, subjected to a load of 20 N for 900 s. The hardness was 6.7 times that of the substrate, and the mass loss was reduced by up to 32.7%. As shown in Figure 8, the morphology of the TiB_2/Ni-based composite coating revealed that the powder was not fully melted, resulting in significant pore defects between particles.

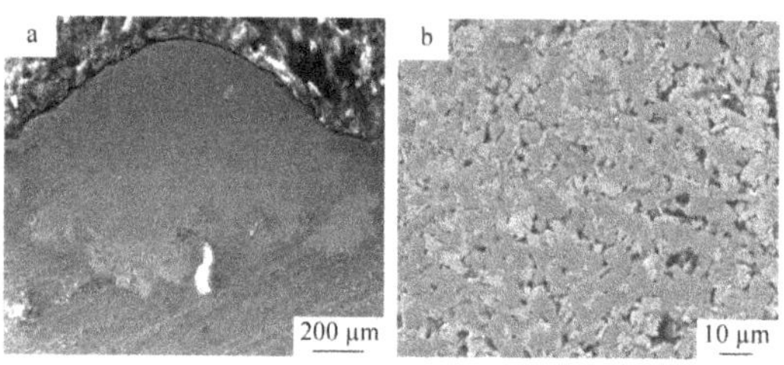

Figure 8. SEM of a cross-section of coating, (**a**) is the macroscopic morphology and (**b**) is the microstructure in the middle of the coating [69].

Liang et al. [70] prepared a NiCrBSi cladding layer on an Al-Si alloy surface using a CO_2 laser. The cladding layer contained a large amount of Ni_3Al, with the highest microhardness reaching 1200 HV. In addition, by analyzing the aluminum–nickel eutectic system composition, they discovered that the solid solution formed by Cr, Si, and B with Al helped improve the resistance of the aluminum alloy to gas corrosion.

These studies suggest that forming aluminum–nickel intermetallic compounds can significantly enhance the hardness and wear resistance of coatings. However, the aluminum alloy substrate's and nickel-based cladding materials' physical and chemical properties differ greatly. As a result, defects such as cracks and pores can quickly form in the cladding layer. Relevant studies have shown that adding rare earth elements can alleviate this issue.

Wang et al. [71] added 5% rare earth elements to the Ni60 alloy during laser cladding on the surface of a 6063 aluminum alloy. They found that adding rare earth elements could improve the fluidity of the melt pool during the solidification process of the cladding layer and purify the melt pool. As shown in Figure 9, when only Ni60 alloy powder was used, many pores and cracks appeared in the cladding layer. After adding a certain amount of rare earth elements, the defects in the cladding layer, such as pores and cracks, were noticeably improved. After adding rare earth elements, the cladding layer's structure was refined, with no apparent pores or segregation.

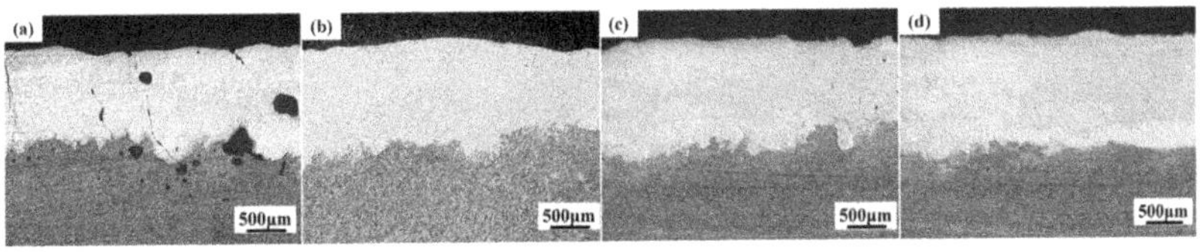

Figure 9. SEM of a cross-section of coatings, (**a**) is Ni60, (**b**) is La_2O_3 + Ni60, (**c**) is Y_2O_3 + Ni60, and (**d**) is CeO_2 + Ni60 [71].

Zhang et al. [72] prepared a nickel-based cladding layer with a nickel–copper ratio of 4:1 on the surface of a 6061 aluminum alloy. They added a certain proportion of rare earth CeO_2, Si, and Co powders to the cladding powder. The top of the coating was primarily composed of equiaxed crystals, dendrites, and tiny equiaxed crystals, while the bottom of the cladding layer was mainly composed of long rod-shaped crystals, dendrites, and some particles. After adding CeO_2, Si, and Co powders into the cladding materials, the cladding layer's hardness increased, and the average friction factor decreased, thanks to the grain refinement of CeO_2 and the dispersion strengthening effect of Si and Co in the coating. Figure 10 portrays the microstructural variances across the top and bottom regions of Ni-Cu, CeO_2/Ni-Cu, Si/Ni-Cu, and Co/Ni-Cu composite coatings. Specifically, Figure 10a,b demonstrate the microstructure of the top and bottom regions in a 100 wt.% M coating, wherein the top area is primarily populated by grey rod-like and equiaxed grains. The bottom zone, conversely, presents a different structure. Figure 10j,k extend this analysis to the Co/M coating, revealing that the top region predominantly comprises short dendritic phases. Figure 10d–f illustrate the microstructure when the base metal is alloyed with 0.9 wt.% of cerium oxide (CeO_2), resulting in a composition that is 99.1 wt.% M and 0.9 wt.% CeO_2. This minor addition of CeO_2 is intended to investigate the effects of rare earth oxide inclusion on the microstructure. Figure 10g–i display the coatings where silicon (Si) has been added at 0.9 wt.%, making up a total composition of 99.1 wt.% M with the inclusion of silicon. This variation is aimed at understanding the influence of Si on the microstructural characteristics of the coating. The corresponding XRD results, illustrated in Figure 10c,l, corroborate these structural characterizations. These findings underscore the significant microstructural differences between the top and bottom regions across various composite coatings.

Table 3 presents the key findings for Ni-based alloys. Current domestic and international research shows that with the use of a laser heat source, the temperature field distribution of the melt pool is uneven, leading to different structures at the top and bottom of the cladding layer. Nickel-based alloys, especially when modified with Ni-based materials as the coating, have good compatibility with the aluminum substrate. The intermetallic compounds formed by Ni and Al not only improve the hardness of the cladding layer but also significantly increase its brittleness. Introducing rare earth oxides in a Ni-based coating material can enhance the quality of the formed layers, and this approach has garnered attention as a method for improving the coating's formation quality. However, there is a critical consideration: the melting point of many rare earth oxides surpasses the boiling point of aluminum alloys. In producing superior coatings, the laser heat input must achieve a delicate balance—sufficient to melt the added powder thoroughly but without overly diluting or evaporating the substrate. This challenge narrows the range of acceptable process parameters for laser deposition when incorporating high-melting-point rare earth oxides, increasing the influence of system errors on the coating quality during experimentation. Achieving a uniform microstructure, refined grains, and reasonably controlling the formation of brittle and hard phases is critical to the laser cladding of nickel-based alloys on the surface of aluminum alloys. Simultaneously, while the inclusion of rare earth elements can notably enhance the quality of the formed coatings and diminish their sensitivity to cracking, the elevated costs and high melting point characteristics associated with these elements necessitate technological advancements to mitigate these issues when applying nickel-based coatings in aluminum alloy surface cladding.

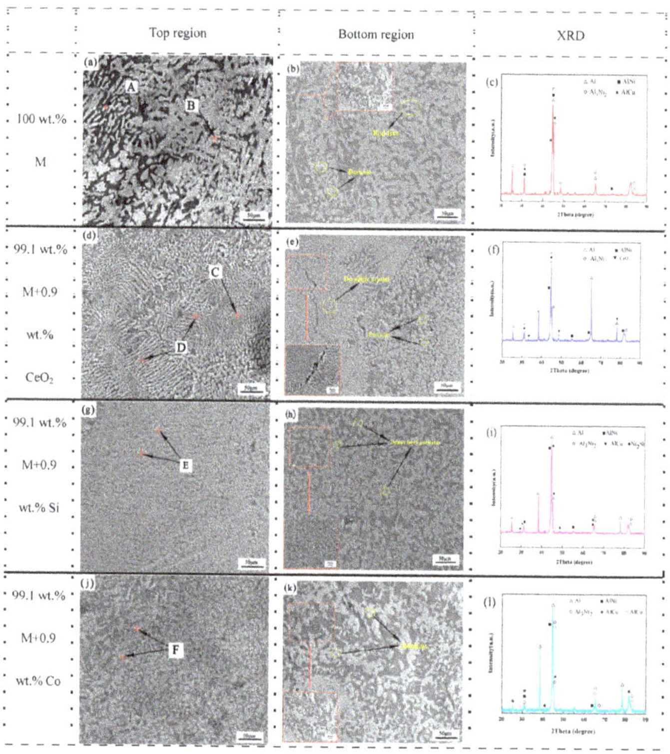

Figure 10. Characteristic microstructures of cladding coatings across different compositions. (**a–c**) is 100% wt. M (**d–f**) is 99.1% wt.%M + 0.9 wt.% CeO_2 (**g–i**) is 99.1% wt.%M + 0.9 wt.% Si (**j–l**) is 99.1% wt.%M + 0.9 wt.% Co [72].

Table 3. Key findings on Ni-based alloy coatings.

Authors	Coating Material	Key Findings
Wu et al. [68]	Nickel-based alloy	Microstructure variations; peak hardness at 8200 MPa; effect of speed on defects
He et al. [69]	TiB_2-reinforced nickel-based composite	Coating hardness 6.7× substrate; mass loss reduced by 32.7%; notable pore defects.
Tan et al. [73]	Nickel-based alloy	Enhanced coating hardness and wear resistance compared to substrate.

Table 3. *Cont.*

Authors	Coating Material	Key Findings
Liang et al. [70]	NiCrBSi	Cladding contains Ni_3Al with a peak microhardness of 1200 HV.
Wang et al. [71]	Ni60 alloy with 5% rare earth elements	Rare earth elements improved coating structure and reduced defects.
Zhang et al. [72]	Nickel-based (Ni-Cu ratio of 4:1) with CeO_2, Si, and Co	Additives (CeO_2, Si, Co) improved coating hardness and reduced friction.

3. Ceramic-Reinforced Composite Coating

Ceramics, with their high hardness, high melting point, and excellent wear resistance, are widely acknowledged as suitable materials for laser cladding. Adding ceramic particles to metal-based laser cladding materials can exploit the non-linear superimposed effects of their physical properties between the ceramic reinforcement phase and the metal matrix. Combining the excellent toughness and machinability of the metal matrix with the high hardness and wear resistance of the ceramic reinforcement phase can enhance the surface of the base material. In laser cladding on the surface of aluminum alloys, ceramic phases are commonly used as reinforcement, added to the metal-based composite materials. These ceramic phases include carbide ceramics, such as WC, SiC, TiC, oxide, nitride, and boride.

Sun et al. [74] via laser cladding, prepared a SiC/Al-12Si composite coating on the surface of an Al-12Si aluminum alloy substrate. Studies have found that the layer and the substrate are divided, and the organization of the bonding area is mainly composed of α-Al dendrites, a small amount of Al-Si eutectic, and a small number of SiC particles. In contrast to the base material, the hardness of the cladding layer increased by about two times, reaching approximately 260 $HV_{0.2}$.

Anandkumar et al. [75] prepared a 12 wt.%TiB_2/Al-based composite coating on the surface of a 7 wt.%Si/Al casting aluminum alloy via laser cladding. To assess the wear performance of the prepared cladding, microscale abrasive wear tests were conducted using a ball cratering device. The abrasive medium for these tests was a suspension containing 35 wt.% SiC particles (with an average diameter of 4.25 μm) in water. The counter body, which rotates against the sample's surface during testing, was a 19 mm diameter sphere made of quenched and tempered AISI 440C tool steel, boasting a hardness of 800 HV. The tests employed a uniform sliding distance of 150 m and exerted a regular load of 0.1 N. The cladding layer mainly consisted of α-Al dendrites and α-Al-Si eutectic, with a micro-hardness of 156 $HV_{0.2}$, an approximately 1.7 times increase compared to the base material. Figure 11 illustrates that the microstructure of the deposited material is highly dependent on the laser processing parameters. A cladding prepared with a power density of 330 W/m^2 and a brief interaction time of 0.08 s—corresponding to a specific energy of 26 J/m^2—exhibits a microstructure primarily composed of SiC particles dispersed within an aluminum alloy matrix, as shown in Figure 11a. This matrix comprises primarily α-Al dendrites and an α-Al+Si eutectic, as shown in Figure 11d. The volume fraction of SiC stands at 28 ± 3%. The TiB_2 reinforcement phase in the coating effectively protected it, enhancing its wear resistance and reducing its sliding friction coefficient to 2.65×10^{-5} $mm^3/(N{\cdot}m)$, approximately 2.4 times lower than the base material.

They [76] also laser clad a mixture of Al-12%Si and TiC ceramic particles on the surface of Al-7%Si casting aluminum alloy, successfully obtaining a cladding layer with no apparent defects. The cladding layer mainly contains uniformly distributed TiC particles, interdendritic α-Al+Si eutectic, dendritic α-Al, and a small amount of Ti_3SiC_2, indicating the partial dissolution of TiC particles. The cladding layer's friction coefficient decreased 2–3 times compared to the base material. Only a small amount of Al and Si were found in the worn-off material from the base material's surface, with no TiC particles detected.

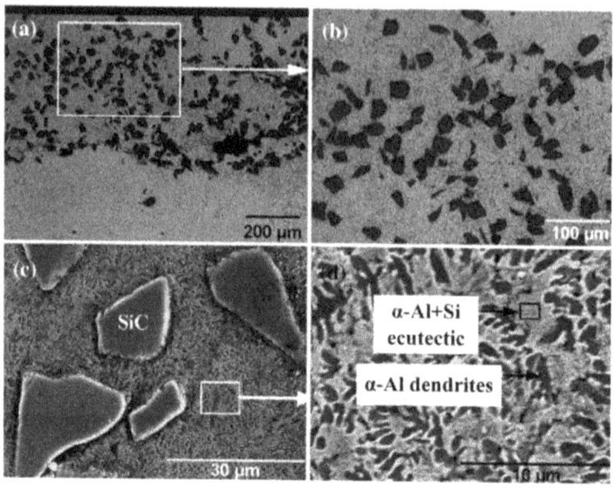

Figure 11. Microscopic illustrations of a coating formulated utilizing a power density of 330 MW/m^2 and a contact period of 0.08 s. (**a**) is a transversal representation of the layer; (**b**,**c**) are SiC particles scattered in the matrix; and (**d**) is the intricate design of the matrix at a microscopic level [75].

Liu et al. [77] prepared an Al-Si-based composite layer reinforced with SiC particles on a 4032 aluminum alloy. Chemical copper plating improved the wettability between SiC and the substrate, while increasing the SiC content within a specific range can reduce the cladding layer's wear.

Li et al. [78] prepared a Ti/TiBCN coating on the surface of a 7075 aluminum alloy using laser cladding. To evaluate the wear resistance of the cladding coatings, tests were conducted using an MFT-R4000 straight-line-reciprocating dry sliding wear tester at room temperature without any lubrication. The test parameters were set at a normal load of 5 N, a friction distance of 5 mm, a frequency of 2 Hz, and a sliding duration of 20 min. The counterpart material for the friction test was GCr15 steel (5 mm in diameter), with a macro-hardness of HRC65. Figure 12 presents a detailed look at the microstructures in various zones. Figure 12a,b reveal the microstructure of the coating zone, composed primarily of equiaxed grains and uniformly dispersed white lath-like crystals. Figure 12c,d portray the transition zone's microstructure, notable for its sudden shift from equiaxed grains to an elongated structure filled with small white particles. Figure 12e,f depict the heat-affected zone's microstructure, characterized by a transverse equiaxed crystal structure resulting from the effects of rapid heating and cooling. The average hardness of the coating is 519.4 $HV_{0.2}$, 4.3 times that of the base material; the average friction coefficient of the layer is 0.208, half that of the base material. Compared with solely Ti powder cladding, adding a certain proportion of TiBCN ceramics to the cladding material can further enhance the coating's wear resistance and corrosion resistance. When the TiBCN content is 15%, the wear and corrosion opposition reach their best.

He et al. [69] prepared a Ni-based alloy composite coating reinforced with TiB_2 ceramic particles on the surface of a 7005 aluminum alloy to enhance the wear resistance of aluminum alloy friction components. Tribological assessments were conducted using an HT-500 ball-on-disk tribometer, where specimens slid against a GCr15 ball (U = 4 mm) under varied loads of 3, 6, 9, and 12 N at a steady speed of 0.5 $m \cdot s^{-1}$ for 500 m, conducted at 20 °C. The micro-hardness of this ceramic-reinforced composite material cladding layer is 855.8 $HV_{0.5}$, 15.4% higher than the Ni-based alloy coating and 6.7 times that of the aluminum alloy substrate.

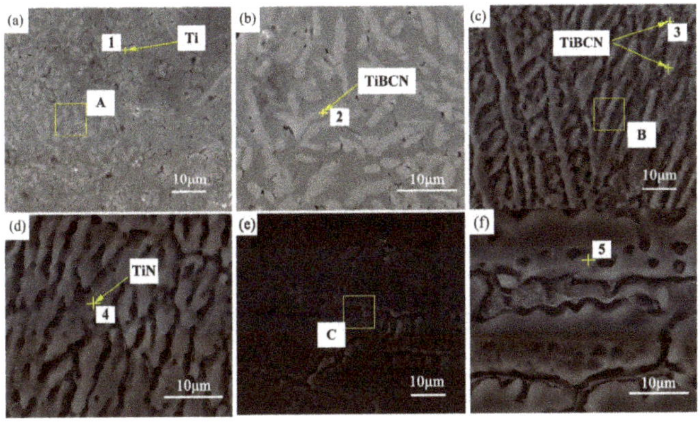

Figure 12. Distinctive cross-sectional microstructures within different zones of the cladding coating prepared with 15 wt.% TiBCN powder. (**a**) is the uppermost layer of the cladding coating. (**b**) is a magnified snapshot of quadrangle [A] in (**a**). (**c**) is the transition zone. (**d**) is a magnified view of the quadrangle in (**c**). (**e**) is the heat-affected zone. (**f**) is an enlarged depiction of the quadrangle in (**e**) [78].

The poor ductility, small elastic modulus, low thermal conductivity, and high melting point of ceramic materials can lead to cracks, deformation, and delamination in the cladding layer. Therefore, research on cladding a single ceramic onto the surface of an aluminum alloy is relatively scarce, and composite cladding layers of ceramic-reinforced materials mixed with metal materials are typically prepared. To address the challenges of laser cladding ceramic materials on aluminum alloys, solutions can be sought from the perspectives of optimizing process parameters, controlling the composition of cladding materials, improving powder properties, adding appropriate transition layers, and the in situ generation of ceramic reinforcement phases.

Regarding the control of the cladding material composition, Riquelme et al. [79] aimed to address the issue of harmful Al_4C_3 compound formation during the laser cladding of a SiCp-reinforced Al-based composite coating on an aluminum alloy surface. They manipulated the reaction equilibrium of Al_4C_3 by adding different alloy elements, thereby preventing the formation of Al_4C_3. When the Si content was low, needle-shaped Al_4C_3 appeared in the cladding layer. When the Si content exceeded 40 wt.%, the appearance of Al_4C_3 was entirely inhibited, and more SiC particles were observed, significantly improving the hardness of the cladding layer.

In terms of the improvement of powder properties, Kamaal et al. [80] addressed the issue of limited cladding efficiency due to the high reflectivity of aluminum alloys and the difficulty with ceramic material cladding. They studied the impact of powder satellite technology on the cladding of Al-TiC powder on a 6082-T6 aluminum alloy substrate. The hardness and wear performance of the specimens were then assessed. The wear performance of these samples was evaluated using a CETR Universal Micro-Tribometer 3, following the ASTM G133-05 standard [81]. Tests were conducted over 10 min, using a 5 mm displacement amplitude at 1 Hz frequency. An AISI steel ball (grade 440 C, 800 HV, 6 mm diameter) was the counter body under a 5 N load. The top surfaces of the claddings, post grinding to their center, were the focus of these tests.

In the satellite technology employed in that study, the Al-TiC powder mixture underwent both dry and wet mixing processes. A binder was sprayed over the mixed powder to achieve a uniform distribution. Subsequent mixing steps ensured the homogeneous distribution of the components, breaking down any potential agglomerations or clusters. After mixing, the satellited powder was dried, evaporating the water content and leaving behind only a minuscule percentage of the binder. This method aims to obtain a better homogenized powder feedstock, ensuring the consistent distribution of its components.

Compared with the directly mixed cladding material, the cladding efficiency and TiC content in the cladding layer increased by 29% and 113%, respectively, after using satellite technology. Figure 13 shows the cross-sectional images of the directly mixed powder and satellite powder cladding layer. As can be seen, the cladding material after using satellite technology resulted in better formation of the cladding layer, with more TiC content and a more uniform distribution. The microhardness measurements for blended and satellited claddings were created at different laser densities. The average microhardness for the combined class lies between 37.7 and 51.6 $HV_{0.2}$. Claddings produced at 104 and 122 J/mm^2 have similar hardness values. A slight increase in hardness from 104 to 122 J/mm^2 is linked to denser structures and an 83% decrease in porosity at the higher laser density. The average micro-hardness of the cladding layer improved by 60%, and the wear rate decreased by 64%.

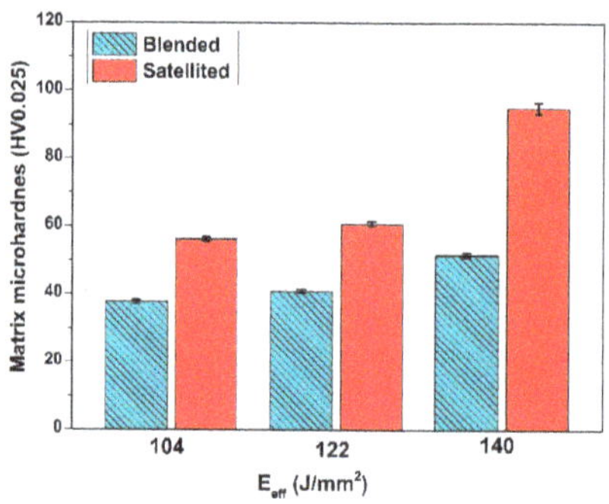

Figure 13. The microhardness measurements of blended and satellited cladding matrices generated at assorted Eeff values [80].

Table 4 presents the key findings for ceramic-reinforced composites. In general, due to their inherent high hardness and exceptional wear resistance, ceramics significantly enhance the cladding layer's hardness when applied to aluminum alloy surfaces. Introducing ceramic particles into metal-based laser cladding materials can produce a synergistic effect between the ceramic reinforcing phase and the metal matrix. This synergy results in a composite coating that combines high strength with flexibility. Regarding material selection, various ceramics such as WC, SiC, TiC, and others like oxides, nitrides, and borides can be used as reinforcing materials, providing a diverse array of choices tailored to specific applications. For instance, coatings enriched with TiBCN have showcased superior corrosion resistance. However, ceramic laser cladding on aluminum alloys comes with its challenges. Given the high hardness and brittleness of ceramics, coating cracking is a potential risk. Ensuring a strong metallurgical bond between the ceramic phase and the aluminum alloy substrate is paramount, especially considering the non-metallic nature of ceramics. Additionally, the microstructure of the ceramic material can be highly influenced by laser processing parameters, emphasizing the importance of optimal parameter selection to achieve the desired outcome. Finally, from a financial standpoint, the cost of employing high-quality ceramics should be taken into account, as it might have implications on the overall economic feasibility of a project.

Table 4. Key findings on ceramic-reinforced composite coatings.

Authors	Coating Material	Key Findings
Sun et al. [74]	SiC/Al-12Si composite	Doubled the hardness of the cladding layer to approximately 260 $HV_{0.2}$.
YANG et al. [77]	Al-Si-based composite with SiC	Wear resistance of cladding layer increased.
Li et al. [78]	Ti/TiBCN	Coating's hardness 4.3 times higher than base material; reduced friction coefficient.
Kamaal et al. [80]	SiC-reinforced Al-based	Achieved improved hardness by preventing Al_4C_3 formation.

4. Emerging Coatings

Currently, the emerging material coatings studied for laser cladding on the surface of aluminum alloys primarily include amorphous alloys and high-entropy alloys (HEAs).

4.1. Amorphous Materials

The traditional alloy cladding materials significantly differ from aluminum substrates in terms of properties, resulting in an increased difficulty with and reduced effectiveness of laser cladding on aluminum alloys. Table 5 shows the mixing enthalpy, melting point, and specific gravity of Al elements and some metal cladding materials [81]. It can be seen that the mixing enthalpy of aluminum elements with other metal elements is relatively small. The aluminum substrate has a low melting point and high dilution rate, and the lower-density aluminum elements in the substrate can easily transition to the cladding layer and form brittle compound phases with features in the cladding layer, increasing the tendency for cracking in the cladding layer. If the design of cladding materials could consider the characteristics of rapid heating and cooling in laser cladding and prepare cladding layers that are less prone to brittle phases, that would be the first new attempt and exploration in the field of laser cladding in recent years. Currently, the emerging material coatings studied for laser cladding on the surface of aluminum alloys primarily include amorphous alloys and high-entropy alloys

Table 5. Thermal properties of typical transition metals: mixing enthalpy, melting point, and specific weight [81].

	Fe	Cr	Ni	Co	Cu	Al
Mixing Enthalpy with Al (kJ/mol)	−11	−10	−22	−19	−1	-
Melting Point (°C)	1538	1857	1455	1495	1083	660
Specific Weight	7.89	7.19	8.9	8.9	8.96	2.6

Amorphous alloys, which have long-range disordered and short-range ordered structures similar to glass, are formed by inhibiting nucleation during the solidification of molten materials. They lack crystal defects such as vacancies, dislocations, and stacking faults and therefore possess advantages such as high hardness, wear resistance, and corrosion resistance. The fast cooling rate of laser cladding makes it possible to apply amorphous materials to laser cladding [82–86].

Sohrabi et al. [87] prepared Zr-based amorphous cladding layers on an AA2011-T6 aluminum alloy surface and studied the relationship between the coating crystallization rate and thermal input. The cladding layers produced under higher laser power yielded more diffraction peaks than other samples. Reducing the energy density can inhibit crystal crystallization. For samples with a laser power of 35 W and overlapping distance of 95 μm, the obtained cladding layer appears amorphous under the detection resolution of XRD tests. The wear resistance of the Zr-based amorphous cladding layer is 20 times that of the aluminum alloy substrate. Figure 14a displays a SEM examination of the wear track for a sample created with optimized parameters, leading to 1% crystallization. This wear evaluation was conducted under specific test conditions to ensure a comprehensive understanding of the material's behavior. Wear tests were conducted on the surface of the samples, which were polished down to a sandpaper with a 2500 mesh size, using a reciprocating tribometer in a ball-on-flat configuration. Al_2O_3 balls with a diameter of 6 mm served as the counterpart. During the test, an applied normal force of 2 N was maintained at an average speed of 6 mm/s and a 1 Hz frequency. Each trial lasted 30 min, with a stroke length of 3 mm, culminating in a total sliding distance of 10.8 m. Precautions were taken to clean the sample surfaces and the counterpart balls in ultrasound baths with acetone, followed by ethanol, and they were then dried with oil-free compressed air.

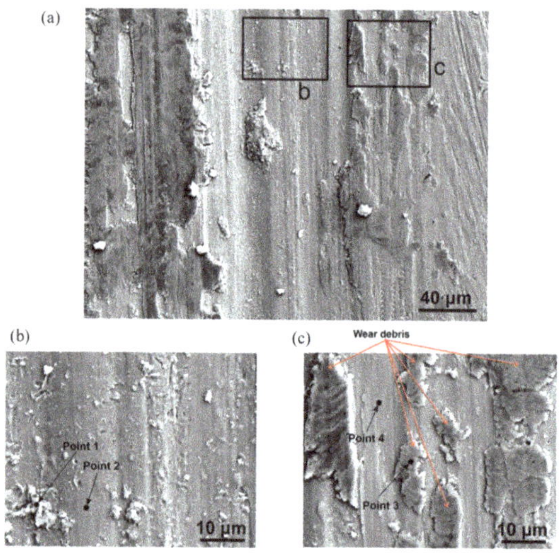

Figure 14. (**a**) SEM micrograph of the wear track (**b**) magnified image of region B in (**a**), and (**c**) magnified image of region C in (**a**) [87].

Figure 14b,c give detailed views of regions b and c. Wear debris is spotted at Points 1 and 3, while Points 2 and 4 represent the areas with preserved coating. The oxygen at Points 1 and 3 is likely due to the wear debris interacting with air. The observed wear patterns and debris suggest a wear mechanism that is a combination of abrasive and oxidation wear.

The formation of amorphous materials is related to the ratio of material components, and obtaining high-quality amorphous cladding layers requires preventing segregation in the coating. Wang et al. [88] used a fiber laser to clad an Al-Ni-Y ternary alloy powder onto a ZL114 aluminum alloy surface to study its ability to form amorphous coatings on aluminum alloys via laser cladding. The cladding layer mainly comprised blocky dark areas and fewer net-like bright areas. The bright regions contained more Ni and Y elements, while the dark spots contained less Ni and Y. The atomic ratio of elements in the bright grain boundary areas is close to $Al_{85.8}Ni_{9.1}Y_{5.1}$, which has the strongest ability to form amorphous alloys in the Al-Ni-Y system. However, XRD results did not show the apparent complete reflection of the amorphous alloy, indicating that there are still some difficulties in preparing amorphous coatings on the aluminum alloy surface via laser cladding under these experimental conditions. The Ni and Y transition issue in the cladding layer must be resolved.

Table 6 presents the key findings for amorphous coatings. As a novel coating material, current research on applying amorphous coatings to aluminum alloy surfaces remains limited. However, such coatings can significantly enhance the substrate's corrosion and wear resistance. Precise control over processing parameters and material composition ratios is vital to prevent segregation within the coating. Delving deeper into optimizing these amorphous coatings' structures, organization, and performances through laser cladding is still a topic warranting further investigation.

Table 6. Key findings on amorphous coatings.

Authors	Coating Material	Key Findings
Sohrabi et al. [87]	Zr-based amorphous	Coating's wear resistance is 20× that of the substrate; reduced energy density inhibits crystallization.
Wang et al. [88]	Al-Ni-Y ternary alloy powder	Bright regions of the layer contain more Ni and Y; challenges in achieving fully amorphous coatings.

4.2. High-Entropy Alloys

In traditional metallurgy, one or two elements are usually selected as the main components of alloy design, with a small number of other elements added to adjust alloy properties. Only a tiny fraction of the given material can be changed. This imposes a significant constraint on the combination of elements, making it challenging to improve traditional materials or develop new ones. Many application fields are not adequately catered for, especially where hardness and wear resistance are paramount. Greer's "Confusion Principle" suggests that an excess of elements would make it difficult for an alloy to maintain its crystalline state, leaning more towards an amorphous state. Due to this, traditional alloy design struggles to break free from the shackles of single-element design. The emergence of HEA, a new material system, breaks this classic alloy design pattern, opening up a new direction for alloy design in materials science [89–95].

HEAs comprise five or more elements in equal atomic ratios. In recent years, scholars have expanded the concept of HEA to include 3 or 4, but fewer than 13, primary elements, with each element's concentration ranging from 5 at% to 35 at% [96–99]. According to traditional metallurgy theories, multi-principal element alloys such as HEAs would generate complex intermetallic compounds (IMCs). However, due to the high configurational entropy of HEAs, simple solid-solution phases become the main phase structures of HEAs, such as Face-Centered Cubic (FCC), Body-Centered Cubic (BCC), and Hexagonal Close-Packed (HCP) structures [64,98,100–107]. The difficulty in preparing aluminum alloy coatings lies in inhibiting the formation of hard and brittle phases caused by substrate dilution behavior and ensuring excellent coating performance. The four core effects of high-entropy alloys provide a theoretical basis for solving this problem.

Shon et al. [108] used a two-step thermal spray–laser cladding process to prepare AlCrFeCoNi HEAs on the surface of a 1100 aluminum alloy in situ, having found that the high-energy density multi-layer cladding process could reduce defects like coating porosity and cracking. They discovered that the high-energy density multi-layer cladding process could significantly minimize defects such as coating porosity and cracking. An essential factor in this process is the energy density, given by the formula $E = P/(v \times w)$, where E is the energy density (J/mm^2), P is the laser power (W), v is the scanning speed (mm/s), and w is the beam width (mm). This energy density is pivotal as it governs the melting behavior and the dilution between the coating and substrate. The energy density plays a crucial role in determining the dilution rate of the coating. When the energy density is high, the substrate elements are more diffused into the coating, leading to a deviation from the coating's initial compositional ratio. Conversely, at lower energy densities, the diffusion from the substrate is limited, allowing the coating to maintain its original compositional balance.

Figure 15a represents various samples' corrosion rates (color) and corrosion potentials (Ecorr) extracted from potentiodynamic polarization curves. Although the Ecorr values remain similar across models, there are significant differences in the icorr values. Figure 15b reveals the extensive pitting corrosion in the untreated Al1100 sample. Figure 15c shows the corroded surface of the single-layered coating processed at 21 J/mm^2. It exhibits substantial corrosion, predominantly in pitting and surface cracking. There are also signs of micro-galvanic corrosion. Figure 15d illustrates the corroded surface of a double-layered sample processed at 21 J/mm^2. Although minor pitting is observed, it is highly localized, indicating a homogenous composition and microstructure. Figure 15e shows a double-layered sample processed at 25 J/mm^2, which experienced minimal or no pitting corrosion. This demonstrates the benefits of minimized compositional dilution and uniform distribution of the high-entropy alloy phase. Different processing conditions affected the microstructure and subsequent corrosion behavior of the materials. Single-layer samples at 21 J/mm^2 exhibited lower corrosion resistance than double-layered samples due to a more heterogeneous structure and substantial substrate dilution. This caused an increase in corrosion rate.

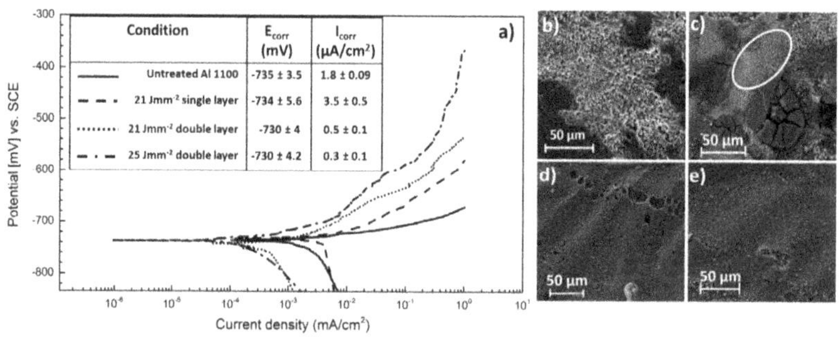

Figure 15. The corrosion results for the HEA coatings (**a**) is a cyclic polarization diagram for the coatings and Al1100 substrate, (**b**) is the corroded surface micrographs for untreated Al1100 substrate, (**c**) is 21 J/mm^2 single HEA layer, (**d**) is 21 J/mm^2 double HEA layer and (**e**) is 25 J/mm^2 double layer [108].

In contrast, double-layered samples showed improved corrosion resistance, particularly at a processing intensity of 25 J/mm^2, which exhibited minimal or no pitting corrosion. This was attributed to reduced compositional dilution and a uniform high-entropy alloy phase. The system benefited from corrosion-resistant elements and a lower aluminum content.

Siddiqui et al. [109] surface-modified an aluminum substrate with mixed powders of high-purity Cu, Fe, Ni, and Ti, and the $Al_xCu_{0.5}FeNiTi$ HEA coating displayed a mixture of two FCC solid solutions and disordered BCC, with hardness reaching 18 times that of the substrate.

Ye et al. [110] laser-clad an $Al_xFeCoNiCuCr$ HEA coating on the surface of an aluminum alloy. As the Al content increased from 1 to 2, the hardness increased from 390 to 687 $HV_{0.2}$. Li et al. laser-clad a $Ti_xCrFeCoNiCu$ HEA coating on the surface of an aluminum alloy. Wear tests were conducted under ambient conditions, utilizing a GCr15 steel ring with a hardness of 61 HRC as the counter material. This ring measured 50 mm in diameter and 10 mm in thickness. The test samples had dimensions of 7 × 14 × 10 mm. Under a 98 N load and at 40 r/min, the experiments lasted for 30 min. Subsequently, the wear rate was determined from the evaluated wear volume. As the Ti content increased, the phase structure of the $Ti_xCrFeCoNiCu$ coating changed from a single FCC to FCC + B2 phase and FCC + Laves phase, and the hardness increased from 215 $HV_{0.2}$ to 585 $HV_{0.2}$. The addition of Ti significantly improved the wear resistance of the coating. Ma et al. [111] added equal moles of Cu and Ti to FeCoNiCr HEA powder to improve the performance of a 6061 aluminum alloy. The coating had high hardness and wear resistance due to the synergistic effect of solid solution strengthening, dispersion strengthening, and grain refinement. Shi et al. [27,112,113] prepared an AlCrFeNiCuCoCu system HEA coating on a 5083 aluminum alloy. The coating consisted of FCC + BCC and Al-rich phases, with higher hardness and corrosion resistance than the substrate.

Li et al. [114] explored the use of $Al_{0.8}FeCoNiCrCu0.5Si_x$ (x ranges from 0 to 0.5) HEAs as cladding materials. They harnessed the unique high-entropy impact of HEAs to limit the development of hard, brittle intermetallic compounds, enhancing the coating's quality and thereby upgrading the surface attributes of the Al alloy. The wear tests were conducted at room temperature, using a GCr15 steel ring as the counter material, which has a hardness of 61 HRC. The steel ring had a diameter of 50 mm and a thickness of 10 mm, while the wear sample dimensions were 7 × 14 × 10 mm. The tests were performed under a load of 98 N and at a speed of 40 r/min for 30 min. The wear rate of the material was deduced from the calculated wear volume. These specific testing conditions ensure that the wear behavior of the HEA claddings is understood within a defined operational environment.

As the Si content increased, the structure of the $Al_{0.8}CrFeCoNiCu_{0.5}Si_x$ coating shifted from FCC + BCC1 + BCC2 to a solely BCC1 + BCC2 formation. The hardness of the $Al_{0.8}FeCoNiCrCu_{0.5}Si_x$ coating initially rose, then fell as the Si content increased. The most and least hard coatings were those composed of $Al_{0.8}FeCoNiCrCu_{0.5}Si_{0.4}$ (592 $HV_{0.2}$) and $Al_{0.8}FeCoNiCrCu_{0.5}$ (412 $HV_{0.2}$), respectively, representing hardness levels five to seven times that of the substrate. The Si content influenced the wear resistance of the coating similarly to its effect on hardness. With different Si contents, the wear rates ranged from 1.19×10^{-6} mm^3/Nm to 8.99×10^{-7} mm^3/Nm, constituting only 0.25% to 0.34% of the substrate's rate. Figure 16a presents the wear surface of an Al alloy, which is characterized by severe plastic deformation along the wear direction, elongated dimples, and a stepped fracture. Figure 16b shows the Si_0 coating's wear surface, displaying a peeling surface, furrow, and fine abrasive particles. The $Si_{0.2}$ coating's wear surface in Figure 16c shows similar features to the Si_0 coating. The wear surfaces of $Si_{0.3}$ and $Si_{0.4}$ coatings in Figure 16d,e, respectively, exhibit a disappearance of the peeling surface and a predominance of abrasive particles. Among them, the $Si_{0.4}$ coating presents fewer abrasive particles and a shallower furrow, suggesting better wear resistance. Finally, in Figure 16f, the wear surface of the $Si_{0.5}$ coating is featured, showing the distribution of an adhesion layer and furrow interval.

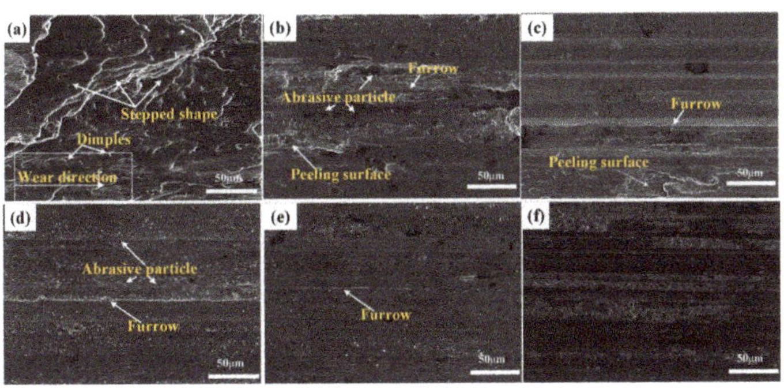

Figure 16. Wear surface morphology of samples; (**a**) is 5083 aluminum alloy; (**b**) is $Al_{0.8}FeCoNiCrCu_{0.5}Si$; (**c**) is $Al_{0.8}FeCoNiCrCu_{0.5}Si_{0.2}$; (**d**) is $Al_{0.8}FeCoNiCrCu_{0.5}Si_{0.3}$; and (**e**) is $Al_{0.8}FeCoNiCrCu_{0.5}Si_{0.4}$; (**f**) is $Al_{0.8}FeCoNiCrCu_{0.5}Si_{0.5}$ [114].

Table 7 presents the key findings for HEA coatings. Due to their high mixing entropy, HEAs have high degrees of disorder at room temperature, which increases with temperature. According to Gibbs's free energy, the larger the entropy, the smaller the system's free energy and the more stable the system. The high mixing entropy of HEAs significantly reduces their free energy, thereby reducing the trend of order during solidification, inhibiting the generation of intergranular compounds, promoting the formation of solid solutions, and resulting in coatings with more stable crystal structures. When applied to the surface of aluminum alloys, these coatings exhibit excellent mechanical properties and corrosion resistance. However, the industrial application of high-entropy alloys with high melting points and high costs still needs further exploration. However, given aluminum's low melting point, current laser cladding HEA coatings on aluminum alloy surfaces predominantly focus on the FeCoNiCrCu system. The area of immiscible HEAs is yet to be thoroughly researched, and the feasibility of this particular system in laser cladding on aluminum alloy surfaces remains to be validated. Further confirmation is also needed regarding incorporating ceramic reinforcing phases into HEAs. Furthermore, from an industrial perspective, implementing high-entropy alloys, which often come with high melting points and costs, still requires more in-depth exploration.

Table 7. Key findings on HEA coatings.

Authors	Coating Material	Key Findings
Shon et al. [108]	AlCrFeCoNi HEA	Reduced defects like coating porosity and cracking through high-energy density multi-layer cladding.
Siddiqui et al. [109]	Mixed powders of Cu, Fe, Ni, Ti	Hardness reached 18 times that of substrate.
Ye et al. [110]	AlxFeCoNiCuCr HEA	As Al content increased, hardness increased from 390 to 687 $HV_{0.2}$.
Ma et al. [111]	Al, Cu, Ti in FeCoNiCr HEA	Synergistic effect led to high hardness and wear resistance.
Li et al. [114]	$Al_{0.8}FeCoNiCrCu0.5Si_x$ HEAs	Increasing Si content shifted structure; highest hardness achieved with $Al_{0.8}FeCoNiCrCu_{0.5}Si_{0.4}$, with wear resistance being five to seven times the substrate's.

5. Discussion and Suggestions

In terms of metal-based coatings, aluminum-based coatings have demonstrated their reparative prowess in laser cladding on aluminum alloy surfaces. For instance, researchers such as Meinert [57] applied the 4000 series of aluminum alloys to the 6061 and 7075 aluminum alloys. Given the ability of laser cladding to target damaged areas with precision and carry out repairs swiftly, this method circumvents the excessive heating and enlarged heat-affected zones that traditional repair methods might induce. The thermal property compatibility between aluminum-based coatings and the base material renders the laser cladding of aluminum alloy surfaces with aluminum-based coatings an efficient and cost-effective strategy. Research by scholars like Corbin [59] indicates that heat treatments can enhance coating hardness. However, this process might also escalate residual stresses within the coating, amplifying the risks of cracks and other potential defects. Since both the coating and the base material are aluminum-centric, enhancements in mechanical properties compared to the original material are relatively limited.

Fe-based coatings have notably augmented the mechanical properties of the substrate. The FeCrMoVC coating studied by Jeyaprakash et al. [63], and the Fe alloy with Wt.%Al explored by Tomida et al. [63], showcased remarkable hardness enhancement. Specifically, in the study by Tomida et al. [63], the alloy achieved a significant boost in hardness values, ranging from 600–1000 HV, by introducing 10–40 wt.% Al content. However, a common issue with Fe-based coatings is the formation of cracks. Research by Ye et al. [64] pinpointed that the cracks appearing at the bond interface between the coating and the base material are attributed to Fe-Al intermetallic compounds. These cracks are associated with the internal stresses within these intermetallic compounds and their mismatch with the substrate. The works of Mei et al. [66] and Carroll et al. [67] further substantiated this viewpoint; the presence of multiple brittle intermetallic compounds in the coating leads to its susceptibility to cracking. A solution to the cracking issue might be controlling the composition of these brittle intermetallic compounds prone to inducing cracks. Moreover, adjusting the coating fabrication process might also be method aiming to minimize the formation of intermetallic compounds or alter their morphology and distribution in the coating, thereby optimizing its performance.

Nickel-based alloy coatings are the most widely studied in the realm of metal-based coatings. The TiB_2-reinforced nickel-based composite researched by He et al. [69], and the NiCrBSi alloy explored by Liang et al. [70], exhibited significant hardness enhancements. The coating developed by He et al. [69] achieved a hardness that was 6.7 times that of the substrate. At the same time, studies by Tan et al. [73], as well as Wang et al. [71], confirmed that, compared to the substrate, nickel-based coatings boast superior wear resistance. Testing wear resistance on aluminum alloy surfaces suggests that nickel-based coatings can be multi-layered on aluminum alloy surfaces, with the bond strength between the coating and substrate exceeding that of Fe-based alloys. However, due to the pronounced physical and chemical differences between the coatings and the substrate, defects such as cracks and porosities can easily form in the coatings. The incorporation of rare-earth elements has been shown to alleviate these issues. However, many rare earth oxides have melting points exceeding the boiling point of aluminum alloys. This implies that during coating preparation, a balance must be struck between ensuring complete powder melting and avoiding excessive dilution or evaporation of the substrate. This necessitates precise control of the laser's thermal input.

Incorporating ceramics into metal bases can notably enhance the substrate's hardness and wear resistance. For instance, in the study by Sun et al. [74], the SiC/Al-12Si composite led to a doubling of the coating layer's hardness, reaching approximately 260 $HV_{0.2}$. Similarly, the Al-Si-based composite containing SiC researched by Yang et al. [77] also showcased the augmented wear resistance of the coating. The extremely high melting points of ceramics pose challenges for their application. In preparing ceramic-inclusive coatings, there is a need to control the thermal input of the laser precisely.

Research on laser cladding for aluminum alloy surfaces is primarily focused on two emerging types of coatings: amorphous alloys and high-entropy alloys. Amorphous alloys, characterized by long-range disorder and short-range order similar to glass, are formed by suppressing nucleation during the solidification of molten materials. Due to their absence of crystallographic defects such as vacancies, dislocations, and stacking faults, these materials theoretically possess superior properties such as high hardness, wear resistance, and corrosion resistance. Studies by Sohrabi et al. [87] and Wang et al. [88] highlight the potential of amorphous coatings when laser clad onto aluminum alloy surfaces, significantly enhancing the substrate's wear resistance. Future research should focus on reducing or eliminating segregation in the coating during the laser cladding process to achieve a uniform and defect-free amorphous layer.

Laser-clad HEA coatings have been proven to enhance aluminum alloy surfaces' hardness, wear resistance, and corrosion resistance. Researchers like Siddiqui [109], Ye [110], and Ma et al. [111], through their respective studies, have revealed the advantages of HEAs. By introducing elements like Al and Si, we can enhance the coating's performance. However, HEA research is still in its early stages and lacks a solid theoretical basis. The high-temperature attributes of laser-clad HEAs are primarily limited to basic properties such as hardness and resistance to oxidation. Comprehensive studies on their high-temperature behaviors, like creep and thermal fatigue, are absent. The rapid cooling in the laser cladding of HEAs occurs under non-equilibrium conditions, and the underlying mechanisms are not well understood. The industrial applications of laser-clad HEAs are limited, with many process parameters being experimentally optimized. Future research can aim to develop a design model for laser-clad HEAs, explore the underlying scientific theories of component design, and expand on current alloy theories to understand the component–structure–performance interplay better. Additionally, in-depth studies into these HEAs' solidification dynamics and complex structures under non-equilibrium conditions are vital.

Consolidating the research findings above, Table 8 provides an integrated overview of various coating types, their recommended application environments, expected performance, and pertinent precautions.

Table 8. Comparative analysis of coating types, their application recommendations, expected outcomes, and implementation precautions.

Coating Type	Recommended Application Environment/Conditions	Expected Performance	Precautions
Aluminum-based coating	Component repair sectors; areas with low surface performance demands	Hardness and corrosion resistance are similar to the base material	Heat treatment can enhance mechanical properties and ensure process control to prevent coating from cracking.
Iron-based coating	Low-cost coating needs; high surface hardness domains	High surface hardness	Limited improvement in corrosion resistance; the coating has high crack sensitivity.
Nickel-based coating	Cost-insensitive areas; high hardness and wear resistance domains	High surface hardness, wear resistance, and moderate corrosion resistance	Add rare earth elements to reduce cracks; monitor laser heat closely.
Ceramic-based Coating	High hardness and wear resistance domains	High surface hardness and wear resistance	Strong crack sensitivity; complex reaction products. Process control is essential.
Amorphous alloy Coating	Wear-resistant domains; cost-insensitive areas	Decent hardness and wear resistance	Emerging coating with limited research; the practical application needs verification.
High-Entropy alloy Coating	Cost-insensitive areas; high hardness, wear, and corrosion resistance domains	High surface hardness, wear and corrosion resistance	Coating is expensive. Utilize the "cocktail effect" of high-entropy alloys; regulate Al, Si, Ti, and Cr components and composition for performance tuning.

6. Conclusions

The failure and scrapping of mechanical equipment under harsh conditions cause substantial economic losses to industrial production. Laser cladding technology provides a rapid and environmentally friendly approach to the surface strengthening and remanufacturing of workpieces. It effectively improves the surface properties of aluminum alloys or serves as a surface repair technology for these alloys. At present, the cladding materials for aluminum alloy laser cladding research cover metal-based alloys, ceramic-reinforced composite materials, and some emerging materials, focusing on enhancing the surface hardness, wear resistance, or corrosion resistance of the aluminum alloy based on the different characteristics of the materials.

Although significant achievements and developments have been made in research on laser cladding coatings for aluminum alloy surfaces domestically and abroad in recent years, many key issues are still worthy of further in-depth study and improvement. Due to the inherent difficulties in aluminum alloy laser cladding, such as high laser reflectivity, the ease of forming a high-melting-point oxide film on the aluminum alloy surface during cladding, the high dilution rate of the cladding layer, deviations in the coating composition from the designed nominal composition, and the occurrence of cracks, pores, and other defects, the alloy elements undergo oxidation and burning, leaving aluminum alloy laser cladding primarily in the experimental stage, unable to be applied on a large scale in actual industrial production.

To address the problems in aluminum alloy laser cladding, current solutions mainly focus on the following areas:

(1) Composition design and control of the cladding layer material. Based on traditional composition design, and considering the process characteristics of laser cladding technology, cladding powder systems are designed for aluminum alloy laser cladding conditions. This includes developing emerging cladding layers, adding absorbers to enhance laser

absorption capacity, adding rare earth elements to improve the flowability of the molten pool, the in situ generation of ceramic materials, and the research and development of gradient coatings.

(2) Regulation of laser process parameters and the research and development of other auxiliary technologies. Reasonable control of laser power, scanning speed, powder feed rate, and other laser process parameters is required to establish an appropriate aluminum alloy laser cladding process parameter window. Simultaneously, the introduction of auxiliary technologies such as electric fields, magnetic fields, ultrasonic vibrations, and the combination of laser cladding with laser shock, micro-forging, etc., can effectively improve the distribution of residual stress, suppress the problems of cracks and pores in the cladding layer, and enhance the fatigue resistance and other comprehensive properties of the samples.

(3) Pre-treatment and heat treatment of the aluminum alloy surface. The use of physical or chemical methods to remove the oxide film on the aluminum alloy surface before cladding, preheating of the substrate, and the possible remelting or annealing of the coating after cladding, or other methods, can reduce the temperature difference between the cladding layer material and the aluminum alloy substrate during the laser cladding process, thereby reducing thermal stress and achieving the aim of lowering the amount of cracking of the cladding layer.

To address the current challenges in this area, future endeavors could be channeled into the following areas:

(1) The design of novel cladding material composition and gradient structures: Current efforts have explored the potential of amorphous alloys and HEAs as coatings. However, due to the large linear expansion coefficient, high thermal conductivity, and relatively low elastic modulus of aluminum alloys, there is significant difficulty in matching the thermophysical properties of the surface reinforcement layer material to the alloy. To circumvent the pronounced thermophysical disparities between the coating and the substrate, the design of cladding layer materials can be approached by achieving a gradient change in the powder layer composition and in situ generation on the surface. Incorporating emerging materials, like gradient amorphous coatings and gradient HEA coatings, may pave the way for coatings that combine high toughness at the interface with the substrate and hardness at the surface, fulfilling the promise of innovative coatings.

(2) The optimization and control of the interface between reinforcement particles and the matrix: The rapid cooling and solidification process inherent to laser cladding results in structural and performance variations at the interface between the particles and the matrix, as compared to those obtained through comprehensive reaction processes such as sintering and casting. Given that the interfacial performance significantly influences the composite coating's overall attributes, harnessing techniques like process optimization and material compatibility assessments is imperative to achieve superior interface properties.

(3) The development of advanced laser cladding systems: This entails the creation of high-powered, short-wavelength lasers to augment the energy density and cladding efficiency of aluminum alloy laser cladding. Additionally, there is a need to set up a comprehensive database dedicated to aluminum alloy laser cladding. Such a database would aim to establish benchmarks and repositories for accurately evaluating cladding materials, processing parameters, and structural attributes, aligning with the emerging trends of digitalization and informatization.

Author Contributions: Conceptualization, P.Z.; methodology, Z.S.; validation, Y.L.; formal analysis, Y.L.; investigation, Y.L.; resources, Z.S.; data curation, X.W.; writing—original draft preparation, Y.L.; writing—review and editing, Y.L.; visualization, Y.L. and M.Z.; supervision, Z.C.; project administration, X.W.; and funding acquisition, M.Z. and J.L. All authors have read and agreed to the published version of the manuscript.

Funding: This study was supported by the Natural Science Research Project of Anhui Provincial Department of Education (Grant Nos. KJ2021A0947), Anhui Provincial Major Science and Technology Project (202003c08020002), the University-level natural science research project of West Anhui University (Grant Nos. WXZR202116, WGKQ2021068), the High-level Talents Research Project of West Anhui University (Grant No. WGKQ2021068, WGKQ 201802004). The Key Science and Technology Program of Anhui Province of China (2021001), the Key generic technology research and development project of Hefei City (2021GJ053), the Natural Science Foundation of Anhui Province (2108085ME142), the National Natural Science Foundation of China [52001304], the Science and Technology Program of Inner Mongolia Autonomous Region (2022YFDZ0004).

Data Availability Statement: The data that support the findings in this study are available from the author Y.L. upon reasonable request.

Conflicts of Interest: The authors declare no conflict of interest.

References

1. Dixit, A.; Kumar, K. Optimization of Mechanical Properties of Silica Gel Reinforced Aluminium MMC by using Taguchi Method. *Mater. Today-Proc.* **2015**, *2*, 2359–2366. [CrossRef]
2. Doshi, S.J.; Gohil, A.V.; Mehta, N.; Vaghasiya, S. Challenges in Fusion Welding of Al alloy for Body in White. *Mater. Today-Proc.* **2018**, *5*, 6370–6375. [CrossRef]
3. Feng, M.N.; Xie, Y.; Zhao, C.F.; Luo, Z. Microstructure and mechanical performance of ultrasonic spot welded open-cell Cu foam/Al joint. *J. Manuf. Process.* **2018**, *33*, 86–95. [CrossRef]
4. Sun, X.; Jie, J.; Peng, B.; Dong, G.; Liu, J.; Liu, S.; Li, T. Numerical Study and Experimental Verification on Solidification Characteristics in Commercial Purity Aluminum under Mechanical Vibration. *J. Mater. Eng. Perform.* **2023**, 1–8. [CrossRef]
5. Wang, B.; An, X.; Xue, P.; Liu, F.; Ni, D.; Xiao, B.; Liu, Y.; Ma, Z. Grain size effects on high cycle fatigue behaviors of pure aluminum. *Int. J. Fatigue* **2023**, *170*, 107556. [CrossRef]
6. Yamagishi, H. Cu/Al Dissimilar Cold Spot Forge Welding: Effects of Bonding Temperature and Reduction Ratio on Joint Strength and Reaction Layer Growth. *Met. Mater. Trans. A Phys. Metall. Mater. Sci.* **2023**, *54*, 3519–3536. [CrossRef]
7. Li, J.; Wang, G.; Zhang, M.; Li, J.; Fang, X.; Ma, X. Strengthening mechanisms of a heterostructured pure aluminum with extraordinary mechanical properties. *Mater. Charact.* **2023**, *202*, 113049. [CrossRef]
8. Singh, B.; Kumar, I.; Saxena, K.K.; Mohammed, K.A.; Khan, M.I.; Ben Moussa, S.; Abdullaev, S.S. A future prospects and current scenario of aluminium metal matrix composites characteristics. *Alex. Eng. J.* **2023**, *76*, 1–17. [CrossRef]
9. Liu, M.; Xu, Z.; Yang, F. Characterization of cyclic dynamic and creep responses of pure aluminum by instrumented indentation. *Sci. Sin. Phys. Mech. Astron.* **2023**, *53*, 214605. [CrossRef]
10. Scharifi, E.; Yardley, V.A.; Weidig, U.; Szegda, D.; Lin, J.; Steinhoff, K. Hot Sheet Metal Forming Strategies for High-Strength Aluminum Alloys: A Review—Fundamentals and Applications. *Adv. Eng. Mater.* **2023**, *25*, 2300141. [CrossRef]
11. Chinh, N.Q.; Olasz, D.; Ahmed, A.Q.; Bobruk, E.V.; Valiev, R.Z. Review on Grain Size- and Grain Boundary Phenomenon in Unusual Mechanical Behavior of Ultra fine-Grained Al Alloys. *Mater. Trans.* **2023**, *64*, 1844–1855. [CrossRef]
12. Kumar, P.K.D.; Gnanaraj, S.D. Aluminium-Silicon based Metal Matrix Composites for brake rotor applications: A review. *Eng. Res. Express* **2023**, *5*, 022002. [CrossRef]
13. Eskandarzade, M.; Masalehdan, T.; Tutunchi, A.; Osouli-Bostanabad, K.; Hildyard, R.; Bewsher, S.R.; Mohammadpour, M. Study of Tribological Properties of Bulk Nanostructured Aluminum and Copper Samples Applicable in Automotive Bearing Application. *J. Mater. Eng. Perform.* **2023**, *32*, 8807–8817. [CrossRef]
14. Yuan, L.; Zeng, X.; Zhao, X.; Xie, Y.; Gandra, J.; Guan, D. Microstructure evolution and tensile behaviour of fine-grained 6082 Al wire with high ultimate strength and high work hardening by friction stir extrusion of bulk Al sheet. *Mater. Sci. Eng. A* **2023**, *864*, 144589. [CrossRef]
15. Ashebir, D.A.; Mengesha, G.A.; Sinha, D.K.; Bereda, Y.B. Multi-response optimization of process and reinforcement parameters of hybrid reinforced Al matrix composites using Taguchi- Grey relational analysis. *Eng. Res. Express* **2022**, *4*, 045038. [CrossRef]
16. Shahri, S.E.E.; Gholami, M.; Rakhshkhorshid, M. Numerical-experimental study of die geometry in the constrained groove pressing of 6061 aluminum sheets. *Proc. Inst. Mech. Eng. Part B J. Eng. Manuf.* **2022**, *237*, 1836–1846. [CrossRef]
17. Pungaiah, S.S.; Sahu, K.K.; Yuvaraja, R.; Prabhu, G.J.; Sudharsan, R.R.; Thamizhvalavan, P.; Ramaswamy, K. Investigation on Mechanical Behaviour of LM6 Aluminum Alloy Hybrid Composites Processed Using Stir Casting Process. *Adv. Mater. Sci. Eng.* **2022**, *2022*, 7539546.
18. Mahmoud, E.R.I.; Khan, S.Z.; Aljabri, A.; Almohamadi, H.; Elkotb, M.A.; Gepreel, M.A.; Ebied, S. Free Intermetallic Cladding Interface between Aluminum and Steel through Friction Stir Processing. *Crystals* **2022**, *12*, 1413. [CrossRef]
19. Niu, G.; Wang, J.; Ye, J.; Mao, J. Enhancing Fe content tolerance in A356 alloys for achieving low carbon footprint aluminum structure castings. *J. Mater. Sci. Technol.* **2023**, *161*, 180–191. [CrossRef]
20. Wang, J.; Li, F. Effect of Sm + Er and Heat Treatment on As-Cast Microstructure and Mechanical Properties of 7055 Aluminum Alloy. *Materials* **2023**, *16*, 4846. [CrossRef]

21. Thanabumrungkul, S.; Jumpol, W.; Meemongkol, N.; Wannasin, J. Investment casting of semi-solid 6063 aluminum alloy using the GISS process. *Mater. Res. Express* **2023**, *10*, 076501. [CrossRef]
22. Wang, L.X.; Yin, K.; Deng, Q.W.; Huang, Q.Q.; He, J.; Duan, J.A. Wetting Ridge-Guided Directional Water Self-Transport. *Adv. Sci.* **2022**, *9*, 2204891. [CrossRef]
23. Yang, D.; Tang, S.; Wu, Z.; Qin, K.; Zhang, H.; Cui, J.; Nagaumi, H. Research Status and Development Trend of Alloying and Process Technology of High-zinc Aluminum Alloy. *Mater. Rev.* **2023**, *37*, 134–139.
24. Li, Q.; Sun, J.; Ren, X.; Hu, M. Effects of High Temperature RRA on Microstructure and Properties of 7075 Aluminum Alloy. *Hot Work. Technol.* **2018**, *47*, 129–132, 137.
25. Mironov, A.E.; Gershman, I.S.; Gershman, E.I.; Zheleznov, M.M. Relationship between the Tribological Properties of Experimental Aluminum Alloys and Their Chemical Composition. *J. Frict. Wear* **2017**, *38*, 87–91. [CrossRef]
26. Li, Y.Z.; Shi, Y.; Olugbade, E. Microstructure, mechanical, and corrosion resistance properties of $Al_{0.8}CrFeCoNiCu_x$ high-entropy alloy coatings on aluminum by laser cladding. *Mater. Res. Express* **2020**, *7*, 026504. [CrossRef]
27. Li, Y.Z.; Shi, Y. Microstructure and wear resistance of the laser-cladded $Al_{0.8}CrFeCoNiCu_{0.5}B_x$ high-entropy alloy coating on aluminum. *Mater. Res. Express* **2020**, *7*, 026517. [CrossRef]
28. Li, Y.; Shi, Y. Microhardness, wear resistance, and corrosion resistance of $Al_xCrFeCoNiCu$ high-entropy alloy coatings on aluminum by laser cladding. *Opt. Laser Technol.* **2021**, *134*, 106632. [CrossRef]
29. Zhang, Y.; Gao, M.; Chen, T.; Pan, L.; Wu, M. Study on the Corrosion Resistance of Thermal Spraying Zinc Aluminum Coating with High Aluminum Content. *Mater. Prot.* **2021**, *54*, 28–33.
30. Riah, B.; Ayad, A.; Camus, J.; Rammal, M.; Boukari, F.; Chekour, L.; Djouadi, M.; Rouag, N. Textured hexagonal and cubic phases of AlN films deposited on Si (100) by DC magnetron sputtering and high power impulse magnetron sputtering. *Thin Solid Films* **2018**, *655*, 34–40. [CrossRef]
31. Zhang, J.; Song, B.; Wei, Q.; Bourell, D.; Shi, Y. A review of selective laser melting of aluminum alloys: Processing, microstructure, property and developing trends. *J. Mater. Sci. Technol.* **2019**, *35*, 270–284. [CrossRef]
32. Wang, C.; Wu, G.; Zhang, Q.; Zhang, Y.; Xiu, Z.; Chen, G. A Review for Corrosion Resistance and Protection Methods for Aluminum Matrix Composites. *J. Chin. Soc. Corros. Prot.* **2008**, *28*, 59–64.
33. Chen, B.; Li, Y.; Nie, J.; Wei, C.; Dang, L.; Ma, Q. Microstructure and mechanical properties of NiCoCrAlY laser cladding coating after high-current pulsed electron beam irradiation. *Philos. Mag. Lett.* **2023**, *103*, 2191223. [CrossRef]
34. Jing, Z.; Xu, P.; Liu, Q.; Yu, C. Residual stress release during laser cladding process: A review. *J. Laser Appl.* **2023**, *35*, 031201. [CrossRef]
35. Jiang, L.; Cui, X.; Jin, G.; Tian, Z.; Wen, X.; Tian, H.; Liu, E. Design and characterization of a novel $Cu_{2.3}Al_{1.3}Ni_{1.7}SnCr_{0.3}$ multi-principal element alloy coating on magnesium alloy by laser cladding. *J. Mater. Sci. Technol.* **2023**, *152*, 220–236. [CrossRef]
36. Arif, Z.U.; Khalid, M.Y.; Rehman, E.U.; Ullah, S.; Atif, M.; Tariq, A. A review on laser cladding of high-entropy alloys, their recent trends and potential applications. *J. Manuf. Process.* **2021**, *68*, 225–273. [CrossRef]
37. Liu, Y.; Ding, Y.; Yang, L.; Sun, R.; Zhang, T.; Yang, X. Research and progress of laser cladding on engineering alloys: A review. *J. Manuf. Process.* **2021**, *66*, 341–363. [CrossRef]
38. Siddiqui, A.A.; Dubey, A.K. Recent trends in laser cladding and surface alloying. *Opt. Laser Technol.* **2021**, *134*, 106619. [CrossRef]
39. Liu, M.; Duan, C.; Li, G.; Cai, Y.; Li, L.; Wang, F. Multi-indicator evaluation and material selection of hybrid additive-subtractive manufacturing to repair automobile panel dies and molds. *Int. J. Adv. Manuf. Technol.* **2023**, *127*, 1675–1690. [CrossRef]
40. Izumi, T.; Arai, M. Numerical simulation of the 3D propeller repair process by laser cladding of SUS316L on SUS304. *J. Manuf. Process.* **2023**, *98*, 234–253. [CrossRef]
41. Gao, Y.; Lu, P.; Bai, S.; Qin, B.; Zhang, D. Influence of Laser Power on Microstructure and Properties of Al-Si+Y_2O_3 Coating. *Coatings* **2023**, *13*, 1289. [CrossRef]
42. Huang, Q.; Yin, K.; Wang, L.; Deng, Q.; Arnusch, C.J. Femtosecond laser-scribed superhydrophilic/superhydrophobic self-splitting patterns for one droplet multi-detection. *Nanoscale* **2023**, *15*, 11247–11254. [CrossRef] [PubMed]
43. Zuo, T.C.; Chen, K. Laser manufacturing technology and its application on the surface engineering. In *Contributions of Surface Engineering to Modern Manufacturing and Remanufacturing*; Southwest Jiaotong University Press: Chengdu, China, 2002.
44. Ozdemir, M.; Sadikoglu, H. A new and emerging technology: Laser-induced surface modification of polymers. *Trends Food Sci. Technol.* **1998**, *9*, 159–167. [CrossRef]
45. Folkes, J. Surface modification and coating with lasers. In *Surface Coatings for Advanced Materials*; Agarwala, R.P., Ed.; Universities in Chicago: Chicago, America, 1997; Volume 246, pp. 261–277.
46. Chen, J.F.; Li, X.C.; Wang, J.T.; Lei, W.N. Effect of CO_2 Laser Surface Melting on the Microstructure of AZ91D Magnesium Alloy. *Key Eng. Mater.* **2011**, *464*, 461–464. [CrossRef]
47. Han, S.W.; Joo, B.D.; Moon, Y.H. Selective surface hardening by laser melting of alloying powder. *Mater. Res. Innov.* **2014**, *18*, S2-902. [CrossRef]
48. Lee, J.H.; Jang, J.H.; Joo, B.D.; Son, Y.M.; Moon, Y.H. Laser surface hardening of AISI H13 tool steel. *Trans. Nonferrous Met. Soc. China* **2009**, *19*, 917–920. [CrossRef]
49. Huang, X.B. Research and application of laser shock processing. In Proceedings of the ICOSM 2020: Optoelectronic Science and Materials, Hefei, China, 25–27 September 2020; Volume 11606, pp. 119–123.

50. Zhou, M.; Zhang, Y.; Cai, L. Laser shock forming on coated metal sheets characterized by ultrahigh-strain-rate plastic deformation. *J. Appl. Phys.* **2002**, *91*, 5501–5503. [CrossRef]
51. Lai, Z.L.; Wang, C.; He, W.F.; Zhou, L.C.; Bin An, Z. Application of High-Energy Laser in Strain-Hardening of Material. *Appl. Mech. Mater.* **2013**, *321–324*, 141–145. [CrossRef]
52. Zhou, Y.; Zhang, K.; Liang, Y.; Cheng, J.; Dai, Y. Selective Laser Melted Magnesium Alloys: Fabrication, Microstructure and Property. *Materials* **2022**, *15*, 7049. [CrossRef]
53. Knyzeva, A.G.; Sharkeev, Y.P. Two-Dimensional Model of Laser Alloying of Binary Alloy Powder with Interval of Melting Temperature. In Proceedings of the XXV Conference on High-Energy Processes in Condensed Matter (HEPCM 2017), Novosibirsk, Russia, 5–9 June 2017.
54. John, M.; Kuruveri, U.B.; Menezes, P.L. Laser Cladding-Based Surface Modification of Carbon Steel and High-Alloy Steel for Extreme Condition Applications. *Coatings* **2022**, *12*, 1444. [CrossRef]
55. Cheng, J.; Xing, Y.; Dong, E.; Zhao, L.; Liu, H.; Chang, T.; Chen, M.; Wang, J.; Lu, J.; Wan, J. An Overview of Laser Metal Deposition for Cladding: Defect Formation Mechanisms, Defect Suppression Methods and Performance Improvements of Laser-Cladded Layers. *Materials* **2022**, *15*, 5522. [CrossRef] [PubMed]
56. Weng, F.; Chen, C.; Yu, H. Research status of laser cladding on titanium and its alloys: A review. *Mater. Des.* **2014**, *58*, 412–425. [CrossRef]
57. Meinert, K.C.; Bergan, P. Refurbishment of aluminum alloys by laser cladding. In Proceedings of the ICALEO'99: Proceedings of the Laser Materials Processing Conference, San Diego, CA USA, 15–18 November 1999.
58. Cottam, R.; Luzin, V.; Liu, Q.; Wong, Y.C.; Wang, J.; Brandt, M. Investigation into Heat Treatment and Residual Stress in Laser Clad AA7075 Powder on AA7075 Substrate. *Met. Microstruct. Anal.* **2013**, *2*, 205–212. [CrossRef]
59. Grohol, C.M.; Shin, Y.C.; Frank, A. Laser cladding of aluminum alloy 6061 via off-axis powder injection. *Surf. Coat. Technol.* **2021**, *415*, 127099. [CrossRef]
60. Yang, Z.; Wang, A.; Weng, Z.; Xiong, D.; Ye, B.; Qi, X. Porosity elimination and heat treatment of diode laser-clad homogeneous coating on cast aluminum-copper alloy. *Surf. Coatings Technol.* **2017**, *321*, 26–35. [CrossRef]
61. Song, M.; Wu, L.; Liu, J.; Hu, Y. Effects of laser cladding on crack resistance improvement for aluminum alloy used in aircraft skin. *Opt. Laser Technol.* **2021**, *133*, 106531. [CrossRef]
62. Dubourg, L.; Pelletier, H.; Vaissiere, D.; Hlawka, F.; Cornet, A. Mechanical characterisation of laser surface alloyed aluminium–copper systems. *Wear* **2002**, *253*, 1077–1085. [CrossRef]
63. Jeyaprakash, N.; Yang, C.-H.; Sivasankaran, S. Formation of FeCrMoVC Layers on AA6061 by Laser Cladding Process: Microstructure and Wear Characteristics. *Trans. Indian Inst. Met.* **2020**, *73*, 1611–1617. [CrossRef]
64. Sathiyamoorthi, P.; Kim, H.S. High-entropy alloys with heterogeneous microstructure: Processing and mechanical properties. *Prog. Mater. Sci.* **2022**, *123*, 100709. [CrossRef]
65. Tomida, S.; Nakata, K. Fe–Al composite layers on aluminum alloy formed by laser surface alloying with iron powder. *Surf. Coatings Technol.* **2003**, *174–175*, 559–563. [CrossRef]
66. Mei, Z.; Wang, W.; Wang, A. Transmission electron microscopy characterization of laser-clad iron–based alloy on Al–Si alloy. *Mater. Charact.* **2006**, *56*, 185–191. [CrossRef]
67. Carroll, J.W.; Liu, Y.; Mazumder, J.; Perry, T.A. Laser surface alloying of aluminum 319 alloy with iron. In Proceedings of the 20th International Congress on Applications of Lasers and Electro-Optics (ICALEO 2001), Jacksonville, FL, USA, 14–18 October 2001.
68. Wu, X.Q.; Yan, H.; Xin, Y.; Yu, B.B.; Hu, Z.; Sun, Y.H. Microstructure and Wear Properties of Ni-based Composite Coatings on Aluminum Alloy Prepared by Laser Cladding. *Rare Met. Mater. Eng.* **2020**, *49*, 2574–2582.
69. He, L.; Tan, Y.F.; Wang, X.L.; Jing, Q.F.; Hong, X. Tribological properties of laser cladding TiB_2 particles reinforced Ni-base alloy composite coatings on aluminum alloy. *Rare Met.* **2015**, *34*, 789–796. [CrossRef]
70. Liang, G.; Su, J. The microstructure and tribological characteristics of laser-clad Ni–Cr–Al coatings on aluminium alloy. *Mater. Sci. Eng. A Struct. Mater. Prop. Microstruct. Process.* **2000**, *290*, 207–212. [CrossRef]
71. Wang, C.; Gao, Y.; Zeng, Z.; Fu, Y. Effect of rare-earth on friction and wear properties of laser cladding Ni-based coatings on 6063Al. *J. Alloys Compd.* **2017**, *727*, 278–285. [CrossRef]
72. Zhang, P.F.; Li, Y.X. Effects of CeO_2, Si and Co on the microstructure and properties of Ni-Cu composite coating on 6061 aluminium alloy by laser cladding. *Philos. Mag. Lett.* **2021**, *101*, 303–311. [CrossRef]
73. Tan, Y.F.; He, L.; Wang, X.L.; Tan, H.; Zhou, C.H. Tribological properties of laser surface cladding Ni-base alloy coatings under dry friction and seawater conditions. *Int. J. Surf. Sci. Eng.* **2016**, *10*, 147–161. [CrossRef]
74. Sun, R.; Lei, Y. Microstructure and hardness of laser clad SiCp–Al composite coatings on Al alloys. *Mater. Lett.* **2008**, *62*, 3272–3275. [CrossRef]
75. Anandkumar, R.; Almeida, A.; Colaco, R.; Vilar, R.; Ocelik, V.; De Hosson, J.T.M. Microstructure and wear studies of laser clad Al-Si/SiC(p) composite coatings. *Surf. Coat. Technol.* **2007**, *201*, 9497–9505. [CrossRef]
76. Anandkumar, R.; Almeida, A.; Vilar, R. Microstructure and sliding wear resistance of an Al-12 wt.% Si/TiC laser clad coating. *Wear* **2012**, *282*, 31–39. [CrossRef]
77. Liu, Y.; Li, G.D.; Jiang, W.T. Effects of Cu-Coated SiC Content on Microstructure and Properties of Laser Cladding SiCp/Al-Si Composite Coatings. *Materials* **2019**, *12*, 1537. [CrossRef]

78. Li, Y.; Zhang, P.; Bai, P.; Wu, L.; Liu, B.; Zhao, Z. Microstructure and properties of Ti/TiBCN coating on 7075 aluminum alloy by laser cladding. *Surf. Coatings Technol.* **2018**, *334*, 142–149. [CrossRef]
79. Riquelme, A.; Escalera-Rodríguez, M.D.; Rodrigo, P.; Otero, E.; Rams, J. Effect of alloy elements added on microstructure and hardening of Al/SiC laser clad coatings. *J. Alloys Compd.* **2017**, *727*, 671–682. [CrossRef]
80. Al-Hamdani, K.S.; Murray, J.W.; Hussain, T.; Clare, A.T. Controlling ceramic-reinforcement distribution in laser cladding of MMCs. *Surf. Coatings Technol.* **2019**, *381*, 125128. [CrossRef]
81. Ayerdi, J.J.; Aginagalde, A.; Llavori, I.; Bonse, J.; Zabala, A.J.W. Ball-on-flat linear reciprocating tests: Critical assessment of wear volume determination methods and suggested improvements for ASTM D7755 standard. *Wear* **2021**, *470–471*, 203620. [CrossRef]
82. Yu, J.; Qiao, C.; Zhang, S.; Liu, Z.; Wróblewski, P. Tribological properties of laser-cladded Fe-based amorphous composite coatings under dry and lubricated sliding. *Opt. Laser Technol.* **2023**, *166*, 109583. [CrossRef]
83. Zhang, Z.; Yuan, J.; Jing, Z.; Cheng, Y.; Liang, X. Research Progress and Development Trend of Amorphous and Nanocrystalline Composite Coatings: A Review. *JOM* **2022**, *74*, 4597–4611. [CrossRef]
84. Zhang, Y.H.; Xu, J.F.; Hu, Y.Q.; Ding, S.H.; Xia, R. Cold welding of ultrathin metallic glass nanowires with side-to-side contact using molecular dynamics simulations. *Mater. Today Commun.* **2022**, *32*, 103937. [CrossRef]
85. Song, S.; Laurell, F.; Meehan, B.; Hawkins, T.W.; Ballato, J.; Gibson, U.J. Localised structuring of metal-semiconductor cores in silica clad fibres using laser-driven thermal gradients. *Nat. Commun.* **2022**, *13*, 2680. [CrossRef]
86. Ibrahim, M.Z.; Halilu, A.; Sarhan, A.A.; Kuo, T.; Yusuf, F.; Shaikh, M.; Hamdi, M. In-vitro viability of laser cladded Fe-based metallic glass as a promising bioactive material for improved osseointegration of orthopedic implants. *Med. Eng. Phys.* **2022**, *102*, 103782. [CrossRef]
87. Sohrabi, N.; Panikar, R.S.; Jhabvala, J.; Buch, A.R.; Mischler, S.; Logé, R.E. Laser coating of a Zr-based metallic glass on an aluminum substrate. *Surf. Coatings Technol.* **2020**, *400*, 126223. [CrossRef]
88. Wang, Y.; Ostrowska, M.; Cacciamani, G. Thermodynamic modeling of selected ternary systems containing Y and CALPHAD simulation of CoNiCrAlY metallic coatings. *Calphad* **2021**, *72*, 102214. [CrossRef]
89. Yan, J.; Zhang, P.; Liu, J.; Yu, H.; Hu, Q.; Yang, J.; Zhang, Z. Design and optimization of the composition and mechanical properties for non-equiatomic CoCrNi medium-entropy alloys. *J. Mater. Sci. Technol.* **2023**, *139*, 232–244. [CrossRef]
90. Huang, A.; Fensin, S.J.; Meyers, M.A. Strain-rate effects and dynamic behavior of high entropy alloys. *J. Mater. Res. Technol.* **2023**, *22*, 307–347. [CrossRef]
91. Laleh, M.; Sadeghi, E.; Revilla, R.I.; Chao, Q.; Haghdadi, N.; Hughes, A.E.; Xu, W.; De Graeve, I.; Qian, M.; Gibson, I.; et al. Heat treatment for metal additive manufacturing. *Prog. Mater. Sci.* **2023**, *133*, 101051. [CrossRef]
92. Meena, N.; Rao, A.G.; Dommeti, S.G.; Prabhu, N. Effect of Multi-Pass Friction Stir Processing on Microstructure and Mechanical Properties of a Metastable Dual-Phase High Entropy Alloy. *Lubricants* **2023**, *11*, 2. [CrossRef]
93. Li, Y.; Shi, Y.; Wang, H.; Zhou, B.; Li, D.; Lin, H.; Wang, J. Effect of Y on Microstructure and Properties of $Al_{0.8}FeCrCoNiCu_{0.5}$ High Entropy Alloy Coating on 5083 Aluminum by Laser Cladding. *Lubricants* **2023**, *11*, 50. [CrossRef]
94. Reverte, E.; Keller, C.; Calvo-Dahlborg, M.; Alcalá, G.; Campos, M.; Cornide, J. Effect of Y_2O_3 addition on the microstructure and mechanical properties of an $Al_{1.8}CoCrCu_{0.5}FeNi$ BCC HEA. *J. Alloys Compd.* **2023**, *960*, 170647. [CrossRef]
95. Wen, J.; Liu, Y.; Huang, Y.; Zhao, Y. Effects of electromagnetic stirring and subsequent homogenization treatment on the microstructure and mechanical properties of $Al_{70}Zn_{10}Mg_{10}Cu_5Si_5$ multi-component alloy. *J. Alloys Compd.* **2023**, *960*, 170725. [CrossRef]
96. Yang, B.; Fang, J.; Xu, C.; Cao, H.; Zhang, R.; Zhao, B.; Huang, M.; Wang, X.; Lv, H.; Che, R. One-Dimensional Magnetic FeCoNi Alloy Toward Low-Frequency Electromagnetic Wave Absorption. *Nano-Micro Lett.* **2022**, *14*, 170. [CrossRef]
97. Xiong, W.; Guo, A.X.; Zhan, S.; Liu, C.-T.; Cao, S.C. Refractory high-entropy alloys: A focused review of preparation methods and properties. *J. Mater. Sci. Technol.* **2023**, *142*, 196–215. [CrossRef]
98. Li, H.; Han, Y.; Feng, H.; Zhou, G.; Jiang, Z.; Cai, M.; Li, Y.; Huang, M. Enhanced strength-ductility synergy via high dislocation density-induced strain hardening in nitrogen interstitial CrMnFeCoNi high-entropy alloy. *J. Mater. Sci. Technol.* **2023**, *141*, 184–192. [CrossRef]
99. Wang, X.; Zhai, W.; Wang, J.; Wei, B. Strength and ductility enhancement of high-entropy $FeCoNi_2Al_{0.9}$ alloy by ultrasonically refining eutectic structures. *Scr. Mater.* **2023**, *225*, 115154. [CrossRef]
100. Alamiery, A.A.; Isahak, W.N.R.W.; Takriff, M.S. Inhibition of Mild Steel Corrosion by 4-benzyl-1-(4-oxo-4-phenylbutanoyl)thiosemi carbazide: Gravimetrical, Adsorption and Theoretical Studies. *Lubricants* **2021**, *9*, 93. [CrossRef]
101. Zhang, D.; Du, D.; Liu, G.; Pu, Z.; Xue, S.; Chang, B. Microstructure and Wear Resistance of FeCuNiTiAl High-Entropy Alloy Coating on Ti6Al4V Substrate Fabricated by Laser Metal Deposition. *Lubricants* **2022**, *10*, 263. [CrossRef]
102. Zhang, X.; Tong, Y.; Hu, Y.; Liang, X.; Chen, Y.; Wang, K.; Zhang, M.; Xu, J. Microstructure and Performance of $Fe_{50}Mn_{30}Cr_{10}Ni_{10}$ High-Entropy Alloy Produced by High-Efficiency and Low-Cost Wire Arc Additive Manufacturing. *Lubricants* **2022**, *10*, 344. [CrossRef]
103. Cantor, B. Multicomponent high-entropy Cantor alloys. *Prog. Mater. Sci.* **2021**, *120*, 100754. [CrossRef]
104. Jin, C.; Li, X.L.; Kang, J.H.; Li, H.Z.; Wang, H.F. Effect of interstitial oxygen/nitrogen on mechanical and wear properties of TiZrHfNb refractory high-entropy alloy. *J. Alloys Compd.* **2023**, *960*, 170863. [CrossRef]
105. Liu, Q.; Fu, H.; Wu, W.; Wang, L.; Cheng, J.; Xia, P.; Gan, B.; Xia, M.; Zhao, F. Influence of annealing temperatures on the microstructure and deformation behavior of a CrCoNi based medium-entropy alloy. *J. Alloys Compd.* **2023**, *960*, 170877. [CrossRef]

106. Xu, T.; Chen, Q.; Ji, L.; Zheng, Z.; Wang, K.; Liu, H. BCC/B2 structure and dislocation strengthening behavior in high Ti content TiAlVCrNb high-entropy alloys. *J. Alloys Compd.* **2023**, *956*, 170179. [CrossRef]
107. Wang, Y.; Chen, Y.; Xie, J.; Ni, J.; Zhang, T.; Wang, S.; Yin, L. Microstructure and mechanical properties of CrMnFeCoNi high entropy alloy/Al composite with different reinforcement content. *J. Alloys Compd.* **2023**, *960*, 170882. [CrossRef]
108. Shon, Y.; Joshi, S.S.; Katakam, S.; Rajamure, R.S.; Dahotre, N.B. Laser additive synthesis of high entropy alloy coating on aluminum: Corrosion behavior. *Mater. Lett.* **2015**, *142*, 122–125. [CrossRef]
109. Siddiqui, A.A.; Dubey, A.K.; Paul, C.P. A study of metallurgy and erosion in laser surface alloying of $AlxCu_{0.5}FeNiTi$ high entropy alloy. *Surf. Coatings Technol.* **2019**, *361*, 27–34. [CrossRef]
110. Ye, X.; Ma, M.; Liu, W.; Li, L.; Zhong, M.; Liu, Y.; Wu, Q. Synthesis and Characterization of High-Entropy Alloy FeCoNiCuCr by Laser Cladding. *Adv. Mater. Sci. Eng.* **2011**, *2011*, 485942. [CrossRef]
111. Ma, Z.; Xia, C.; Zhong, H.; Yang, T.; Liu, N.; Liang, C.; Li, Q. Microstructure, mechanical property and corrosion resistance of FeCoCrNi-M high-entropy alloy coatings on 6061 aluminum alloy prepared by laser cladding. *Surf. Coat. Technol.* **2023**, *455*, 129217. [CrossRef]
112. Shi, Y.; Ni, C.; Liu, J.; Huang, G. Microstructure and properties of laser clad high-entropy alloy coating on aluminium. *Mater. Sci. Technol.* **2018**, *34*, 1239–1245. [CrossRef]
113. Ni, C.; Shi, Y.; Liu, J.; Huang, G. Characterization of $Al_{0.5}FeCu_{0.7}NiCoCr$ high-entropy alloy coating on aluminum alloy by laser cladding. *Opt. Laser Technol.* **2018**, *105*, 257–263. [CrossRef]
114. Li, Y.Z.; Shi, Y. Phase assemblage and wear resistance of laser-cladding $Al_{0.8}FeCoNiCrCu_{0.5}Si_x$ high-entropy alloys on aluminum. *Mater. Res. Express* **2020**, *7*, 086504. [CrossRef]

lubricants

MDPI

Article

Effect of Nb Addition on the Corrosion and Wear Resistance of Laser Clad AlCr2FeCoNi High-Entropy Alloy Coatings

Xiulin Ji [1,*], Kunpeng Guan [1], Yayun Bao [2], Zhongfa Mao [1], Fengtao Wang [1] and Houfu Dai [1,*]

[1] Department of Mechanical Engineering, Shantou University, Shantou 515063, China; 22kpguan@stu.edu.cn (K.G.); zfmao@stu.edu.cn (Z.M.); ftwang@stu.edu.cn (F.W.)
[2] Bomag (Changzhou) Construction Machinery Co., Ltd., Changzhou 213125, China; paulyayun@163.com
* Correspondence: xiulinji@gmail.com (X.J.); houfudai@stu.edu.cn (H.D.)

Abstract: Laser clad AlCr2FeCoNiNbx (x = 0, 0.5, 1.0, 1.5, 2.0, with x values in molar ratio) high-entropy alloy (HEA) coatings were fabricated on Q345 carbon steel. This study delves into the impact of Nb incorporation on the reciprocating sliding wear resistance of these laser clad coatings against a Φ6 mm silicon nitride ball. The microstructure of the as-clad AlCr2FeCoNiNbx coatings transformed from a single Face-Centered Cubic (FCC) solid solution (when x = 0) to the hypoeutectic state (when x = 0.5) and progressed to the hypereutectic state (when $x \geq 1.0$). This evolution was marked by an increase in the Laves phase and a decrease in FCC. Consequently, the HEA coatings exhibited a gradually increasing Vickers hardness, reaching a peak at HV 820. Despite a decline in corrosion resistance, there was a notable enhancement in wear resistance, and the friction of the HEA coating could be reduced by Nb addition. The phase evolution induced by Nb addition led to a shift in the predominant wear mechanism from delamination wear to abrasive wear. The wear rate of Nb0.5 was impressively low, at 6.2×10^{-6} mm N^{-1} m^{-1} when reciprocating sliding under 20 N in air. In comparison to Nb0, Nb0.5 showcased 3.6, 7.2, and 6.5 times higher wear resistance at 5 N, 10 N, and 20 N, respectively. Under all applied loads, Nb1.5 has the lowest wear rate among all HEA coatings. This substantiates that the subtle introduction of Laves phase-forming elements to modulate hardness and oxidation ability proves to be an effective strategy for improving the wear resistance of HEA coatings.

Keywords: wear resistance; laser cladding; microstructure; high-entropy alloy; corrosion; micro-alloying

Citation: Ji, X.; Guan, K.; Bao, Y.; Mao, Z.; Wang, F.; Dai, H. Effect of Nb Addition on the Corrosion and Wear Resistance of Laser Clad AlCr2FeCoNi High-Entropy Alloy Coatings. *Lubricants* **2024**, *12*, 5. https://doi.org/10.3390/lubricants12010005

Received: 8 November 2023
Revised: 10 December 2023
Accepted: 21 December 2023
Published: 24 December 2023

1. Introduction

Wear is a basic form of component failure, and improving the wear resistance of materials can improve the service life of components. The industries of metallurgy, mining, electricity, water conservancy, and agricultural machinery have high requirements for the wear resistance of mechanical equipment. It is generally believed that the hardness of materials is closely related to their wear resistance. High-entropy alloys (HEAs), as newly developed alloys, have exhibited lots of outstanding properties, including high hardness, good corrosion resistance, high toughness along with high strength, excellent wear resistance, and exceptional oxidation resistance [1–3]. The wear resistance of HEAs and their coatings is attractive and has been widely investigated [4], since they have great potential for engineering applications such as turbine blades and heat exchangers [5]. Compared to traditional alloys, some HEAs have a relatively higher material hardness. HEA with high hardness is expected to have correspondingly high wear resistance. Previous studies on the wear resistance of HEA have shown that, compared to traditional alloy materials, HEA exhibits better wear resistance. For example, the wear resistance of NiCrAlCoMo is higher than that of AISI1050 steel [6]; the wear resistance of Al2CrFeCoxCuNiTi (x = 0.5, 1, 1.5, and 2) is more than three times that of Q235 steel [7]; the wear resistance of Al3CrFeCoNiCu is about four times that of bearing steel [8], and so on. Therefore, HEA has potential value as a wear-resistant coating material. The in-depth study of the influencing factors on the wear resistance of HEA has also become one of the current research hotspots in the field of HEA.

The choice of metal elements and their ratio will directly determine the properties of the HEA itself. The HEA, composed of several transition metals such as Fe, Ni, and Co, along with several other elements such as Al, Cr, Cu, Mn, Ti, etc., first attracted research interest. At present, high-entropy alloys mainly composed of FeCoNi have become the main component series in current HEA research due to their excellent mechanical properties. For the good corrosion resistance comparable to SS304 stainless steel [9] and the high compressive strength of up to 3920 MPa [10], Al-Co-Cr-Fe-Ni high entropy alloy was investigated as a wear-resistant coating. However, the wear resistance of laser clad AlCr2FeCoNi was not as high as expected.

In addition, due to the excellent corrosion resistance and biocompatibility of IIIB and IVB group elements such as Ti and Zr, the TiZr series alloy composition has become a research hotspot for another type of HEA wear-resistant material. Changes in the element content of HEAs cause phase transitions, which have been noted, especially when adding those elements with high mixing enthalpy to other elements. For example, adding Mo to CoCrFeNiMox (x = 0–1.5) leads to the optimum combination of intermetallic compounds and FCC phases, promoting the optimization of wear resistance [11]. Adding Nb to AlCoCrFeNi HEA also resulted in phase evolution and Laves phase formation [12]. With increasing the Nb content in CoCrFeNi HEA, the yield strength is improved accordingly [13]. Furthermore, the eutectic microstructure was found in CoCrFeNi HEA with Nb addition [14]. Unfortunately, how the addition of Nb affects wear resistance is uncertain. The Nb element can directly combine with C at high temperatures to form NbC, which has extremely high hardness. Therefore, adding Nb elements to high-entropy alloys is expected to have a dispersion-strengthening effect, thereby improving wear resistance. Cheng et al. [10] prepared CoNiCuFeCr and CoNiCuFeCrNb coatings using the plasma transfer arc melting method and found that under the same wear test conditions, the addition of Nb can improve the wear resistance of the coating by about 1.5 times. Guo et al. [15] prepared AlCrFeMoNbxTiW high-entropy alloy coatings using laser cladding technology and conducted friction and wear experiments at a load of 200 N and a speed of 200 r·min^{-1}. The results indicate that the average friction coefficient and wear rate of the coating decrease with the increase in Nb content, and the delamination phenomenon caused by adhesive wear of the coating disappears, enhancing the anti-adhesive wearability of the coating. The AlCrFeNi2W0.2Nbx HEA coatings were synthesized on the 304 stainless steel by laser cladding, and the AlCrFeNi2W0.2Nbx (x = 1.5, 2.0) high-entropy alloy coatings exhibit an order of magnitude lower wear than 304 stainless steel [16]. Lin et al. [17] prepared FeCoCrNiAlNbx HEA coatings and found that the friction coefficient and mass loss of the HEA coating first decreased and then increased with the increase in Nb content, and both reached their minimum values at x = 0.5. Therefore, the addition of an appropriate amount of Nb could be an effective way to improve the wear resistance of the HEA coating.

The main preparation methods for HEAs and their coatings include vacuum melting, powder metallurgy, mechanical alloying, laser cladding, thermal spraying, cold spraying, magnetron sputtering, plasma-based ion implantation, electrochemical deposition, etc. Among them, laser cladding is a widely used surface technology, and Ni-based, Fe-based, and Co-based alloys are commonly used as laser clad coatings to improve the surface performance of conventional metal parts [18,19]. Compared with arc melting, laser cladding reduces the economic cost significantly by melting a thin layer of the substrate, which ensures superior metallurgical properties such as supersaturated solid solutions, fast heating and cooling rates, and dense microstructure [19–21]. So, fabricating HEA coatings with laser cladding has a wide potential application in strengthening the protective coating market. The former research has proved that a suitable amount of intermetallic compound in the solid solution phases presents a positive effect on the mechanical performance, although many HEAs are designed to prevent their formation [20,22,23]. Consequently, the wear resistance of HEAs may also be strongly influenced by adjusting the phase transformation with a minor addition.

In current work, laser clad AlCr2FeCoNiNb*x* (x = 0, 0.5, 1.0, 1.5, and 2.0) HEAs coatings were manufactured on low carbon steel, and Nb was added to form an intermetallic compound because it has a very negative mixing entropy with other elements. Although our former work on the slurry erosion resistance of AlCr2FeCoNiNb*x* HEA coatings displayed that the erosion resistance was enhanced by Nb addition [24], the reciprocating sliding wear of this HEA coating has not been studied. Consequently, for the requirement of a protective coating of hydraulic cylinder piston rods, the sliding wear resistance of AlCr2FeCoNiNb*x* coatings is investigated carefully, as are their microstructure, phase evolution, hardness, and corrosion resistance.

2. Materials and Methods

Laser clad $AlCr_2FeCoNiNb_x$ (x = 0, 0.5, 1.0, 1.5, and 2.0) coatings with different x values in the molar ratio are referred to as Nb0, Nb0.5, Nb1.0, Nb1.5, and Nb2.0, respectively. The nominal alloy compositions of AlCr2FeNiCoNb*x* HEA coatings are listed in Table 1. The pure metal powders with a purity >99.5 wt.% and an average particle size of 53 μm were mixed by dry ball milling. The ball milling process adopts an ND7-type planetary (Tencan Powder Technology Co., Ltd., Changsha, China) ball mill, an alumina ball milling tank (same manufacturer), and alumina ceramic balls (same manufacturer) with a diameter range of 5–20 mm. The relevant parameters used are frequency 16 Hz (450 r min^{-1}), each running time of 3 min, the following pausing time of 7 min, and the actual ball milling time of 20 h. Then, the mixed powders were paved on the substrate with a 4.5 wt% PVC solution. The thickness of the pre-coating layer is about 0.8 mm. The substrate is low carbon steel (Q345 with 0.12–0.20 wt.% C, 0.20–0.55 wt.% Si, 1.20–1.60 wt.% Mn, <0.045 wt.% P, and <0.045 wt.% S), which is cleaned by sandpaper before pre-coating. The pre-coated layer was dried for 3 h in a vacuum oven. The pulsed laser processing machine (GD-ECYW300, Shenzhen Gd Laser Technology Co., Ltd., Shenzhen, China) was used for laser cladding. The specific processing parameters were: spot diameter D = 1 mm, peak power P = 3.5 kW, pulse frequency F = 19 Hz, pulse width W = 3.0 ms, scanning speed V = 140 mm/min, and an overlap ratio of 50%. For the following tests, the HEA coating samples were cut to a size of 10 × 10 × 10 mm.

Table 1. Chemical compositions of AlCr2FeNiCoNbx HEA coating (wt.%).

x	Al	Cr	Fe	Ni	Co	Nb
0.00	8.87	34.16	18.34	19.28	19.35	/
0.50	7.69	29.63	15.91	16.72	16.79	13.26
1.00	6.79	26.17	14.05	14.77	14.83	23.39
1.50	6.08	24.43	12.58	13.22	13.28	30.41
2.00	5.51	21.21	11.39	11.97	12.02	37.90

The phase compositions were analyzed by X-ray diffraction (XRD, Rigaku D/max 2500, Akishima, Japan)-using Cu Kα radiation. The microstructure was observed by a scanning electron microscope (SEM, FEI Quanta 200, Thermo Fisher Scientific, Waltham, MA, USA). Through SEM energy dispersive spectrometry (EDS), the chemical compositions of different microareas were studied. The hardness of HEA coatings was tested by a Vickers hardness tester (HXD-1000TC, Shanghai Caikang Optical Instrument Co., Ltd., Shanghai, China) under a loading of 200 g with a holding time of 15 s. Each hardness value was obtained from the average value of five measurements.

After polishing with 240 to 1200 grit SiC paper, the HEA coating samples obtained the same average surface roughness (R_a = 0.1 μm). The anodic polarization curves were performed for the study of corrosion resistance. The measurement starts from −1000 mV (vs. SCE) to 200 mV (vs. SCE) with a sweep rate of 1 mV·s^{-1}. A platinum wire was applied as the auxiliary electrode. The working electrode was the HEA coating sample, which was covered by epoxy resin except for one side surface (1 cm^2). The reference electrode was a saturated calomel electrode (SCE).

The sliding wear resistance of the laser clad coating samples was characterized by reciprocating sliding using a tribological testing machine (CFT-I style, Lanzhou Zhongke Kaihua Technology Development Co., Ltd., Lanzhou, China). The sliding speed was 20 $mm \cdot s^{-1}$, the sliding stroke was 5 mm, and the total sliding duration was 60 min. During sliding against a Φ6 mm silicon nitride ball, the Coefficient of Friction (COF) was continuously recorded. The contact loads were 5 N, 10 N, and 20 N, respectively. After each test, the surface profiles were detected along the wear track. The average values of the cross-sectional area obtained from at least 5 positions along the wear track were used for the calculation of volume loss. The wear rate can be calculated from the following expression: $\omega = V/(F \times L)$, where V means the wear volume loss (mm^3), F represents the applied normal load (N), and L is the total distance of sliding (m).

3. Results and Discussion

3.1. Microstructure and Hardness

3.1.1. Microstructure Evolution

As-clad $AlCr_2FeCoNiNb_x$ (x = 0, 0.5, 1.0, 1.5, and 2.0) HEA coatings were characterized by XRD. The corresponding XRD patterns are presented in Figure 1. Based on the XRD pattern of Nb0, it can be seen that it contains a single FCC solid solution phase. The addition of Nb in the HEA coating Nb0 leads to the emergence of the BCC and Laves phases. Meanwhile, the main crystal planes of the FCC phase transform from (200) to (111). By continuously increasing the content of Nb, the diffraction peak intensity of the FCC phase gradually decreases. When $x \geq 1.0$, FCC phase peaks are almost invisible. At the same time, the diffraction peak intensities of BCC and Laves phases gradually increase with the increase in Nb. Therefore, Nb addition is beneficial to the phase evolution from FCC to BCC and the Laves phase. Such phase evolution may greatly enhance the HEA's wear resistance [25]. However, further addition of Nb causes a decrease in BCC peak intensity and an increase in Laves phase peak intensity. So, when $x \geq 1.0$, more Nb content favors the transition from BCC to the Laves phase.

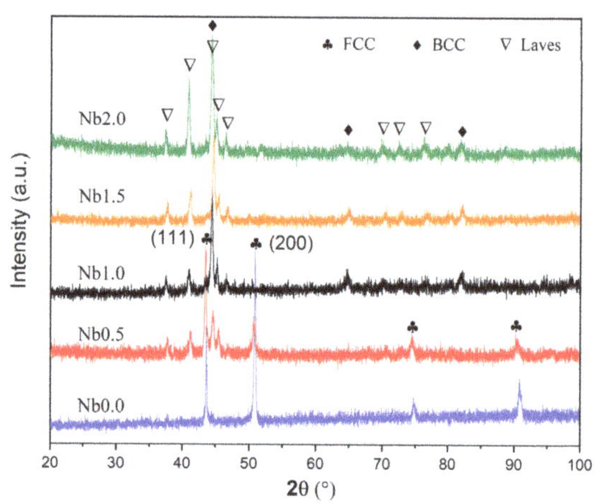

Figure 1. XRD patterns of as-clad AlCr2FeCoNiNbx (x = 0, 0.5, 1.0, 1.5, and 2.0) coatings.

The phase formation of multi-component alloys mainly depends on the competitive outcomes of mixing enthalpy (ΔH_{mix}), mixing entropy (ΔS), the atom-size difference (ΔR), valence electron concentration (VEC), and electro-negativity difference (ΔX) [26]. Here, the Nb element possesses the largest atomic size in the alloying system. Adding Nb will cause severe lattice distortion, which promotes the transition of FCC to higher-density

BCC. It results in the FCC being reduced and the BCC phase increasing gradually. On the other hand, following the Miedema model, the ΔH_{mix} in an alloy system is defined as the following [27–29]:

$$\Delta H_{mix} = \sum_{i=1, i \neq j}^{n} \Omega_{ij} C_i C_j, \tag{1}$$

$$\Omega_{ij} = 4\Delta H_{AB}^{mix}, \tag{2}$$

where n is the number of components in an alloy system, C_i and C_j are the atomic percentages of the ith and jth elements, and ΔH_{mix} is the mixing enthalpy of binary AB alloys, which can be found in Refs. [30,31]. The calculated values of ΔH_{mix} are −11.00, −14.25, −16.49, −18.03, and −19.06 kJ/mol, respectively, when x = 0, 0.5, 1.0, 1.5, and 2.0, revealing that the increase in Nb content leads to the decrease of ΔH_{mix}. It is well known that the more negative ΔH_{mix} indicates a higher binding energy between atoms. As a result, it favors the formation of the Laves phase. More Nb addition will lead to the disintegration of the solid solution, the failure of the high-entropy effect, and the formation of intermetallic compounds. In other words, due to the significant difference between Nb and other atoms (the very low negative value of ΔH_{mix}), the addition of Nb promotes dense atomic structure and the formation of the Laves phase. In addition, lattice distortion is common in high-entropy alloys with multiple principal components. The strong binding between certain atoms in the lattice may exacerbate lattice distortion and lead to the formation of relatively loose atomic structures, causing the phase transition from FCC to BCC. Therefore, for the greater ΔH_{mix}, the addition of Nb promotes the formation of the Laves phase, and the content of the Laves phase increases with the addition of Nb.

Furthermore, the chemical compositions of the zones pointed out in Figure 2 are characterized by EDS, as listed in Table 2. As can be seen, all coatings contain a large amount of Fe content. Meanwhile, Al and Ni content is lower than the nominal composition, indicating the surface of Q345 steel is melted partially into the HEA coating during laser melting, resulting in dilution. In addition, Al and Ni may be seriously burned in part by the high-energy laser beam and discharged in the form of scum eventually. The EDS (Table 2) analysis indicates that the FCC contains more Fe, Cr, and Ni, which is consistent with Refs. [13,14]. While the Laves phase and the dendrite are enriched with Nb because of its largest atomic radius. The degree of the supersaturation increased with the addition of Nb, resulting in more Nb-rich BCC and Laves phases being formed.

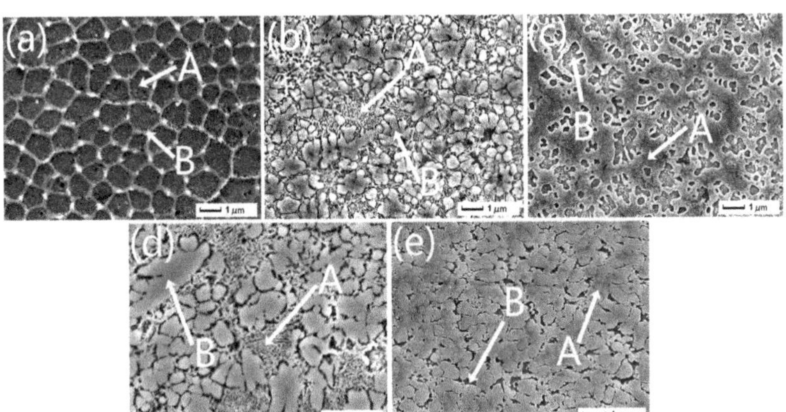

Figure 2. Microstructure observations of as-clad HEA coatings: (**a**) Nb0, (**b**) Nb0.5, (**c**) Nb1.0, (**d**) Nb1.5, and (**e**) Nb2.0. The zones marked as "A" and "B" in each subfigure are measured with EDS, as the results shown in Table 2.

Table 2. EDS results of as-clad $AlCr_2FeCoNiNb_x$ (x = 0, 0.5, 1.0, 1.5, and 2.0) HEA coatings (at. %).

x	Zone	Al	Cr	Fe	Co	Ni	Nb
0	A	4.91	19.52	51.67	12.68	11.21	/
	B	9.02	21.33	45.77	13.95	9.93	/
0.5	A	3.52	14.46	52.75	8.23	9.92	11.11
	B	8.20	15.36	55.01	8.73	7.21	5.49
1.0	A	3.10	8.86	56.15	6.23	5.82	19.84
	B	6.09	7.96	63.73	5.95	5.73	10.53
1.5	A	2.78	7.78	58.99	4.43	5.19	20.83
	B	0.72	6.72	51.78	4.74	5.39	30.65
2.0	A	2.14	8.15	46.74	3.37	2.18	37.41
	B	1.11	8.00	61.11	5.59	5.28	18.90

The microstructure of the as-clad HEA alloy coatings with a different Nb content is presented in Figure 2. The microstructure of Nb0 is relatively simple and exhibits equiaxed crystal morphology (Figure 2a), suggesting the FCC solid solution structure based on the XRD results shown in Figure 1a. A typical hypoeutectic structure with numerous coarse FCC-equiaxed crystals is present in Nb0.5, meaning this HEA alloy belongs to a eutectic high-entropy alloy (EHEA) [32,33]. The eutectic structure is composed of BCC and a network of Laves phases. FCC decreases obviously with the Nb addition. Compared to Nb0 and Nb0.5, Nb1.0 displays a hypereutectic structure, as shown in Figure 2c. The main phase composition of Nb1.0 is the BCC phase, and the eutectic mixture is composed of the BCC and Laves phases. With the addition of Nb, the BCC phase increases continually, becoming a primary phase, as shown in Figure 2d. Meanwhile, the lamellar eutectic structure decreases. Further adding Nb, the coarse BCC phase occupies the majority of the space of Nb2.0, as shown in Figure 2e, and the eutectic structure and Laves phase almost disappear. So, the eutectic point may be located at a chemical composition between Nb0.5 and Nb1.0. The eutectic reaction during solidification may be elucidated as L→FCC + BCC + Laves. Similarly, by adding the Nb element, Ma et al. [12] reported that the binary eutectic point of the $AlCoCrFeNb_xNi$ alloy ranges from 0.5 to 0.75, which is slightly different from the ternary eutectic point found in this work.

3.1.2. Hardness Evolution

Figure 3 exhibits the cross-sectional hardness variation curve in laser clad AlCr2FeCoNiNbx (x = 0, 0.5, 1.0, 1.5, and 2.0) HEA coatings. Based on the relationship between the hardness values and the distance from the coating surface, it can be inferred that the thickness of the HEA coatings is about 200 μm and the Vickers hardness of HEA coatings with Nb addition is higher than that of Nb0. With the increasing Nb content, the HEA coating's hardness gradually increases, reaching a maximum of around HV 820. On the other hand, the hardness of Nb0.5 increases by about 135% compared to Nb0, while the hardness of Nb1.0 increases by about 14% compared to Nb0.5. However, the hardness increases in Nb1.5 compared to Nb1.0 and Nb2.0 compared to Nb1.5 is lower, with values around 9%. The increasing hardness of HEA coating can be seen as the result of the increase in BCC and Laves phases. Adding Nb also results in a decrease in the softer FCC content. At the same time, the lattice distortion of the HEA coating is also strengthened by the addition of Nb with larger atomic radii, which severely hinders dislocation movement.

3.2. Corrosion Resistance

The corrosion resistance of laser clad AlCr2FeCoNiNbx (x = 0, 0.5, 1.0, 1.5, and 2.0) HEA coatings was analyzed by their polarization curves detected in a 3.5% NaCl salt solution, as shown in Figure 4. By using cathode Tafel line extrapolation to intersect with corrosion potential, relevant data on corrosion potential E_{corr} and corrosion current density I_{corr} can be obtained, as shown in Table 3. Nb0 has the highest corrosion potential and the minimum corrosion current, indicating that Nb0 possesses the best corrosion resistance.

Adding the Nb element causes a negative shift in corrosion potential. Meanwhile, the corrosion current significantly increases. Compared with other HEAs, Nb0 only has an FCC phase, which is beneficial for reducing galvanic corrosion between phases. The corrosion potential of Nb0.5 is higher than that of Nb1.0, Nb1.5, and Nb2.0 because the microstructure of Nb0.5 is still dominated by FCC and there is no obvious primary cell formed between phases, resulting in a higher corrosion potential. Due to its most complex multiphase structure containing BCC, Laves, and a small amount of FCC simultaneously, Nb1.0 has the most negative E_{corr} and the highest i_{corr}, indicating its worst corrosion resistance. As Nb further increases, the microstructure is simplified to similar BCC and Laves phases, resulting in similar corrosion potentials and corrosion current density. So, a uniform single microstructure is beneficial for improving corrosion resistance, and adding Nb is harmful to corrosion resistance.

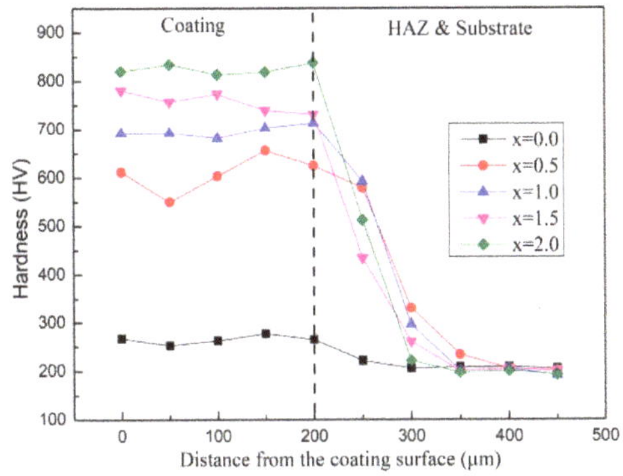

Figure 3. Cross-sectional hardness values of as-clad AlCr2FeCoNiNb*x* (*x* = 0, 0.5, 1.0, 1.5, and 2.0) coatings.

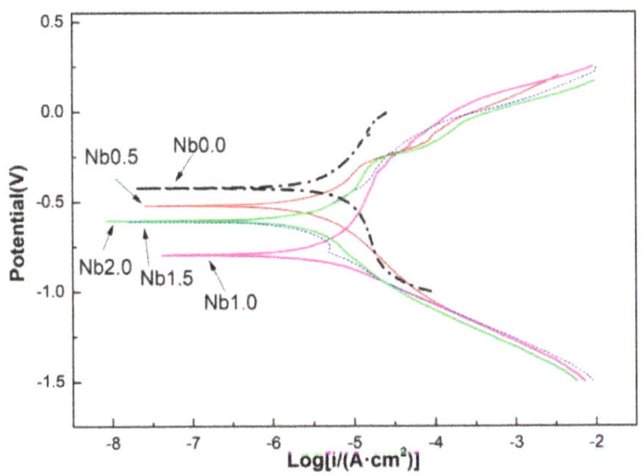

Figure 4. Polarization curves for as-clad $AlCr_2FeCoNiNb_x$ HEA coatings in a 3.5% NaCl solution.

Table 3. Polarization parameters for as-clad AlCr2FeCoNiNb*x* HEA coatings in a 3.5% NaCl solution.

HEA Coatings	Nb0	Nb0.5	Nb1.0	Nb1.5	Nb2.0
E_{corr}(V)	−0.423	−0.520	−0.795	−0.608	−0.602
i_{corr}(μA cm^{-2})	0.857	2.990	3.293	2.427	2.708

Due to defects such as cracks and pores, the corrosion resistance of the coating is strongly influenced by both the material and the preparation process. Based on the electrochemical kinetic parameters estimated from the polarization curves after exposure to a 3.5 wt.% NaCl solution, a high-velocity oxy-fuel (HVOF)-prepared 316 L stainless steel coating displays −610 mV corrosion potential and 14.3 $\mu A \cdot cm^{-2}$ corrosion current density [34]. Compared with the HVOF 316L coating, Nb0.5 has a nobler corrosion potential and a much lower corrosion current density. Therefore, although the addition of Nb has slightly reduced the corrosion resistance, the HEA coatings studied in this work exhibit considerable corrosion resistance.

3.3. Sliding Wear Resistance

The effect of Nb addition on the wear resistance of the laser clad AlCr2FeCoNiNb*x* HEA coating was characterized by reciprocating sliding wear against a silicon nitride ball. The detected average Coefficient of Friction (COF) and average wear rate are shown in Figure 5. Except for Nb1.0, the COFs of the HEA coatings generally decreased with the increase in applied load. A very low COF of Nb1.0 coating at 5 N is attributed to its long wear-in stage, which has a COF lower than 0.2. The average COF of Nb1.0 at the stable stage is around 0.74, which is the lowest COF in all the HEA coatings at 5 N. Consequently, compared to Nb0, the average COF of Nb1.0 decreased by 22.7%. However, at 10 N or 20 N loads, the HEA coating Nb1.0 presented the highest COF values, while Nb2.0 possessed the lowest COFs. So, by adding appropriate Nb content, the friction of the HEA coating can be regulated or reduced.

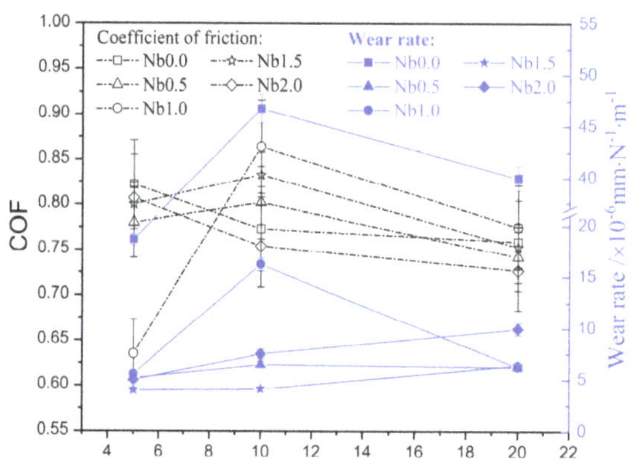

Figure 5. Average COFs and average volume loss rates of laser clad $AlCr_2FeCoNiNb_x$ HEA coatings after reciprocating sliding wear against silicon nitride balls under 5, 10, and 20 N, respectively.

The average volume loss rates of laser clad AlCr2FeCoNiNb*x* HEA coatings significantly decreased after Nb addition. At 5 N load, the average wear rate of Nb0 was 18.7×10^{-6} mm N^{-1} m^{-1}, and the corresponding value of Nb0.5 was 5.2×10^{-6} mm N^{-1} m^{-1}, which is about a 72% decrease. At high applied loads, the effect of Nb addition is more significant on the reduction of wear rate. At 10 N and 20 N, the average wear rate of Nb0 was 46.8 and 40×10^{-6} mm N^{-1} m^{-1}, respectively. Accordingly, the corresponding values of Nb0.5 were 6.5 and 6.2×10^{-6} mm N^{-1} m^{-1}, respectively. So, Nb addition leads to a more than 84% wear rate decrease in the laser clad HEA coating at high sliding applied loads. Because of Nb addition, the hard Laves phase is synthesized in the HEA coating, leading to an increase in hardness and the ability to resist plastic deformation. However, more Nb additions did not result in an obvious decrease in wear rate. Nb1.5 presented the lowest wear rate of all HEA coatings. At 20 N load, Nb0.5, Nb1.0, and Nb1.5 show almost the same low wear rate. The wear rate in Nb2.0 increased with the applied load, indicating this HEA coating possesses the characterization of high strength materials. Compared with the hardness of the HEA coatings (Figure 3) with and without adding Nb, it can be seen that the high hardness is the decisive factor for high wear resistance. However, too high a level of hardness is harmful to wear resistance, especially under high applied loads. Nb1.5 has the lowest wear rate in the applied load range of 5 N to 20 N. So, achieving a strong and tough combination, with the majority of hard phases and a small amount of dispersed soft phases, is an effective way to obtain the lowest wear rate.

3.4. Wear Mechanism

To analyze the wear mechanism, the worn surface morphologies of the HEA coatings were observed, as illustrated in Figure 6. Although the volume loss rate of the HEA coatings can be connected with their hardness, the reason causing the wear rate of Nb1.0 to be much higher than others is not clear. As can be seen in Figure 6a, the worn surface of Nb0 is covered by lots of black tribo-layers, which contain cracks and may be formed from metal oxides and debris. Meanwhile, there are some deep craters with limited deformation resistance, indicating delamination is the primary wear mechanism of Nb0.

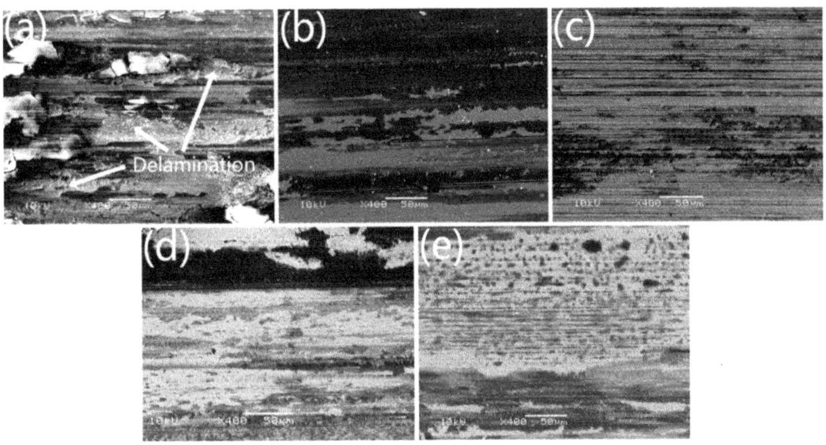

Figure 6. Worn surface morphologies of laser clad $AlCr_2FeCoNiNb_x$ HEA coatings after reciprocating sliding wear against a silicon nitride ball under 10 N: (**a**) Nb0, (**b**) Nb0.5, (**c**) Nb1.0, (**d**) Nb1.5, and (**e**) Nb2.0.

As we all know, oxidation of the metal contact surface is inevitable due to the presence of oxygen and frictional heat when sliding in the atmospheric environment. The worn surface of Nb0.5 presents large-area oxidation layers, as shown in Figure 6b. However, on the worn surface of Nb1.0, the oxidation layers are substituted by many abrasive grooves, as shown in Figure 6c. Oxidation-generated tribo-layer is beneficial to decreasing COF and wear rate because the tribo-layer can inhibit further oxidation and abrasion. Similarly, the low friction and wear of VAlTiCrMo1.6 are attributed to the tribo-oxidation-induced layered oxidic surface [35]. Without the protection of the tribal-layer, Nb1.0 is suffering serious abrasive wear, which leads to a higher COF and a worse wear rate, as shown in Figure 5. With more Nb addition, the HEA coatings Nb1.5 and Nb2.0 present many scratched grooves, as shown in Figure 6d,e, which are shallow when compared with those on the surface of Nb1.0. However, similar to Nb1.0, Nb1.5, and Nb2.0 have limited oxidation layers, which indicates that the FCC phase tends to form an oxidation layer. So, the HEA coating with the Nb addition is mainly controlled by abrasive wear.

4. Conclusions

The microstructure of as-clad AlCr2FeCoNiNb*x* HEA coatings underwent notable changes with increasing Nb content, transitioning from a singular FCC structure (Nb0) to a hypoeutectic microstructure (Nb0.5), and eventually to a hypereutectic configuration.

The eutectic point, potentially situated between Nb0.5 and Nb1.0, manifested a reaction expressed as L→BCC + FCC + Laves. Correspondingly, the volume fraction of BCC and Laves phases increased, paralleled by a rise in hardness. The addition of Nb propelled the Vickers hardness of the HEA coatings to a peak at HV 820, surpassing that of the Nb-free HEA coating by more than threefold. Consequently, the corrosion resistance of the HEA coating declined, while wear resistance exhibited a significant improvement. The reciprocating sliding wear tests showed that the friction of the HEA coating can be reduced by adding appropriate Nb content. Achieving a strong and tough combined microstructure is an effective way to obtain the lowest wear rate. With Nb incorporation, the primary wear mechanism shifted from delamination wear to abrasive wear. Based on the wear rate values of Nb0.5, when an Nb element is added, the wear rate of Nb0 can be reduced by 72% at 5 N and by more than 84% at 10 N or 20 N. This underscores that the adjustment of FCC and BCC contents, coupled with the formation of the Laves phase, significantly enhances the wear resistance of laser clad HEA coatings through Nb addition.

Author Contributions: Conceptualization, X.J. and F.W.; methodology, X.J. and Y.B.; validation, K.G., Y.B. and Z.M.; formal analysis, X.J.; investigation, K.G. and Y.B.; resources, Y.B. and H.D.; data curation, K.G.; writing—original draft preparation, X.J. and Y.B.; writing—review and editing, X.J.; visualization, Z.M. and H.D.; supervision, X.J.; project administration, F.W.; funding acquisition, F.W. All authors have read and agreed to the published version of the manuscript.

Funding: This research was supported by the STU Scientific Research Foundation for Talents (NTF21011). F.W. would like to acknowledge the support from the 2020 Li Ka Shing Foundation Cross-Disciplinary Research (2020LKSFG01D) and Guangdong Provincial University Innovation Team Project (2020KCXTD012).

Data Availability Statement: The authors ensure that the data supporting the results of the research are included in the article.

Acknowledgments: The author, X.J., would like to express his gratitude to MDPI Publishing and the Lubricants editorial department.

Conflicts of Interest: Yayun Bao was employed by Bomag (Changzhou) Construction Machinery Co., Ltd., but he participated in the research when he was a student at Hohai University. The authors declare that the research was conducted in the absence of any commercial or financial relationships that could be construed as a potential conflict of interest.

References

1. Chen, Z.; Chen, W.P.; Wu, B.Y.; Cao, X.Y.; Liu, L.S.; Fu, Z.Q. Effects of Co and Ti on microstructure and mechanical behavior of $Al_{0.75}$FeNiCrCo high entropy alloy prepared by mechanical alloying and spark plasma sintering. *Mater. Sci. Eng. A-Struct.* **2015**, *648*, 217–224. [CrossRef]
2. Ji, X.; Duan, H.; Zhang, H.; Ma, J. Slurry Erosion Resistance of Laser Clad $NiCoCrFeAl_3$ High-Entropy Alloy Coatings. *Tribol. Trans.* **2015**, *58*, 1119–1123. [CrossRef]
3. Ye, Q.F.; Feng, K.; Li, Z.G.; Lu, F.G.; Li, R.F.; Huang, J.; Wu, Y.X. Microstructure and corrosion properties of CrMnFeCoNi high entropy alloy coating. *Appl. Surf. Sci.* **2017**, *396*, 1420–1426. [CrossRef]
4. Kumar, D. Recent advances in tribology of high entropy alloys: A critical review. *Prog. Mater. Sci.* **2023**, *136*, 101106. [CrossRef]
5. Wu, W.H.; Yang, C.C.; Yeh, J.W. Industrial development of high-entropy alloys. In *Annales de Chimie-Science des Materiaux*; Masson: Paris, NY, USA, 2006; Volume 31, pp. 737–747.
6. Lin, Y.C.; Cho, Y.H. Elucidating the microstructural and tribological characteristics of NiCrAlCoCu and NiCrAlCoMo multicomponent alloy clad layers synthesized. *Surf. Coat. Technol.* **2009**, *203*, 1694–1701. [CrossRef]
7. Zhang, H.; Pan, Y.; He, Y.; Jiao, H. Microstructure and properties of Al_2CrFeCoxCuNiTi high-entropy alloy coating prepared by laser cladding. *Mater. Sci. Eng. Powder Met.* **2013**, *18*, 735–740.
8. Ji, X.; Alavi, S.H.; Harimkar, S.P.; Zhang, Y. Sliding Wear of Spark Plasma Sintered CrFeCoNiCu High-Entropy Alloy Coatings: Effect of Aluminum Addition. *J. Mater. Eng. Perform.* **2018**, *27*, 5815–5822. [CrossRef]
9. Kar, S.; Meena, L.K.; Srivastava, V.C.; Mandal, G.K. Microstructural, Mechanical, and Corrosion Behaviour of Ni-Fe-Cr-Al and Ni-Fe-Cr-Al-Co Alloys. *Trans. Indian Inst. Met.* **2023**. [CrossRef]
10. Rogal, L.; Szklarz, Z.; Bobrowski, P.; Kalita, D.; Garzel, G.; Tarasek, A.; Kot, M.; Szlezynger, M. Microstructure and Mechanical Properties of Al-Co-Cr-Fe-Ni Base High Entropy Alloys Obtained Using Powder Metallurgy. *Met. Mater. Int.* **2019**, *25*, 930–945. [CrossRef]

11. Liu, Y.; Xie, Y.X.; Cui, S.G.; Yi, Y.L.; Xing, X.W.; Wang, X.J.; Li, W. Effect of Mo Element on the Mechanical Properties and Tribological Responses of CoCrFeNiMo High-Entropy Alloys. *Metals* **2021**, *11*, 486. [CrossRef]
12. Ma, S.G.; Zhang, Y. Effect of Nb addition on the microstructure and properties of AlCoCrFeNi high-entropy alloy. *Mater. Sci. Eng. A-Struct.* **2012**, *532*, 480–486. [CrossRef]
13. Liu, W.H.; He, J.Y.; Huang, H.L.; Wang, H.; Lu, Z.P.; Liu, C.T. Effects of Nb additions on the microstructure and mechanical property of CoCrFeNi high-entropy alloys. *Intermetallics* **2015**, *60*, 1–8. [CrossRef]
14. Jiang, H.; Jiang, L.; Qiao, D.; Lu, Y.; Wang, T.; Cao, Z.; Li, T. Effect of Niobium on Microstructure and Properties of the CoCrFeNbxNi High Entropy Alloys. *J. Mater. Sci. Technol.* **2017**, *33*, 712–717. [CrossRef]
15. Guo, Y.; Liu, Q.; Zhou, F. Microsturcture and wear resistance of high-melting-point $AlCrFeMoNb_xTiW$ high-entropy alloy coating by laser cladding. *Rare Met.* **2017**, *41*, 1327–1332. [CrossRef]
16. Liang, H.; Yao, H.; Qiao, D.; Nie, S.; Lu, Y.; Deng, D.; Cao, Z.; Wang, T. Microstructures and Wear Resistance of $AlCrFeNi_2W_{0.2}Nb_x$ High-Entropy Alloy Coatings Prepared by Laser Cladding. *J. Therm. Spray Technol.* **2019**, *28*, 1318–1329. [CrossRef]
17. Lin, D.; Zhang, N.; He, B.; Gong, X.; Zhang, Y.; Li, D.; Dong, F. Structural Evolution and Performance Changes in $FeCoCrNiAlNb_x$ High-Entropy Alloy Coatings Cladded by Laser. *J. Therm. Spray Technol.* **2017**, *26*, 2005–2012. [CrossRef]
18. Zhong, M.; Liu, W. Laser surface cladding: The state of the art and challenges. *Proc. Inst. Mech. Eng. Part C J. Mech. Eng. Sci.* **2010**, *224*, 1041–1060. [CrossRef]
19. Zhang, Z.; Yu, T.; Kovacevic, R. Erosion and corrosion resistance of laser cladded AISI 420 stainless steel reinforced with VC. *Appl. Surf. Sci.* **2017**, *410*, 225–240. [CrossRef]
20. Wang, Y.P.; Li, B.S.; Fu, H.Z. Solid Solution or Intermetallics in a High-Entropy Alloy. *Adv. Eng. Mater.* **2009**, *11*, 641–644. [CrossRef]
21. Zhuang, Y.X.; Liu, W.J.; Chen, Z.Y.; Xue, H.D.; He, J.C. Effect of elemental interaction on microstructure and mechanical properties of FeCoNiCuAl alloys. *Mater. Sci. Eng. A-Struct.* **2012**, *556*, 395–399. [CrossRef]
22. Fu, Z.Q.; Chen, W.P.; Chen, Z.; Wen, H.M.; Lavernia, E.J. Influence of Ti addition and sintering method on microstructure and mechanical behavior of a medium-entropy $Al_{0.6}CoNiFe$ alloy. *Mater. Sci. Eng. A-Struct.* **2014**, *619*, 137–145. [CrossRef]
23. Chuang, M.H.; Tsai, M.H.; Wang, W.R.; Lin, S.J.; Yeh, J.W. Microstructure and wear behavior of $Al_xCo_{1.5}CrFeNi_{1.5}Ti_y$ high-entropy alloys. *Acta Mater.* **2011**, *59*, 6308–6317. [CrossRef]
24. Ji, X.; Bao, Y.; Zhao, J.; Gu, P. Erosion wear resistance of laser cladding AlCr2FeCoNiNb*x* high-entropy alloy coatings. In *Asia International Conference on Tribology 2018*; Abdollah, M.F.B., Ed.; Malaysia Tribology Society: Kuching, Malaysia, 2018; pp. 3–5.
25. Yang, W.; Luo, J.; Fu, H.; Cheung, C.F.; Ruan, H.; Yang, X.-S. bcc → hcp phase transition significantly enhancing the wear resistance of metastable refractory high-entropy alloy. *Scr. Mater.* **2022**, *221*, 114966. [CrossRef]
26. Wu, C.L.; Zhang, S.; Zhang, C.H.; Zhang, H.; Dong, S.Y. Phase evolution and cavitation erosion-corrosion behavior of $FeCoCrAlNiTi_x$ high entropy alloy coatings on 304 stainless steel by laser surface alloying. *J. Alloys Compd.* **2017**, *698*, 761–770. [CrossRef]
27. Yeh, J.W.; Chen, S.K.; Lin, S.J.; Gan, J.Y.; Chin, T.S.; Shun, T.T.; Tsau, C.H.; Chang, S.Y. Nanostructured high-entropy alloys with multiple principal elements: Novel alloy design concepts and outcomes. *Adv. Eng. Mater.* **2004**, *6*, 299–303. [CrossRef]
28. Raghavan, R.; Kumar, K.C.H.; Murty, B.S. Analysis of phase formation in multi-component alloys. *J. Alloys Compd.* **2012**, *544*, 152–158. [CrossRef]
29. Pi, J.H.; Pan, Y. Thermodynamic Analysis for Microstructure of High-Entropy Alloys. *Rare Met. Mater. Eng.* **2013**, *42*, 232–237. [CrossRef]
30. Praveen, S.; Murty, B.S.; Kottada, R.S. Alloying behavior in multi-component AlCoCrCuFe and NiCoCrCuFe high entropy alloys. *Mater. Sci. Eng. A-Struct.* **2012**, *534*, 83–89. [CrossRef]
31. Takeuchi, A.; Inoue, A. Classification of bulk metallic glasses by atomic size difference, heat of mixing and period of constituent elements and its application to characterization of the main alloying element. *Mater. Trans.* **2005**, *46*, 2817–2829. [CrossRef]
32. Jiang, H.; Han, K.M.; Gao, X.X.; Lu, Y.P.; Cao, Z.Q.; Gao, M.C.; Hawk, J.A.; Li, T.J. A new strategy to design eutectic high-entropy alloys using simple mixture method. *Mater. Des.* **2018**, *142*, 101–105. [CrossRef]
33. Lu, Y.P.; Jiang, H.; Guo, S.; Wang, T.M.; Cao, Z.Q.; Li, T.J. A new strategy to design eutectic high-entropy alloys using mixing enthalpy. *Intermetallics* **2017**, *91*, 124–128. [CrossRef]
34. Nayak, S.K.; Kumar, A.; Sarkar, K.; Pathak, A.; Banerjee, A.; Laha, T. A Study on the Corrosion Inhibition of Fe-Based Amorphous/Nanocrystalline Coating Synthesized by High-Velocity Oxy-Fuel Spraying in an Extreme Environment. *J. Therm. Spray Technol.* **2019**, *28*, 1433–1447. [CrossRef]
35. Fan, J.; Liu, X.; Pu, J.; Shi, Y. Anti-friction mechanism of VAlTiCrMo high-entropy alloy coatings through tribo-oxidation inducing layered oxidic surface. *Tribol. Int.* **2022**, *171*, 107523. [CrossRef]

lubricants

MDPI

Article

Microstructure and Wear Resistance of Fe_3Al Coating on Grey Cast Iron Prepared via Direct Energy Deposition

Hossein Rajaei [1,*], Sasan Amirabdollahian [2], Cinzia Menapace [1], Giovanni Straffelini [1] and Stefano Gialanella [1]

[1] Department of Industrial Engineering, University of Trento, Via Sommarive 9, 38123 Trento, Italy; cinzia.menapace@unitn.it (C.M.); giovanni.straffelini@unitn.it (G.S.); stefano.gialanella@unitn.it (S.G.)
[2] ProM Facility, Trentino Sviluppo S.p.A., Via Fortunato Zeni, 8, 38068 Rovereto, Italy; sasan.amir@promfacility.eu
* Correspondence: hossein.rajaei@unitn.it

Abstract: In this study, the potential of Fe_3Al coating material as an environmentally friendly alternative to coatings containing critical elements for brake discs was investigated. A buffer layer of Cr–Mo steel (Ferro 55) that was about 500 µm thick was applied on a gray cast iron disc to enhance the coating quality and prevent the formation of hot cracks during solidification. The microstructural analysis of the cross-section of the coating showed that the buffer layer diffused into the Fe_3Al coating, forming a combination of Fe_3Al, Fe, and $Fe_3AlC_{0.5}$ phases. The tribological properties of the Fe_3Al-coated disc were evaluated using pin-on-disc tests against two different copper-free friction materials extracted from commercial brake pads. The wear results show a coefficient of friction comparable to that of an uncoated disc (≈0.55), but with a reduction in particulate matter (PM) emissions, which decreased from 600 to 476 #/cm^3. The last issue is an interesting aspect that is gaining increasing importance in view of the upcoming international standards.

Keywords: DED; Fe_3Al; wear; Friction Coefficient (COF); emissions; brake disc

Citation: Rajaei, H.; Amirabdollahian, S.; Menapace, C.; Straffelini, G.; Gialanella, S. Microstructure and Wear Resistance of Fe_3Al Coating on Grey Cast Iron Prepared via Direct Energy Deposition. *Lubricants* **2023**, *11*, 477. https://doi.org/10.3390/lubricants11110477

Received: 4 September 2023
Revised: 6 October 2023
Accepted: 1 November 2023
Published: 5 November 2023

1. Introduction

Grey cast iron (GCI) has been extensively used in brake rotor applications due to its advantageous properties, which include outstanding castability, high thermal conductivity, good damping capability, cost-effectiveness, and adequate wear resistance. In fact, GCI is widely used as a rotor in brake systems for road vehicles. However, for this application, GCI still exhibits properties that are liable to be improved, like corrosion resistance and that it emits airborne particles [1,2].

In particular, the particles generated via the tribo-oxidation of GCI account for a significant fraction of particles released by braking systems [3] and are indeed the main source of airborne magnetite (Fe_3O_4) particles in urban areas [4]. The contribution of particulate matter emissions from disc brakes to pollution and their potential for emitting toxic components are two of the most serious environmental issues today [5]. New materials and technologies have been developed to address these concerns. For instance, carbon–ceramic composites have been established as an alternative to traditional brake disc materials since they display reduced wear rates and PM emission levels [6–8], However, because of the sophisticated production methods and unique features, these kinds of discs usually come with a higher price and are used for specific luxury cars. The application of coatings and disc surface treatments has also been investigated to reduce the wear and PM emissions of brake discs [2,9–12].

Coating GCI brake discs using direct energy deposition (DED) has proven to be an effective method, resulting in low-wear brake discs with increased durability and less particle emission [2,13,14]. As compared to the reference GCI samples [15], laser-cladded GCI brake discs with a Ni-self fluxing alloy and 60% spheroidized-fused WC exhibit higher wear resistance and lower PM concentrations in PoD testing. However, harmful airborne particles are still released into the environment.

In this regard, intermetallic phase Fe_3Al exhibits promising characteristics. Fe_3Al is known for its low environmental impact, as it is composed of abundant non-toxic elements. Unlike certain coatings containing hazardous elements. like W, Ni, and Co [16–18], Fe_3Al eliminates the concerns associated with the release of harmful airborne particles. Additionally, Fe_3Al exhibits good corrosion resistance, making it an attractive choice for coating applications and for those applications requiring surface durability. These favorable qualities, combined with its relatively low cost and availability, place Fe_3Al as a desirable alternative to stainless steel and other materials in various industrial sectors [19,20]. Recently, the possibility of manufacturing Fe_3Al coatings on materials with lower corrosion resistance, such as low-alloy steel, has been investigated [21]. Comparing the wear resistance of iron aluminide alloys to that of other ceramics and metals, Alman et al.'s [22] research indicated that the resistance of iron aluminides was comparable to that of stainless steel.

Numerous studies have primarily focused on the application of this coating for improving the wear resistance of brake discs [11,12,15], frequently neglecting the critical environmental considerations associated with the selection of coating materials. Considering the promising characteristics of Fe_3Al and its potential as a coating material, this study aims to investigate the microstructure and wear resistance of an Fe_3Al coating on GCI prepared via DED. Using Fe_3Al powder, we intend to minimize the release of toxic elements into the environment and have the more general task of developing environmentally friendly brake systems with an enhanced performance and reduced environmental impact.

2. Materials and Methods

2.1. Materials

Gas-atomized, spherical Fe_3Al powder with a D50 of approximately 90 µm was used as the initial raw material, which was supplied by NANOVAl company, Berlin, Germany. In Figure 1, the scanning electron microscope (SEM) micrographs of the powder and optical microscope (OM) cross-sectional view of the coating are shown. Based on the X-ray diffraction (XRD) analysis conducted on the initial powder (see Figure 2), the only phase detected was Fe_3Al. This is supported by energy dispersive X-ray spectroscopy (EDXS) investigations, which indicated that the composition of the powder lies inside the Fe_3Al phase range (Fe-13.5 wt.% Al). A buffer layer of the Cr–Mo steel coating material was considered for improving the coating quality and adhesion. The composition of the buffer layer material is given in Table 1.

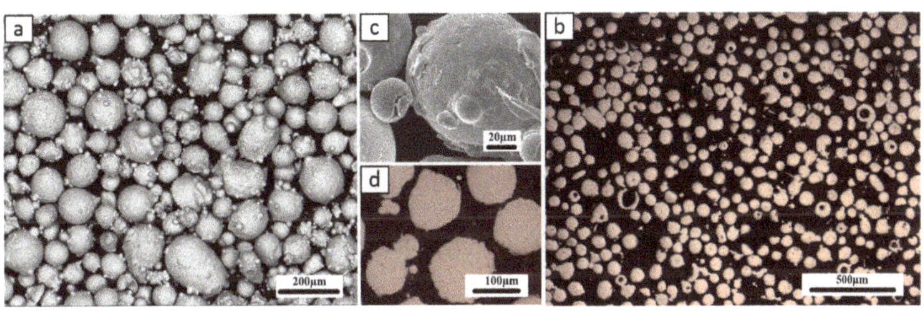

Figure 1. (**a**,**c**) SEM micrographs of gas-atomized Fe_3Al powder. (**b**,**d**) OM view of Fe_3Al powder cross-section.

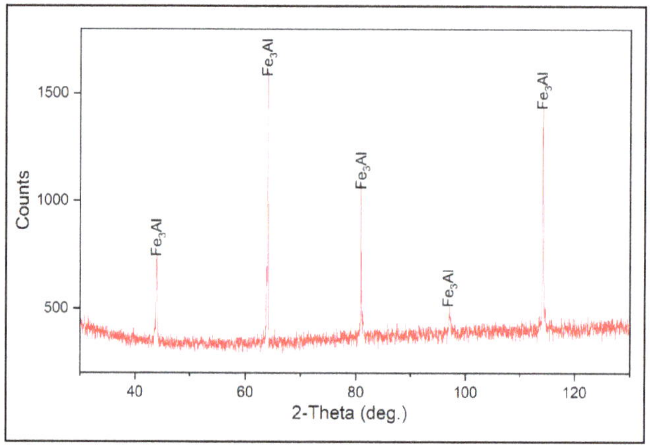

Figure 2. XRD pattern of the gas-atomized Fe_3Al powder.

Table 1. EDXS analysis of the Ferro 55 powder (wt.%).

	Fe	Cr	Mo	Mn	Si	C
Ferro 55	Bal.	7	2.2	1.1	0.3	0.35

2.2. Coating Deposition

For the deposition of the Fe_3Al coating, it is necessary to pre-heat the GCI substrate at temperatures higher than 150 °C in order to prevent defects and cracks generating in the coating structure. The deposition of a buffer layer prior to the deposition of the iron aluminide coating can provide enough energy for pre-heating the substrate and improving the bonding of the coating to the substrate. A buffer layer of Ferro 55 composition was deposited on the GCI disc right before the deposition of Fe_3Al coating (Table 2). After the buffer layer, the Fe_3Al coating was deposited at a scanning speed of 1000 mm/min, with a power of 1000 W and a powder feeding rate of 11 g/min (Figure 3).

An LASERTEC653D hybrid machine (DMGMORIAG, Bielefeld, Germany) with a 2500 W diode laser (=1020 nm) and a Coax14 powder nozzle was used to make for the deposition of the coatings. The laser had a focal length of 13 mm and a pot diameter of 3 mm with a top hat beam profile. The beam profile was shaped like a top hat.

Table 2. Compositions, from EDXS analyses, of the friction materials used in the pin on disc tests. Carbon was not considered in the quantification routine.

Elements (wt%)	Cu-Free	ECOPADS
O	26 ± 1	26 ± 1
Mg	8 ± 1	2.2 ± 0.5
Al	6.1 ± 0.2	1.6 ± 0.5
S	4 ± 1	7 ± 1
Si	3.4 ± 0.6	0.3 ± 0.6
Ca	5.1 ± 0.8	3.2 ± 0.3
Zn	12 ± 1	13.7 ± 0.8
Cr	2.5 ± 0.2	2 ± 0.4
Fe	28 ± 2	21 ± 4
Sn	8.1 ± 0.5	5.2 ± 0.5
Ba	-	21 ± 3

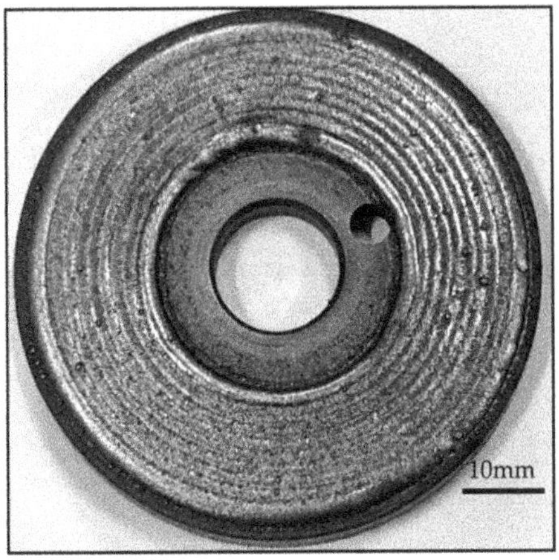

Figure 3. Fe_3Al coating with Ferro 55 buffer layer deposited on GCI substrate.

2.3. Wear Test

To assess the coating's tribological characteristics and to get an insight into the fundamental behavior of the coating sliding against the brake pad friction materials, a dry sliding pin-on-disc (PoD) test was performed. Two friction materials with different compositions were used to investigate the tribological behavior of the coating system. Cu-free and ECOPADS [23] friction materials were chosen as the pin materials (cylindrical shape), and their compositions, as indicated via EDXS analysis, are given in Table 2. For the PoD test, 14,400 m of the total sliding distance, a 2 m/s sliding speed, and 0.6 MPa of applied pressure were used. The selected PoD test conditions were intended to approximately simulate the mild wear condition, which is typical for braking systems [1,9,13]. The PoD test was also carried out on a GCI disc under identical testing conditions in order to provide a better reference point and an understanding of the coating properties comparatively. To avoid the initial surface roughness contributing to the final wear results, the same surface initial roughness Ra was taken into consideration for all the samples (approximately 0.25 μm).

Dry sliding wear tests were conducted using a Biceri PoD tribometer (Biceri, Leeds, UK) at room temperature following the ASTM G99 standard. Three trials were conducted for each combination of disc and friction materials. The assessment of wear resistance was pursued through the measurement of weight loss using a scale with an accuracy of ±0.1 mg. The height deviations of the pins were measured with a digital caliper, and subsequently, these data were used to calculate the wear rate.

2.4. Wear and Emission Measurements

The specific wear rate (K_a) was considered to compare how the different friction materials behaved when they slid against each other. The following formula (Equation (1)) is used to calculate K_a:

$$K_a = \Delta V/(S \times F_n) \quad (1)$$

where ΔV (m^3) is the wear volume measured after the wear test, F_n (N) is the applied load, and S (m) is the sliding distance.

The wear volume of the pin was estimated through the division of its weight reduction by its apparent density. To determine the wear area of individual discs, three measurements were conducted along the radial direction of the wear track using a profilometer instrument (model Hommel Tester T1000) (Hommelwerke GmbH, Villingen-Schwenningen, Germany), featuring a tip with a curvature radius of 5 μm. The scanning length on the disc was 12 mm with a scanning speed of 0.15 mm/s. The average wear volume of the discs was calculated using Equation (2):

$$\Delta V_{disc} = 2\pi r A \quad (2)$$

where r is the average wear radius, and A is the calculated wear area.

An Optical Particle Sizer (OPS) (TSI Incorporated, Shoreview, MN, USA) was used to measure the size of the emitted particles. With a sampling rate of 1Hz, the OPS instrument counts particles between 0.3 m and 10 m in size. As for the PM concentration, the OPS device can sample between 0 and 3000 particles/cm^3.

2.5. Characterization

The cladded layers underwent etching using a Nital solution with a concentration of 2%. The study involved the use of a JEOL IT 300 SEM, (JEOL Ltd., Akishima, Japan) with an EDXS system (EDXS; Bruker, Billerica, MA, USA) for semi-quantitative composition analyses to conduct microstructural observations of the coating and the worn surfaces of the pins and discs. The SEM was operated at an accelerating voltage of 20 kV.

The coated disc underwent XRD analyses in order to assess its phase composition. Data were collected with an Italstructures IPD3000 diffractometer (XRD, GNR Analytical Instruments Group, Novara, Italy), which employed a Co Ka radiation source and an Inel CPS120 detector. The detector was able to simultaneously capture the signal over an angular span ranging from 5° to 120°.

The microhardness of the coatings was measured using a Future-Tech microhardness tester (FM-310) (Future-Tech Corporation, Kawasaki, Japan) on a polished cross-sectional sample. The assessment was conducted from the surface down to the GCI substrate, with a force load of 300 g and a 10 s indentation duration. The study on hardness was conducted in a sequential manner, commencing from the surface layer and progressing downwards, with a series of indentations spaced at regular intervals of 0.15 mm. The objective was to gain a comprehensive understanding of the variations in hardness across the various layers.

3. Results and Discussion

3.1. Coating Characterization

Figure 4 depicts a cross-section of the Fe_3Al coating (Fe_3Al coating with Ferro 55 Buffer layer), where the buffer layer of F55 was first deposited onto the GCI before the deposition of the Fe_3Al coating. The buffer layer was used to improve the coating quality by decreasing defects deriving from the interaction of the Fe_3Al coating with the GCI substrate during deposition and solidification.

Figure 4a shows a low-magnification view of the coating cross-section from the top to the bottom, which shows a defect-free coating microstructure, particularly at the substrate/buffer layer boundary. The microstructure of the Fe_3Al coating, as depicted in the cross-section of the coating near the surface in Figure 4b, reveals the presence of large grains with smaller needle-like structures dispersed throughout the material. These needle-like grains, as supported by the EDXS maps in Figure 5, appear to correspond to Fe-Al carbides. Figure 4c depicts the interface between the top coating and the buffer layer, revealing the significant diffusion of the buffer layer into the Fe_3Al coating, mainly along the grain boundaries. The outer parts of the coating are affected by diffusion from the substrate to a lower extent, showing smaller and less numerous needle-like grains.

Figure 5 shows X-ray element distribution maps acquired at the boundary region between the buffer layer and the Fe_3Al coating (magnified area in Figure 4c). This reveals that the Fe_3Al coating contains large carbon-enriched needle-like grains. These grains are Fe-Al carbides resulting from the diffusion of carbon from the buffer layer. Interestingly, the grain boundary of the coating has a higher concentration of chromium, confirming that the main pathway for the diffusion of carbon is the grain boundary region.

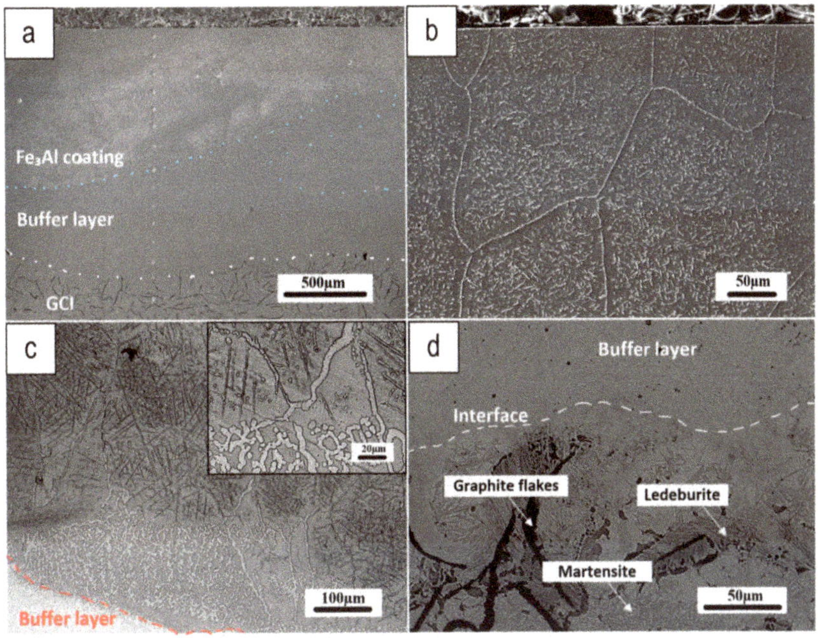

Figure 4. (**a**) Coating cross-section; (**b**) microstructure of the cross-section near the coating surface; (**c**) coating cross-section boundary between the top coating and F55 buffer layer, with a magnified view in the inset; and (**d**) coating cross-section border between the buffer layer and substrate.

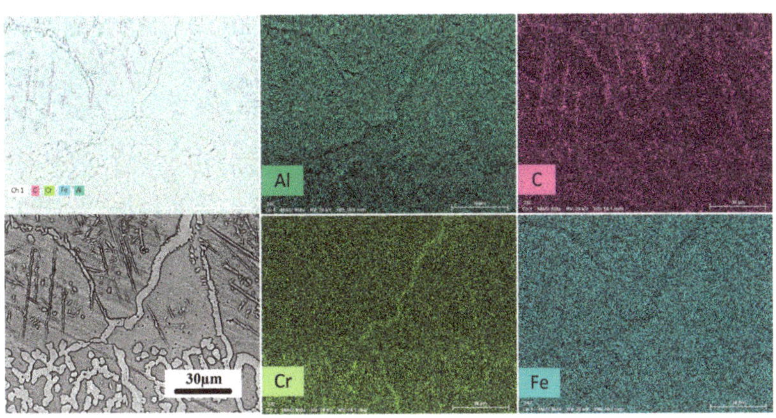

Figure 5. Color X-ray map of the cross section depicting the boundary between the Fe_3Al Coating and F55 buffer layer (refer to Figure 4c).

Figure 4d shows a high-resolution micrograph of the GCI microstructure near the interface with the buffer layer. A laser beam melts the F55 powder during deposition, with the residual heat partially melting the GCI, thus promoting the diffusion of carbon from the graphite flakes in the GCI. As the melt pool passes away, the rapid solidification of the GCI near the deposition boundary produces a martensitic structure. On the edge of the graphite flakes, there are also chilled ledeburite phases.

Grum and Sturm [24] observed the development of ledeburite shells around graphite in cast iron during laser surface remelting. They propose that the carbon-saturated iron surrounding the graphite flakes solidifies near the eutectic point. Because the higher temperatures in the Heat Affected Zone (HAZ) are near the border, carbon diffusion promotes the formation of broad regions of ledeburite along the coating–substrate interface. Figure 4d shows very good metallurgical bonding between the GCI and the buffer layer and the presence of a martensitic structure in the HAZ.

Figure 6 shows SEM images of the coating microstructure. Needle-like particles can be seen dispersed throughout the coating matrix. Figure 6a depicts a mixed area, the edge of which is outlined by a yellow dashed line, where buffer layer components diffused outward into the surface Fe_3Al coating, altering its typical microstructure. Figure 6b shows a higher magnification view of this mixed area close to the interface with Fe_3Al coating, indicating the difference between the grain size in this mixed area and the usual Fe_3Al microstructure (shown in high magnification in Figure 6c). This picture is confirmed by the analytical data. Figure 7a,b shows the EDXS results obtained from the pointed areas on the coating matrix and the needles, respectively (Figure 6), revealing that the elements from the buffer layer, such as Cr, C, and Si, have dissolved into the Fe_3Al coating. Furthermore, by comparing the intensities of the carbon in the matrix and needles (Figure 7a,b), it can be seen that the concentration of carbon in the needles is greater than that in the matrix (as was clearly shown in Figure 5 for the cross-section of this coating). The EDX spectrum in Figure 7c, referring to the mixed area (point C in Figure 6b), reveals the presence of relatively higher quantities of Fe and Cr and a smaller amount of Al in the mixed area. These findings indicate a significant diffusion from the buffer layer, mainly along grain boundaries toward the coating surface. In the literature, other researchers have demonstrated that Cr additions to Fe_3Al may have positive effects on mechanical properties, namely room temperature ductility and fracture toughness, which can be both improved significantly by chromium additions of up to 6% [25–27].

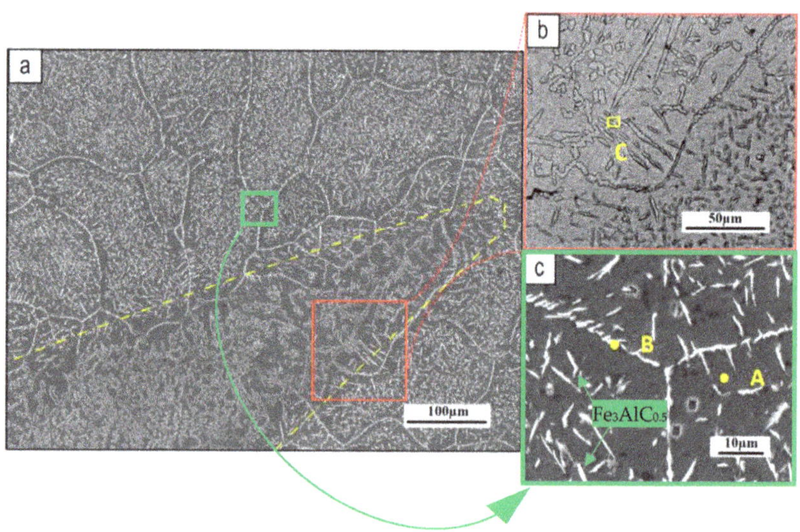

Figure 6. SEM of the Fe_3Al-coated surface: (**a**) general view; (**b**,**c**) higher-magnification images. The dotted line depicts a mixed area, in which the Fe_3Al is modified by the outer diffusion of elements from the buffer layer.

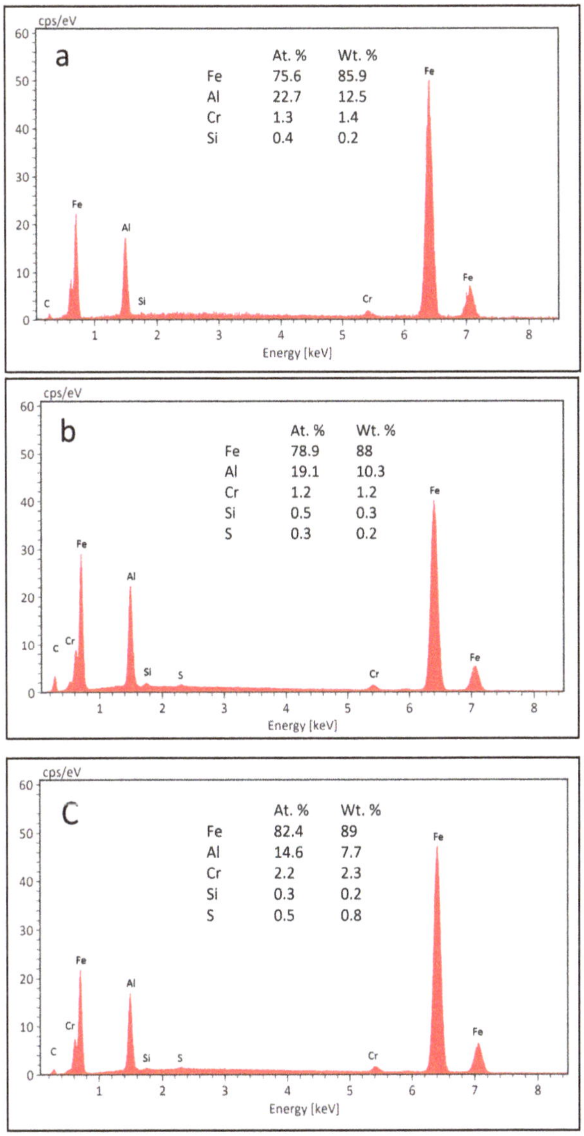

Figure 7. EDXS analysis of the defined areas in Figure 6; (**a**) matrix area "A", (**b**) needle-like grain "B", and (**c**) area "C".

Figure 8 shows the Vickers hardness profile of the Fe_3Al coating. This profile was collected at the center of the sample away from the edges (i.e., the first and last deposition tracks). The hardness of the Fe_3Al coating is between 310 and 350 HV0.3. However, it rises sharply in the F55 buffer layer to roughly 650–690 HV0.3, a typical hardness value for an F55 coating. The hardness of the HAZ also reaches to this range, after passing it through the GCI substrate, the hardness drops below 300 HV0.3.

The existing phases in the Fe_3Al coating were also identified using XRD analysis (Figure 9). The main detected phases are Fe_3Al, $Fe_3AlC_{0.5}$, and Fe. According to the XRD results, the formation of the $Fe_3AlC_{0.5}$ phase is a further consequence of the reaction of the matrix with the diffused and dissolved carbon from the bottom layer.

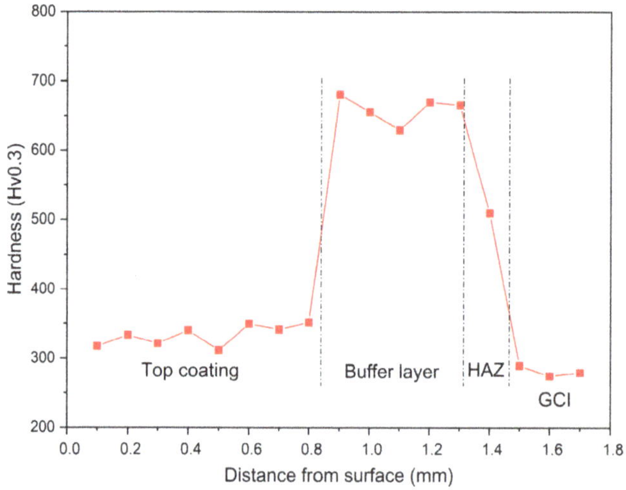

Figure 8. Hardness profile of the Fe_3Al coating from top to down.

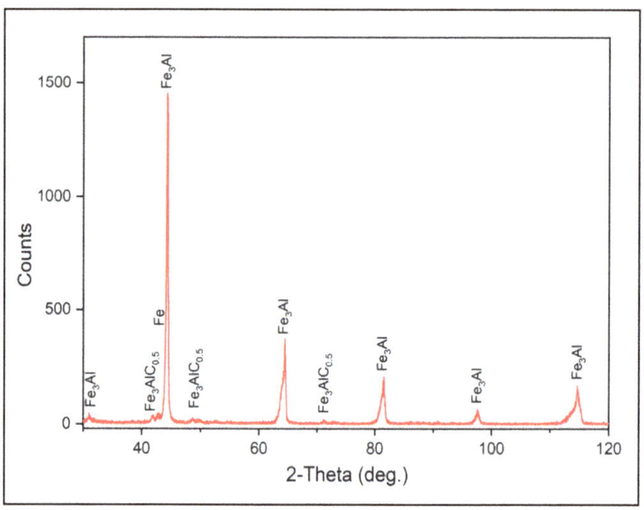

Figure 9. XRD of the Fe_3Al coating on GCI.

The formation of the carbide phase plus the carbon remaining in the solution in the Fe_3Al matrix has a considerable impact on the mechanical properties of iron aluminide alloys, including improvements in the yield strength, creep resistance, and hydrogen-induced cracking resistance [28,29]. Furthermore, recent research has demonstrated that adding carbon to iron aluminides may increase their room-temperature ductility through the formation of perovskite-based $Fe_3AlC_{0.5}$ precipitates in the matrix [30].

3.2. Wear Test Results

Figure 10 shows the trends in the friction coefficient for the Fe_3Al-coated disc, as well as uncoated GCI as a control sample that was compared with the coating. The GCI/Cu-free couple took a longer time to reach a steady state friction coefficient (about 3000 s), whereas the coupling involving the coated disc took place over a relatively short period. The friction coefficient for the GCI/Cu-free and Fe_3Al/ECOPADS sliding couples were approximately the same at a steady state (0.54). The friction coefficient for the Fe_3Al/Cu-free sliding couple reached around 0.57.

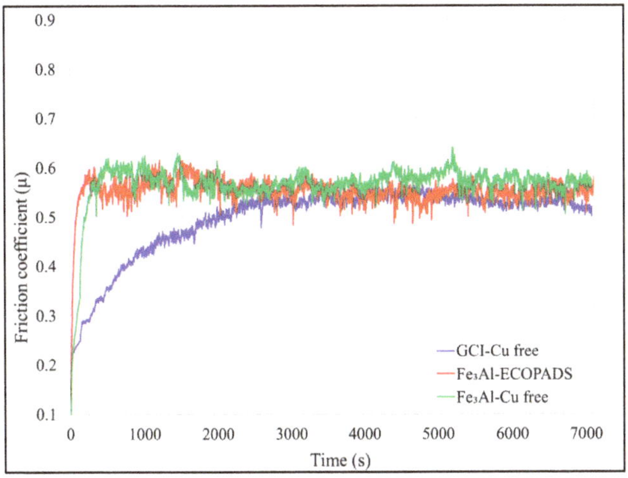

Figure 10. Comparative results for COF of different tribological couplings.

Figure 11 shows the corresponding particulate matter emissions during the pin-on-disc test. The emissions have the same growing tendency as the COF in the case of the uncoated disc, i.e., increasing with an increasing COF during the running-in period before reaching a steady state. Comparing the emissions of different sliding couples, the Fe_3Al/Cu-free couple has considerably fewer emissions (476 #/cm^3) than the GCI/Cu-free couple do (600 #/cm^3). The Fe_3Al /ECOPADS sliding couple, which takes into account the wear dependency in the composition of the friction material, has even fewer emissions (411 #/cm^3) than the Fe_3Al/Cu-free couple does.

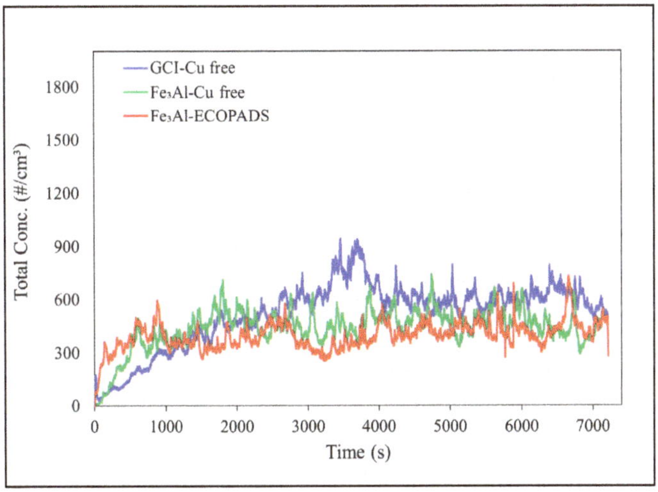

Figure 11. Comparative results for the particulate matter emissions of different tribological couplings.

As stated before, the tribological properties of the Fe_3Al coating were investigated in regard to concerns about its wear behavior when sliding against two different friction materials: Cu-free and ECOPADS. The results are presented in Figure 12, which shows the wear of the pin and counterpart disc under diverse sliding circumstances for all the samples. A comparison of the wear behavior of the coating against Cu-free and ECOPADS friction materials showed that the Fe_3Al/ECOPADS couple has a lower total Ka value than the Cu-free/coating sliding couple does, indicating the better wear resistance of the former. When the coated disc slides against the ECOPADS friction material, the Ka value of the Fe_3Al disc was 2.00×10^{-14} m^2/N, which was slightly lower than that when they slid against Cu-free friction material (2.85×10^{-14} m^2/N). On the other hand, when sliding against Cu-free friction materials, the Ka value of GCI was 2.14×10^{-14} m^2/N, which increased to 2.85×10^{-14} m^2/N when the GCI was replaced with a coated disc (i.e., Fe_3Al/Cu-free sliding pair). Further analysis of the worn surface of the pin and disc will provide an insight into the active mechanisms that caused the observed difference in wear behavior between the coating against the two different friction materials.

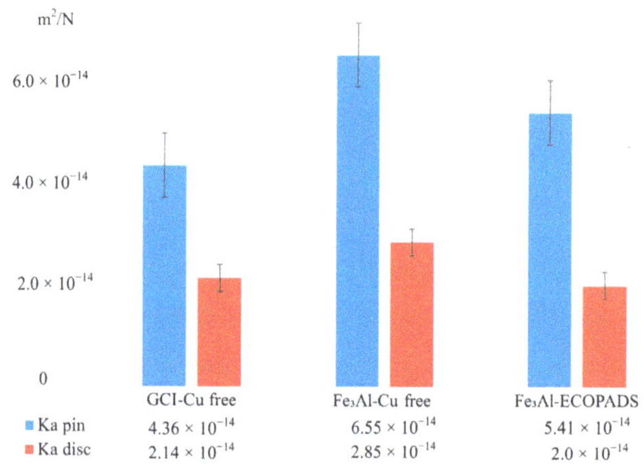

Figure 12. Wear rate, Ka, of pin and disc related to the sliding couples.

Figure 13 shows the typical morphologies of the worn surfaces of the pin and related counterpart discs. In this figure, a compacted secondary plateau can be seen to have formed aside the iron fibers, acting as primary plateaus, emerging from the worn pin surface, thus blocking and favoring the compaction of the wear debris. The X-ray maps in Figure 14 show the spatial distribution of the main elements on the worn pin surface, providing interesting clues regarding the contribution of oxides to the formation and extension of secondary plateaus on the surface of the pin.

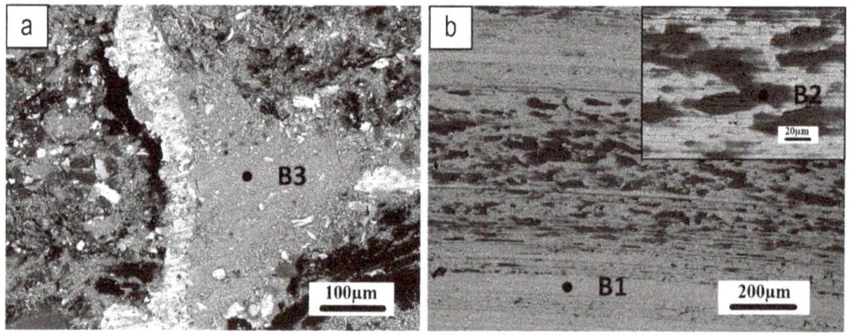

Figure 13. SEM analysis of the Fe_3Al/Cu free sliding couple after PoD test; (**a**) worn pin surface and (**b**) wear track. The spots of the EDXS analyses are indicated (see Table 3 for relevant results).

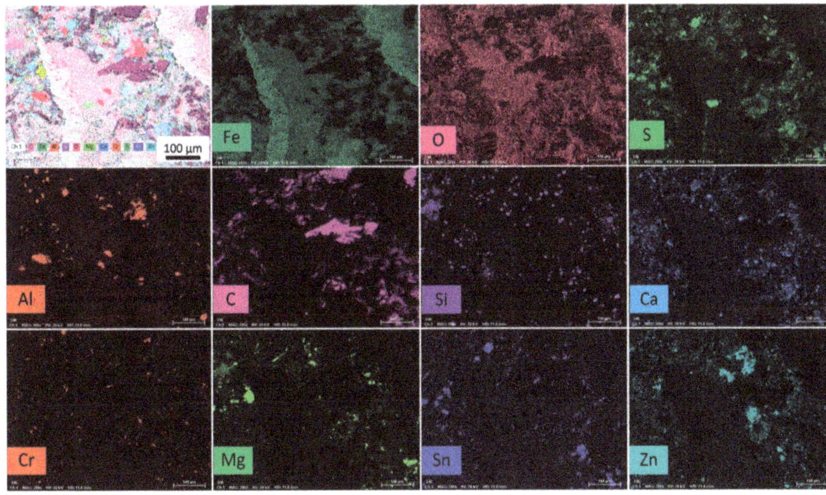

Figure 14. X-ray maps of the Cu-free pin surface for the Fe_3Al-coated/Cu-free sliding couple after the pin-on-disc test.

Table 3. EDXS analysis of the points B1, B2, and B3 marked in Figure 13, which are related to the composition of the friction layers which formed on the Fe_3Al-coated disc and Cu-free pin after the PoD sliding tests. The carbon concentration was not evaluated.

	B1 (wt.%)	B2 (wt.%)	B3 (wt.%)
O	3	19.3	20.8
Al	11.9	7.9	8.8
Cr	1.0	1.1	0.8
Fe	83.3	57.2	55.9
Mg	0.4	3.9	2.8
Si	0.4	1.5	1.8
Zn	-	3.9	3.8
Sn	-	2.3	2.4
S	-	2.2	2.3
Ca	-	0.7	0.6

Figure 13b shows wear tracks on the Fe_3Al-coated disc at two different magnifications. Many grooves can be seen along the sliding direction, indicating the occurrence of abrasion to some extent on the disc. These surface features are typically produced due to abrasion with materials with a higher hardness, such as MgO, Al_2O_3, and other oxides, which contribute to wear during the PoD test. Additionally, darker gray color patches can be seen on the wear tracks, where most of the transferred material is located.

Table 3 shows the results of the EDXS investigations in three different spots, as indicated in Figure 13, selected as follows: a point of the wear track where there is no sign of transferred materials (B1); patches of the transfer layer on the wear track on the disc (B2); a secondary plateau on the worn pin surface (B3). EDXS analysis confirmed the presence of all the elements from the friction material and a relatively large amount of Fe and Al originating from the Fe_3Al coating.

The analyses for the Fe_3Al/Cu-free mating couple indicate a high content of iron oxides in the secondary plateaus on the pin surface and transfer layer on the disc. Iron turned out to be always in association with a relatively large amount of Al, originating from the coating.

Figure 15 depicts the etched surface of the coating after the PoD test against the Cu-free friction material. It was observed that the materials transferred from the pad and coating onto the wear track. The $Fe_3AlC_{0.5}$ carbide phase (Figures 5 and 9) was also observed on the etched surface. The microstructure of the coating indicates that plastic deformation occurred at the surface due to sliding contact with friction materials during the PoD test.

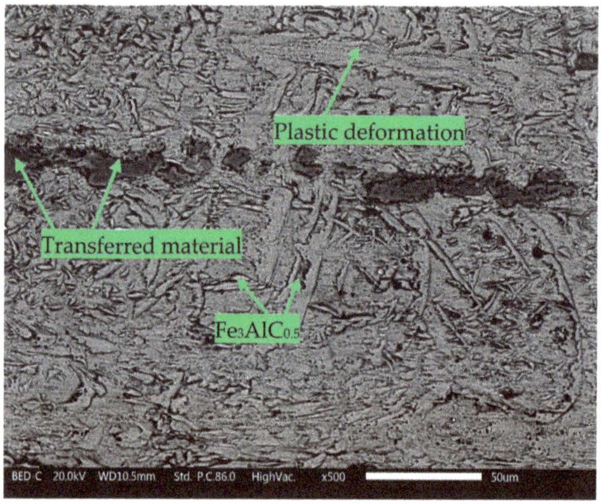

Figure 15. The etched surface of the Fe_3Al coating after PoD test against the Cu-free friction material.

Figure 16 shows SEM micrographs of the worn pin and wear tracks on its counterpart Fe_3Al disc, offering important insights into the tribological behavior of the system. The well-compacted secondary plateaus observed on the worn pin surface (Figure 16a) suggest that the tribo-oxidation mechanism played a role in the wear process. Meanwhile, the parallel scratches and small amount of transferred material observed on the coated disc (Figure 16b) indicate that the abrasion mechanism may have been a significant contributor to wear.

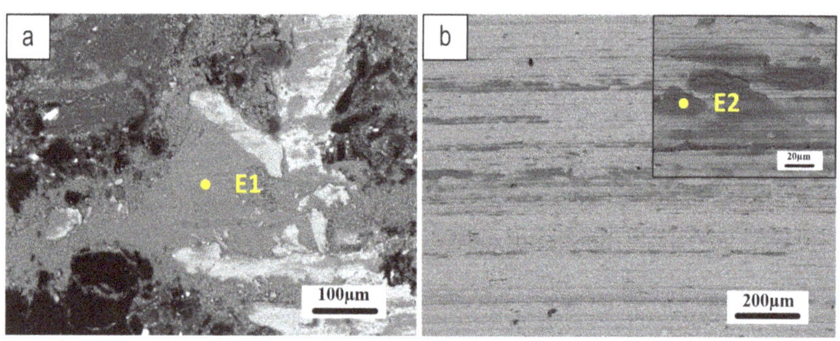

Figure 16. SEM analysis of the Fe_3Al/ECOPADS sliding couple after PoD test; (**a**) worn pin surface and (**b**) wear track.

Table 4 provides the EDXS analysis results of the friction layer on the ECOPADS pin and coated disc, corresponding to the coded points in Figure 16 (E1 and E2, respectively). The elemental compositions reveal contributions from both the pin and coating materials, providing further insight into the underlying mechanisms of the tribological behavior of the system.

Table 4. EDXS analysis of the points (E1 and E2) in Figure 16 related to the friction layer composition formed on the surface of the sliding couples Fe_3Al/ECOPADS. As for the former compositional data, carbon was not quantified.

	E1 (wt.%)	E2 (wt.%)
O	22.6	18.7
Al	7.4	8.5
Cr	0.8	0.7
Fe	52.1	51.2
Ba	6.4	7.0
Mg	2.3	2.7
Si	1	1.0
Zn	2.2	3.2
Sn	1.5	2.1
S	3.1	3.7
Ca	0.6	0.8

In both cases, whether Cu-free or ECOPADS friction materials were used, the formation and extension of secondary plateaus on the surface of the pins were observed (Figures 13a and 16a). These compacted secondary plateaus play a role in providing efficient contact in the interface area, leading to a more consistent and stable friction coefficient. Furthermore, the extension of secondary plateaus on the pin surface facilitates the smooth movement of the sliding bodies [31]. The transferred materials on the coated disc surface also appear to influence the wear behavior [13]. In the case of Cu-free friction materials, a higher amount of Fe was observed in the secondary plateaus and transferred materials onto the disc, which is known to be present on the pad and coating (Figures 13, 15 and 16).

According to a study conducted by Olofsson et al. [13], in the recycling of brake discs using a stainless steel coating, there is a noticeable tendency for materials to transfer onto the disc during the PoD test. This transfer of materials was found to have an adverse effect on the friction coefficient and emissions. Additionally, this resulted in a higher weight loss for both the pin and the disc, which exceeded that of the uncoated GCI disc. For the Fe_3Al coating, the increased Fe content also seemed to enhance adhesive wear by slightly increasing the coefficient of friction. However, the wear results were comparable to those of the uncoated GCI disc.

Figure 17 shows SEM micrographs of the worn pin and wear tracks associated with the GCI/Cu-free sliding couple. Compacted and extended secondary plateaus are observed behind the iron fibers on the surface of the Cu-free pin (Figure 17a), and minor materials transferred to the wear track (Figure 17b). In the wear track on the GCI disc, several grooves running in the sliding direction indicate the abrasive wear mechanism occurring in the GCI disc. The GCI/Cu-free pairing (Figure 17) has a relatively high amount of Fe oxides on the secondary plateaus as well as on the transferred materials (see items F1 and F2 in Table 5), which confirms the contribution of the tribo-oxidation mechanism in the system. This is consistent with the observed low friction coefficient in this scenario. In the case of the Fe_3Al coating, the Fe content decreases, while the Al content originating from the coating takes its place.

Figure 17. SEM analysis of the GCI/Cu free sliding couple after PoD test; (**a**) worn pin surface and (**b**) wear track.

Table 5. EDXS analysis of the points (F1 and F2) in Figure 17 related to the friction layer composition formed on the surface of the GCI/Cu-free sliding couples. As for the former compositional data, carbon was not quantified.

	F1 (wt.%)	F2 (wt.%)
O	28	26.8
Al	1.4	1.3
Cr	0.4	0.6
Fe	61	62.3
Mg	1.7	1.8
Si	1.4	1.4
Zn	2.5	2
Sn	1.4	1.6
S	1.3	1.3
Ca	0.4	0.4
Mn	0.5	0.5

The particles that are released into the PoD chamber can offer valuable insights into the mechanisms of formation and disruption of the friction layer. Figure 17 presents the EDXS semi-quantitative chemical analysis of the particles emitted from the Fe_3Al coating/ECOPADS sliding couple and picked up from the inner walls of the PoD chamber. The full-frame EDXS analysis of the particles (Figure 18a) indicates that their chemical composition is typical of the transfer layer on the worn track and secondary plateaus generated on the pin surface (as shown in Table 3). Moreover, the EDXS analysis of a randomly selected large particle (Figure 18b) that settled inside the chamber revealed high concentrations of Ba, S, and O. This composition is consistent with barium sulfate, which is a typical ingredient of the ECOPADS friction material (see Table 2). Regarding the size of the particles released from the system (Figure 18), larger particulate matter is less harmful to human health than smaller particles is [32]. This is because larger particles, if still airborne, tend to be filtered out by the upper part of the respiratory system, whereas smaller particles can penetrate deeper into the lungs and even enter the bloodstream.

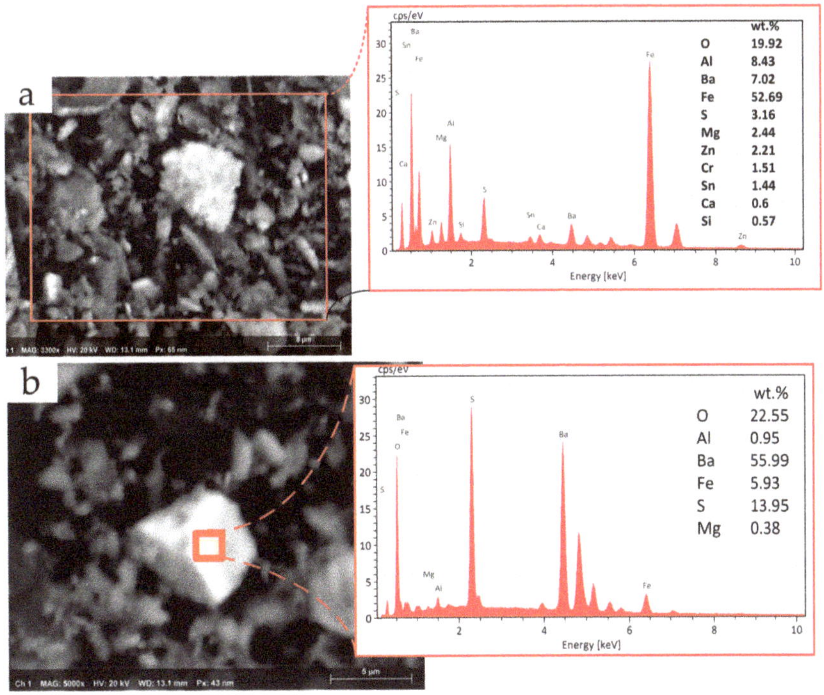

Figure 18. (**a**) SEM and EDXS analyses of the particles collected from the PoD enclosure chamber. (**b**) EDXS analysis of a typical relatively large particle retrieved inside the enclosure chamber of the PoD test rig.

4. Conclusions

The comprehensive analysis presented in this study provides important insights that highlight the application of Fe_3Al coating as an environmentally friendly substitute for coatings containing critical elements for GCI brake discs to mitigate the environmental concerns associated with brake systems.

The microstructure of the Fe_3Al coating reveals the presence of finely dispersed needle-like $Fe_3AlC_{0.5}$ carbides embedded within the alloy grains. These microstructural characteristics influence the tribological performance of the coated disc.

The coating/ECOPADS friction material tribological couple exhibits a reduced wear rate compared to that of the coating/Cu-free sliding couple, underlining better wear resistance in the former configuration.

While the wear results indicated a coefficient of friction similar to that of an uncoated disc (approximately 0.55), the coating/Cu-free sliding couple shows substantially reduced PM emissions compared to those of the GCI/Cu-free pairing, dropping from 600 to 476 #/cm^3. Particularly, even further reductions in PM emissions are achieved with the coating/ECOPADS sliding coupling (411 #/cm^3), underlining the advantageous environmental aspect of the Fe_3Al coating.

Author Contributions: Conceptualization, H.R. and C.M.; methodology, H.R, S.G., C.M. and G.S.; formal analysis, H.R.; investigation, H.R. and S.A.; resources, S.A.; writing—original draft preparation, H.R. and C.M.; writing—review and editing, H.R., S.G., C.M., G.S. and S.A.; supervision, S.G. and G.S. All authors have read and agreed to the published version of the manuscript.

Funding: This research was financed by Fondazione Caritro through the Foundation VRT (Fondazione per la Valorizzazione della Ricerca Trentina); grant "Impact Innovation 2021"; project name: RiFreRiBE.

Data Availability Statement: The data presented in this study are available on request from the corresponding author.

Conflicts of Interest: The authors declare no conflict of interest.

References

1. Lyu, Y.; Leonardi, M.; Mancini, A.; Wahlström, J.; Olofsson, U. Tribology and Airborne Particle Emission of Laser-Cladded Fe-Based Coatings versus Non-Asbestos Organic and Low-Metallic Brake Materials. *Metals* **2021**, *11*, 1703. [CrossRef]
2. Aranke, O.; Algenaid, W.; Awe, S.; Joshi, S. Coatings for Automotive Gray Cast Iron Brake Discs: A Review. *Coatings* **2019**, *9*, 552. [CrossRef]
3. Verma, P.C.; Alemani, M.; Gialanella, S.; Lutterotti, L.; Olofsson, U.; Straffelini, G. Wear debris from brake system materials: A multi-analytical characterization approach. *Tribol. Int.* **2016**, *94*, 249–259. [CrossRef]
4. Zhang, Q.; Lu, D.; Wang, D.; Yang, X.; Zuo, P.; Yang, H.; Fu, Q.; Liu, Q.; Jiang, G. Separation and Tracing of Anthropogenic Magnetite Nanoparticles in the Urban Atmosphere. *Environ. Sci. Technol.* **2020**, *54*, 9274–9284. [CrossRef]
5. Ciudin, R.; Verma, P.C.; Gialanella, S.; Straffelini, G. Wear debris materials from brake systems: Environmental and health issues. *WIT Trans. Ecol. Environ.* **2014**, *191*, 1423–1434. [CrossRef]
6. Li, W.; Yang, X.; Wang, S.; Xiao, J.; Hou, Q. Comprehensive Analysis on the Performance and Material of Automobile Brake Discs. *Metals* **2020**, *10*, 377. [CrossRef]
7. Opel, T.; Langhof, N.; Krenkel, W.; Krenkel, I.W. Development and tribological studies of a novel metal-ceramic hybrid brake disc. *Int. J. Appl. Ceram. Technol.* **2022**, *19*, 62–74. [CrossRef]
8. Goo, B.-C. Development and Characterization of C/C-SiC Brake Disc. *Mater. Manuf. Process.* **2016**, *31*, 979–988. [CrossRef]
9. Rajaei, H.; Menapace, C.; Amirabdollahian, S.; Perini, M.; Straffelini, G.; Gialanella, S. Microstructural and Tribological Evaluation of Brake Disc Refurbishing Using Fe-Based Coating via Directed Energy Deposition. *Metals* **2022**, *12*, 465. [CrossRef]
10. Shamanian, M.; Abarghouie, S.M.; Pour, S.M. Effects of surface alloying on microstructure and wear behavior of ductile iron. *Mater. Des.* **2010**, *31*, 2760–2766. [CrossRef]
11. Chioibasu, D.; Mihai, S.; Cotrut, C.M.; Voiculescu, I.; Popescu, A.C. Tribology and corrosion behavior of gray cast iron brake discs coated with Inconel 718 by direct energy deposition. *Int. J. Adv. Manuf. Technol.* **2022**, *121*, 5091–5107. [CrossRef]
12. Kılıç, H.; Mısırlı, C.; Mutlu, I.; Timur, M. Mechanical and tribological properties of a WC-based HVOF spray coated brake disc. *Mater. Test.* **2022**, *64*, 1150–1161. [CrossRef]
13. Olofsson, U.; Lyu, Y.; Åström, A.H.; Wahlström, J.; Dizdar, S.; Nogueira, A.P.G.; Gialanella, S. Laser Cladding Treatment for Refurbishing Disc Brake Rotors: Environmental and Tribological Analysis. *Tribol. Lett.* **2021**, *69*, 57. [CrossRef]
14. Rettig, M.; Grochowicz, J.; Käsgen, K.; Eaton, R.; Wank, A.; Hitzek, A.; Schmengler, C.; Koss, S.; Voshage, M.; Verpoort, C.; et al. Carbidic Brake Rotor Surface Coating Applied by High-performance-laser Cladding. In Proceedings of the Euro Brake 2020, Online, 16–19 June 2020. [CrossRef]
15. Dizdar, S.; Lyu, Y.; Lampa, C.; Olofsson, U. Grey Cast Iron Brake Discs Laser Cladded with Nickel-Tungsten Carbide—Friction, Wear and Airborne Wear Particle Emission. *Atmosphere* **2020**, *11*, 621. [CrossRef]
16. Bastian, S.; Busch, W.; Kühnel, D.; Springer, A.; Meißner, T.; Holke, R.; Scholz, S.; Iwe, M.; Pompe, W.; Gelinsky, M.; et al. Toxicity of Tungsten Carbide and Cobalt-Doped Tungsten Carbide Nanoparticles in Mammalian Cells in Vitro. *Environ. Heal. Perspect.* **2009**, *117*, 530–536. [CrossRef]
17. US National Toxicology Program (NTP). Available online: https://ntp.niehs.nih.gov/ (accessed on 17 May 2023).
18. European Chemical Agency. Available online: https://echa.europa.eu/home (accessed on 17 May 2023).
19. Liu, Z.; Han, Q.; Guo, Y.; Lang, J.; Shi, D.; Zhang, Y.; Huang, Q.; Deng, H.; Gao, F.; Sun, B.; et al. Development of interatomic potentials for Fe-Cr-Al alloy with the particle swarm optimization method. *J. Alloy. Compd.* **2019**, *780*, 881–887. [CrossRef]
20. Mulyawan, A.; Terai, T.; Fukuda, T. Interpretation of Fe-rich part of Fe–Al phase diagram from magnetic properties of A2-, B2-, and DO3-phases. *J. Alloy. Compd.* **2020**, *834*, 155140. [CrossRef]
21. Luo, X.; Zhang, K.; Cao, J.; Meng, G.; Yu, F.; Zhou, Y.; Zhou, H.; La, P.; Xie, H. Effect of line energy density of the laser beam on the microstructure and wear resistance properties of the obtained Fe_3Al laser cladding coatings. *Optik* **2022**, *261*, 169256. [CrossRef]
22. Alman, D.E.; Hawk, J.A.; Tylczak, J.H.; Doğan, C.P.; Wilson, R.D. Wear of iron-aluminide intermetallic-based alloys and composites by hard particles. *Wear* **2001**, *251*, 875–884. [CrossRef]
23. Menapace, C.; Leonardi, M.; Matějka, V.; Gialanella, S.; Straffelini, G. Dry sliding behavior and friction layer formation in copper-free barite containing friction materials. *Wear* **2018**, *398–399*, 191–200. [CrossRef]
24. Grum, J.; Šturm, R. Comparison of measured and calculated thickness of martensite and ledeburite shells around graphite nodules in the hardened layer of nodular iron after laser surface remelting. *Appl. Surf. Sci.* **2002**, *187*, 116–123. [CrossRef]
25. McKamey, C.; Liu, C. Chromium addition and environmental embrittlement in Fe_3Al. *Scr. Met. Et Mater.* **1990**, *24*, 2119–2122. [CrossRef]
26. Deevi, S.C. Advanced intermetallic iron aluminide coatings for high temperature applications. *Prog. Mater. Sci.* **2021**, *118*, 100769. [CrossRef]
27. Johansson, P.; Uhrenius, B.; Wilson, A.; Ståhlberg, U. Processing, Fabrication, and Mechanical Properties of Fe_3Al Based PM Alloys. *Powder Met.* **1996**, *39*, 53–58. [CrossRef]

28. Parvathavarthini, N.; Prakash, U.; Dayal, R. Effect of carbon addition on hydrogen permeation in an Fe_3Al-based intermetallic alloy. *Intermetallics* **2002**, *10*, 329–332. [CrossRef]
29. Sundar, R.S.; Deevi, S.C. Effect of carbon addition on the strength and creep resistance of FeAl alloys. *Metall Mater Trans A Phys Metall Mater Sci* **2003**, *34*, 2233–2246. [CrossRef]
30. Zhu, S.M.; Guan, X.S.; Shibata, K.; Iwasaki, K. Effect of carbon addition on tribological properties of Fe-Al alloys. *Metall. Mater. Trans. A Phys. Metall. Mater. Sci.* **2002**, *33*, 1292–1295. [CrossRef]
31. Cho, M.H.; Cho, K.H.; Kim, S.J.; Kim, D.H.; Jang, H. The Role of Transfer Layers on Friction Characteristics in the Sliding Interface between Friction Materials against Gray Iron Brake Disks. *Tribol. Lett.* **2005**, *20*, 101–108. [CrossRef]
32. Franck, U.; Odeh, S.; Wiedensohler, A.; Wehner, B.; Herbarth, O. The effect of particle size on cardiovascular disorders—The smaller the worse. *Sci. Total. Environ.* **2011**, *409*, 4217–4221. [CrossRef]

lubricants

Article

Microstructure and Wear Resistance of Ti_5Si_3/Ti_3Al Composite Coatings Prepared by Laser Cladding on TA2 Titanium Alloy

Kaijin Huang [1,2,*] and Wanxia Huang [1]

1 State Key Laboratory of Materials Processing and Die & Mould Technology, Huazhong University of Science and Technology, Wuhan 430074, China; hwxsophia948@163.com
2 State Key Laboratory of Advanced Technology for Materials Synthesis and Processing, Wuhan University of Technology, Wuhan 430070, China
* Correspondence: huangkaijin@hust.edu.cn; Tel.: +86-027-87543676

Abstract: In order to improve the wear resistance of titanium alloy, a Ti_5Si_3/Ti_3Al composite coating with improved wear resistance was successfully prepared by laser cladding TA2 titanium alloy using the double-layer presetting method of Ti-63 wt.% Al mixed powder layer/Si powder layer. The microstructure, phase composition and wear resistance of the coating were studied using X-ray diffraction (XRD), scanning electron microscopy (SEM) and pin-disk friction and wear method. The results show that the coating is mainly composed of the Ti_5Si_3 primary phase and Ti_5Si_3/Ti_3Al eutectic structure. The microhardness of the coating is higher than that of the matrix. The average microhardness of the coating is about 668 $HV_{0.1}$, which is 3.34 times that of the matrix. The coating significantly improves the wear resistance of the TA2 matrix, and the mass wear rate is 1/5.79 of that of the TA2 matrix. The main wear mechanisms of the coating are abrasive wear, adhesive wear and oxidative wear, whereas the main wear mechanisms of the TA2 matrix are adhesive wear and oxidative wear.

Keywords: laser cladding; Ti_5Si_3/Ti_3Al composite coatings; wear resistance; TA2 titanium alloy

Citation: Huang, K.; Huang, W. Microstructure and Wear Resistance of Ti_5Si_3/Ti_3Al Composite Coatings Prepared by Laser Cladding on TA2 Titanium Alloy. *Lubricants* **2023**, *11*, 213. https://doi.org/10.3390/lubricants11050213

Received: 11 April 2023
Revised: 3 May 2023
Accepted: 5 May 2023
Published: 9 May 2023

1. Introduction

Titanium alloy has the advantages of low density, high specific strength and strong corrosion resistance, and has broad application prospects in aerospace, ocean engineering, medical devices and other fields. However, the low hardness and poor wear resistance of titanium alloy significantly restrict its service life under frictional conditions. In order to overcome the weakness of low hardness and poor wear resistance of titanium alloy, surface modification technology is widely used to prepare the coating on titanium alloy to improve the wear resistance of titanium alloy. At present, the commonly used surface modification technologies include ion implantation [1], plasma spraying [2,3], vapor deposition [4,5], micro-arc oxidation [6], laser alloying [7], laser cladding [8,9], etc. Among them, laser cladding is of high significance because of its advantages of metallurgical bonding with the matrix and adjustable coating thickness.

Based on the literature, it is known that two methods, namely, the powder presetting method [10–20] and the powder feeding method [7,9,21–24], are mainly used in the laser surface modification of titanium alloy. Among them, powder presetting methods reported in the literature all used single pure powder or mixed powder for one-time presetting, and there are no reports in the literature on the layered presetting method adopted in this paper. For users who cannot afford a laser cladding powder delivery system but need a laser cladding multi-component powder system, using the layered preset powder layer method can reduce the time of pre-mixing powder and improve work efficiency. Therefore, it is necessary to carry out the feasibility of the layered preset powder layer method. This is the purpose and significance of this paper. The hardness and wear resistance of the coating are greatly improved [8–13,15,17–23]. Ti-Al intermetallic compounds [19] and

Ti_5Si_3 silicide [25,26] formed by laser surface modification of Al and Si elements have excellent wear resistance. The wear mechanisms of the coating and substrate are mainly abrasive wear and adhesive wear, respectively [27]. In addition, Ti_5Si_3 has a high melting point, low density, high hardness and good chemical inertness, whereas Ti_3Al and TA2 titanium alloy have similar thermal expansion coefficients and melting points. Therefore, as long as the ratio of Ti_5Si_3/Ti_3Al is appropriate, it is possible to obtain composite coatings with higher temperature resistance, wear resistance and corrosion resistance than the TA2 titanium alloy.

In this paper, a Ti_5Si_3/Ti_3Al composite coating (which refers to the composite coating mainly composed of Ti_5Si_3 and Ti_3Al phases with high hardness) was prepared on TA2 titanium alloy by laser cladding based on the layered presetting powder method, and the microstructure, phase composition and wear resistance of the coating were studied.

2. Materials and Methods

TA2 titanium alloy was selected as the matrix material, and its specific chemical composition is shown in Table 1. The TA2 titanium alloy with dimensions of 100 mm × 50 mm × 10 mm was selected as the substrate for laser cladding. The substrate was sanded and ultrasonically cleaned with ethanol before the powder was preset.

Table 1. Chemical composition of TA2 (wt.%).

Element	Fe	Si	C	N	H	O	Ti
Content	0.2	0.1	0.1	0.05	0.015	0.1	Bal.

Al powder, Si powder and Ti powder with purity of 99.0 wt.% were selected as laser cladding powder materials in this paper. The powder layer was preset as shown in Figure 1. The thickness of each layer was about 0.4 mm, and the total thickness was about 0.8 mm. The binder used was 4 wt.% polyvinyl alcohol (PVA) aqueous solution. In order to ensure the homogeneity of Ti-63 wt.% Al mixed powder, the process of pre-ball milling was adopted in this paper. The specific parameters of the ball milling were as follows. The XQM-1.2L planetary ball milling produced by Xi'an Liyou Mechanical Equipment Co., Ltd., (Xi'an, China) was used; the weight ratio of the agate ball to powder was 6:1; the rotation speed of the ball milling was 180 r/min; and the ball milling time was 2 h.

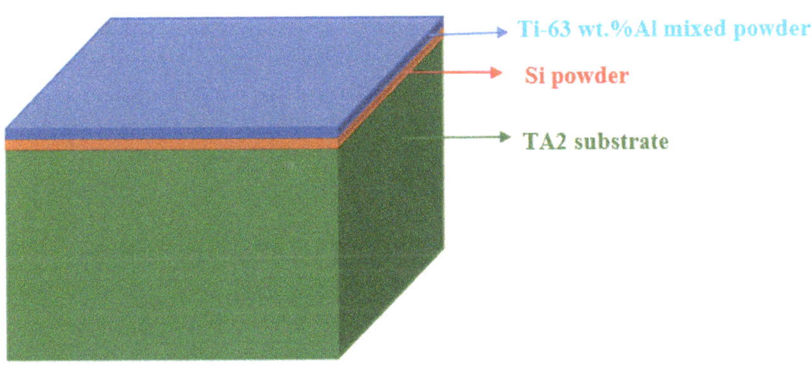

Figure 1. Schematic diagram of presetting powder layer.

Laser cladding was carried out using a Rofin 4 KW fiber laser produced by Rofin Laser Technology Co., Ltd. (Hamburg, Germany). During the laser cladding, argon gas with a purity of 99.999% was used as a protective gas to prevent oxygen in the air from entering the laser molten pool. After several laser cladding parameters were explored, the following optimized laser cladding parameters were finally adopted: laser power P = 1 KW; laser scanning speed V = 10 mm/s; laser spot diameter D = 2 mm; the overlapping rate of the laser spot f = 35%; Ar gas flow rate V_g = 16 L/min.

After laser cladding, the coating and TA2 substrate were cut into 9 mm × 9 mm × 10 mm samples using the wire cutting method, which were mainly used for the XRD phase composition test, SEM microstructure observation, coating microhardness test and wear resistance test. After wire cutting, all samples were cleaned ultrasonically, dried in a drying oven and then placed in a drying dish for later use. An HVS-1000A microhardness tester produced by Laizhou Huayin Test Instrument Co., Ltd., (Laizhou, China) was used to test the microhardness of the cross-section of the coating. The test parameters were as follows: a load of 100 g and a loading time of 15 s. Finally, the relationship curve between the microhardness HV and surface depth h was drawn according to the microhardness values at different depths.

The wear resistance of the coating and TA2 matrix was tested based on the principle of pin-disk friction and wear tests. The size of the wear sample was 9 mm × 9 mm × 10 mm. The wear device model was Ecomet300/Automet300. The parameters of the dry sliding friction and wear (Figure 2) were as follows: the rotary speed was 200 r/min, the applied load was 45 N, the dry sliding speed was 0.921 m/s, the dry sliding wear time was 5 min, the total wear distance was 276.32 m, and the grinding material was 200# SiC sandpaper. The mass of the samples before and after wear was measured using an SOP Quintix124-1CN electronic analysis balance, and the morphology of the wear was observed using SEM to study the wear mechanism of the coating and substrate.

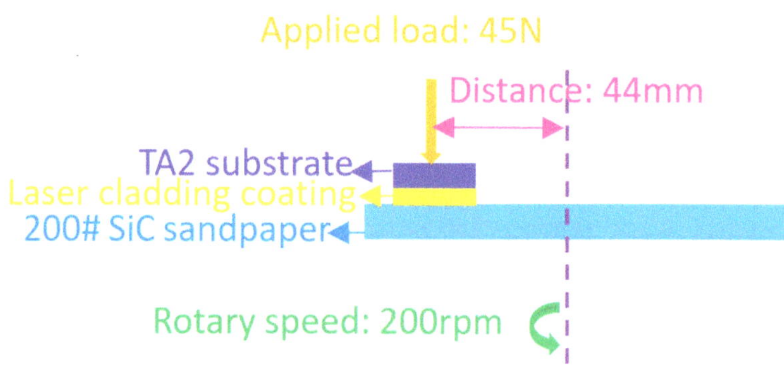

Figure 2. Schematic diagram of dry friction and wear experiment of laser cladding coating.

Phase analysis was performed on the X 'Pert PRO type X-ray diffractometer produced by PANalytical B.V. in the Netherlands (Almelo, The Netherlands). The test conditions were as follows: Cu target, wavelength λ = 0.15406 nm, tube pressure 40 kV, tube flow 40 mA, and 2θ scanning range of 10°~90°. The microstructure of the coating was observed using a ZEISS ΣIGMA HD scanning electron microscope made by Carl Zeiss (Jena, Germany). The micromorphology of the worn coating and TA2 matrix was observed using a JSM-7600F scanning electron microscope produced by JEOL (Akishima, Tokyo). The cross-section of the laser cladding coating should be inlaid, ground, polished and etched before the morphology analysis. The composition and proportion of the etching liquid used were $HF:HNO_3:H_2O$ = 7:43:50 (volume ratio), and the etching time was 5~10 s.

3. Results

3.1. Microstructure of the Laser Cladding Coating

Figure 3 shows the XRD pattern of the laser cladding coating. It can be seen from the figure that the laser cladding coating is mainly composed of Ti_3Al (JCPDS 00-052-0859) and Ti_5Si_3 (JCPDS 03-065-3597). The formation of these two phases is undoubtedly very favorable for improving the wear resistance of laser cladding coating because the hardness of Ti_3Al and Ti_5Si_3 is 5.2 GPa and 11.3 GPa, respectively, both of which are higher than that of Ti (2 GPa). In the laser cladding process, the preset Si powder layer and Ti-63 wt.% Al mixed powder layer melts under the laser and reacts in situ with the partially melted matrix Ti to form Ti_5Si_3 ($5Ti + 3Si = Ti_5Si_3$) and Ti_3Al ($3Ti + Al = Ti_3Al$). According to the calculation results of Zhang [28], the above two reactions in the temperature range of 300~1500 K, ΔG and ΔH are negative. In terms of thermodynamics, a negative ΔH indicates that these two reactions are exothermic and a negative ΔG indicates that these two reactions can occur spontaneously.

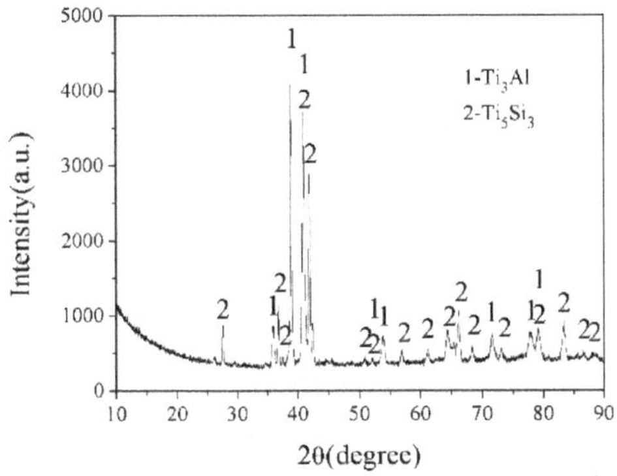

Figure 3. XRD pattern of the laser cladding coating.

Figure 4a shows the overall morphology of the cross-section coating. It can be seen from the figure that the thickness of the laser cladding layer is about 1 mm. There is a metallurgical bonding between the laser cladding layer and matrix, and no obvious porosity or cracks appear in the coating. Figure 4b–d show the amplified SEM morphologies of the different parts in Figure 4a. According to EDS results (Table 2) and XRD calibration results (Figure 3), it can be determined that the large black block in the coating is the primary Ti_5Si_3 phase, and the rest microstructure is a typical Ti_3Al-Ti_5Si_3 eutectic structure. There are more primary Ti_5Si_3 near the top and middle of the coating (point 1 in Figure 4b and point 3 in Figure 4c), whereas the primary Ti_5Si_3 near the bottom of the coating is less and even completely becomes Ti_3Al-Ti_5Si_3 eutectic structure (point 5 in Figure 4d). This type of microstructure is consistent with those reported by Vojtěch et al. [29] and Wu et al. [30]. Moreover, the components of Al, Si and Ti (see Table 1) also fall within the range of hypereutectic components of Ti_3Al-Ti_5Si_3 reported in the literature [30–32].

Table 2. EDS results of the cross-section coating at different points.

Different Points	Ti (at.%)	Al (at.%)	Si (at.%)	Possible Phases
1	60.79	4.83	34.38	Ti_5Si_3
2	70.00	19.82	10.18	Ti_5Si_3/Ti_3Al
3	61.23	4.70	34.07	Ti_5Si_3
4	78.83	12.18	8.99	Ti_5Si_3/Ti_3Al
5	75.06	19.72	5.22	Ti_5Si_3/Ti_3Al

The solidification process is as follows. After a high-power laser is applied to Ti-63 wt.% Al mixed powder layer and Si powder layer, Ti-63 wt.% Al mixed powder, Si powder and part of the Ti matrix melt after absorbing the laser energy. The three kinds of melts are mixed under the action of the stirring force of a molten pool. When the high-power laser leaves, the molten melt in the molten pool begins to solidify. First, the Ti_5Si_3 phase with a high melting point crystallizes, forming the primary Ti_5Si_3 phase. With a decrease in temperature, a high-temperature eutectic reaction occurs in the remaining liquid phase [29]:

$$L \rightarrow \beta\text{-Ti(Al,Si)} + \varepsilon\text{-Ti}_5\text{(Si,Al)}_3 \quad (1)$$

As the temperature continues to decrease, an isomeric transformation occurs in phase β-Ti(Al,Si) [29]:

$$\beta\text{-Ti(Al,Si)} \rightarrow \alpha\text{-Ti(Al,Si)} \quad (2)$$

When the temperature continues to decrease, an ordered reaction occurs in phase α-Ti(Al,Si) [29]:

$$\alpha\text{-Ti(Al,Si)} \rightarrow \alpha_2\text{-Ti}_3\text{(Al,Si)} \quad (3)$$

Thus, when the temperature is lowered to room temperature, the resulting structure is the primary Ti_5Si_3 phase and α_2-Ti_3(Al,Si) + ε-Ti_5(Si,Al)$_3$ is the eutectic structure.

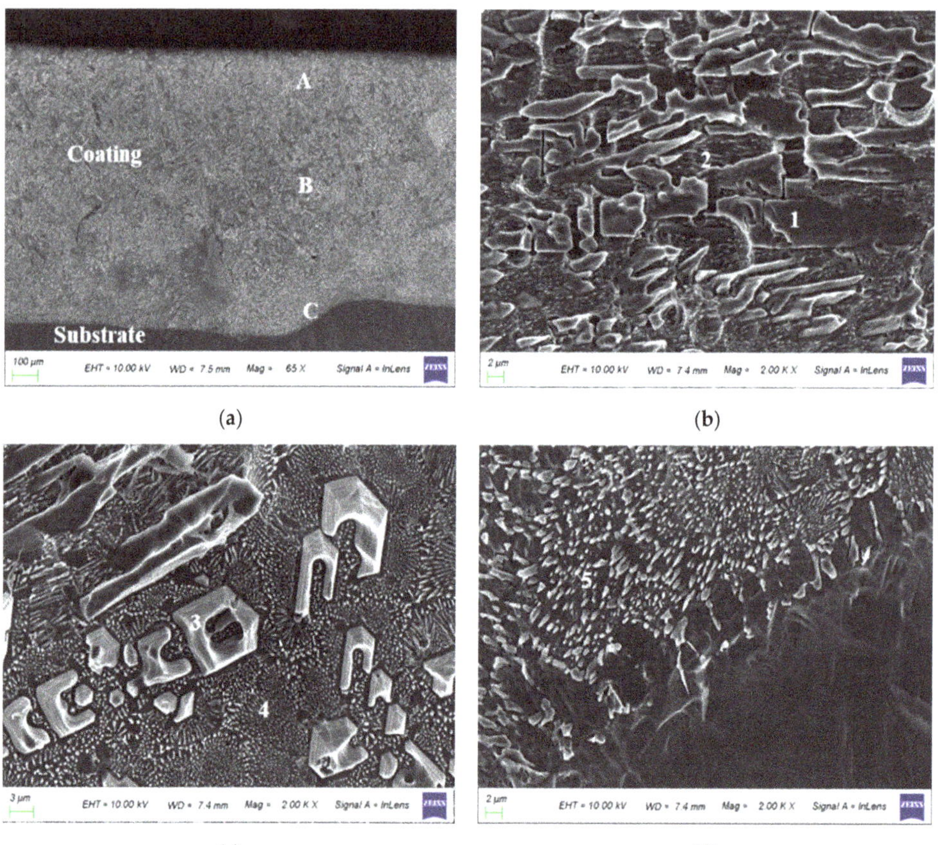

Figure 4. SEM morphology of the cross-section coating: (**a**) whole; (**b**) locally amplified at position A in (**a**); (**c**) locally amplified at position B in (**a**); (**d**) locally amplified at position C in (**a**).1, 2, 3, 4 and 5 in figure represent the location points of the EDS test, respectively.

It should be pointed out that in the literature [28], a single-layer presetting method was adopted to laser cladding Ti, Al and Si mixed powders on Ti-6Al-4V titanium alloy, where the thickness of the presetting powder was 0.6~0.8mm. The prepared coating was mainly composed of Ti_5Si_3, $Ti_7Al_5Si_{12}$, Ti_3AlC_2, Ti_3Al, TiAl, $TiAl_3$ and α-Ti, which is much more complex than the coating prepared by the double-layer presetting method in this paper. This may be related to the flow regularity of Ti, Si and Al powders during the laser action. During the process of laser action, there is a large temperature gradient on the upper and lower surfaces of the molten pool, thus forming a very steep surface tension. The surface tension causes a strong convection in the molten pool, and the convection in the molten pool can produce a strong stirring force so that the liquid metal in the molten pool is mixed evenly. Convection in the molten pool mainly comes from three aspects. First, convection in the molten pool comes from the energy distribution of the laser beam. The Gaussian laser beam used in this experiment can cause severe convection in the molten pool [33]. The second factor is the protective gas blown out in order to prevent the oxidation of the liquid metal at high temperatures. This high-pressure protective gas can produce a certain stirring effect and promote the diffusion of the liquid metal. Third, at high temperatures, the thermal expansion of liquid metal causes a certain density difference in the molten pool, so buoyancy is generated in the molten pool, and the existence of buoyancy causes natural convection of the melt [34]. These three factors that make the layer preset powder can also form a single-layer preset powder structure to reduce the powder mixing time and improve production efficiency.

3.2. Wear Properties of the Laser Cladding Coating

Figure 5 shows the microhardness distribution of the laser cladding coating. As can be seen from the figure, the average microhardness of the coating is about 668 $HV_{0.1}$, which is 3.34 times that of the TA2 matrix (200 $HV_{0.1}$). Since the microhardness of Ti_5Si_3 and Ti_3Al is 1500 HV [28] and 530 $HV_{0.2}$ [35], respectively, the microhardness of the laser cladding coating is mainly due to the existence of these harder intermetallic compounds (Figure 3) and fine-grain strengthening due to rapid solidification during laser cladding. Obviously, an increase in the microhardness of the coating is beneficial to the improvement of the wear resistance of the coating.

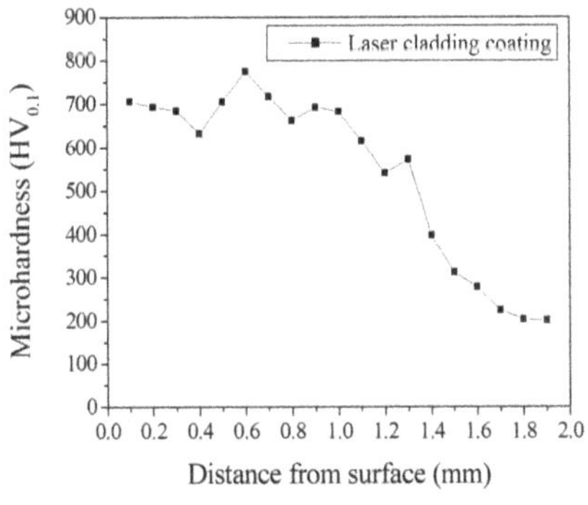

Figure 5. Microhardness distribution of the laser cladding coating.

Figure 6 shows a comparison of the mass wear rates of laser cladding coating and TA2 matrix under dry friction and wear conditions at room temperature. As can be seen from the figure, under the conditions of dry friction loading of 45 N and wear distance of 276.32 m, the mass wear rate of the TA2 matrix is 16.22 mg/min, and that of laser cladding coating is 2.8 mg/min, about 1/5.79 of TA2 matrix. Therefore, the wear resistance of the laser cladding coating is 5.79 times that of the TA2 matrix; that is, the wear resistance of the laser cladding coating is obviously better than that of the TA2 matrix. It is known that the higher the hardness of a material, the better the wear resistance [36]. The average microhardness of the laser cladding coating is 3.34 times that of the TA2 matrix (Figure 5), so the wear resistance of the laser cladding coating is better than that of the TA2 matrix.

It should be pointed out that the improvement factor of wear resistance of the laser cladding coating relative to TA2 matrix is significantly higher than that reported in the literature [37]. In the literature [37], Ti_2SC/TiS composite coating was prepared on TA2 matrix by laser cladding, and the wear rate of TA2 matrix was 3 times that of the composite coating, which was less than 5.79 times that of the coating in this paper.

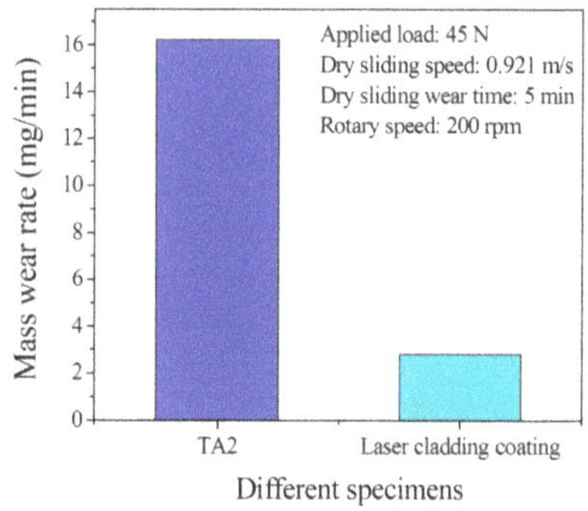

Figure 6. Comparison of mass wear rate between laser cladding coating and TA2 matrix.

Figure 7 shows the SEM morphology of TA2 matrix and laser cladding coating after dry friction and wear. Table 3 shows the EDS results of the worn TA2 matrix and laser cladding coating. According to the SEM morphology of the TA2 matrix after wear in Figure 7a,b, it can be seen that the TA2 matrix after dry friction and wear presents adhesive wear characteristics of being pressed. At the same time, there are deep and wide pear grooves, and there are a large number of fine particles in the grooves. The EDS results in Table 3 show that during dry friction and wear of TA2 matrix, a large number of fine TiO or TiO_x particles are generated due to surface friction heat generation and oxidation. As the abrasive material is SiC with a microhardness of 3200 HV [28], its microhardness is much higher than that of the TA2 matrix. Therefore, under the condition of dry friction and wear, the hard abrasive SiC continuously presses TA2 matrix, causing the surface of the matrix to be scratched and flattened. At the same time, due to the heat generated by dry friction, the Ti in TA2 matrix is oxidized into an incomplete valence TiO_x [38]. Therefore, the main wear mechanisms of TA2 matrix are adhesive wear and oxidative wear.

Table 3. EDS results of worn TA2 matrix and coating.

Different Points	Ti (at.%)	Al (at.%)	Si (at.%)	O (at.%)	Possible Phases
1	91.80	-	-	8.20	Ti + TiO_x
2	51.11	-	-	48.89	TiO
3	100.00	-	-	-	Ti
4	92.11	-	-	7.89	Ti + TiO_x
5	87.00	-	-	13.00	Ti + TiO_x
6	43.34	2.52	24.78	29.36	Ti_5Si_3 + Al_2O_3 + TiO_2
7	64.69	5.27	30.03	-	Ti_5Si_3 + Ti_3Al
8	45.00	2.96	22.99	29.08	Ti_5Si_3 + Al_2O_3 + TiO_2
9	45.54	2.17	29.40	22.88	Ti_5Si_3 + Al_2O_3 + TiO_2
10	40.92	2.31	24.37	32.39	Ti_5Si_3 + Al_2O_3 + TiO_2
11	58.58	14.18	4.80	22.43	Ti_3Al + TiO_2 + SiO_2

Figure 7c,d show the SEM morphologies of the laser cladding coating after dry friction and wear. Combined with the EDS results in Table 3, it can be seen that after dry friction and wear, furrows of different depths and widths and wear particles of different sizes appear on the surface of the coating. Due to the heat generated by dry friction, most oxidation reactions occur and different oxides are generated, but the furrows of all the coatings are shallow and narrow. The different groove depths and widths of the coating should be closely related to the surface microhardness of the coating, and the different sizes of the coating wear particles should be related to the type and volume fraction of the intermetallic compounds in the coating. In general, the larger the volume fraction of Ti_5Si_3 phase with higher microhardness in the coating, the smaller the particles, the more severe the abrasive wear of the coating, the more spalling wear particles, the shallower the groove, the lighter the adhesive wear, and the more severe the oxidation wear. However, because the microhardness of the coating is lower than that of the hard anti-abrasive material SiC, abrasive wear, adhesive wear and oxidation wear coexist.

In summary, the wear mechanisms of TA2 matrix are mainly adhesive wear and oxidative wear, whereas the wear mechanisms of coating are abrasive wear, adhesive wear and oxidative wear that co-exist, which is similar to the results reported in the literature [27,39,40].

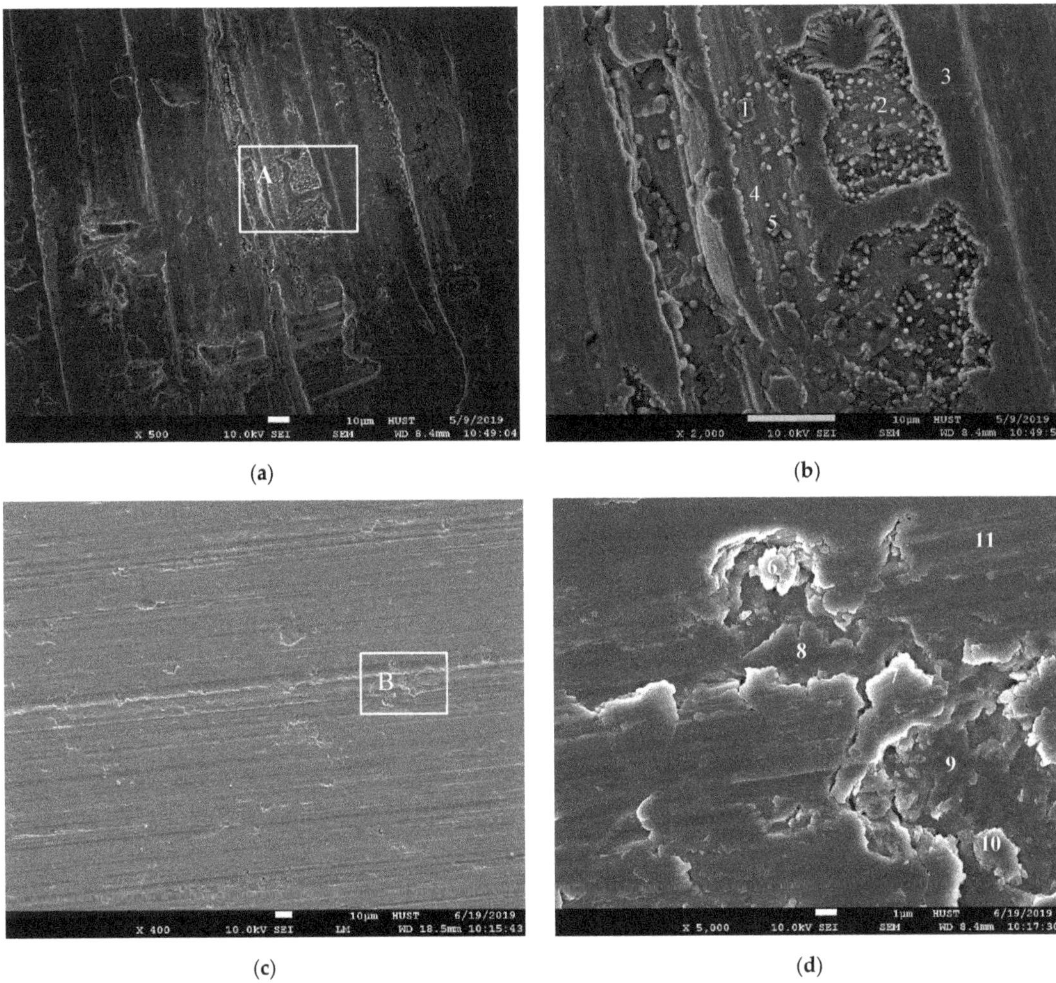

Figure 7. SEM morphology of the worn TA2 matrix and laser cladding coating. (**a**) Whole SEM morphology of the worn TA2 matrix; (**b**) local amplified SEM morphology of the worn TA2 matrix in area A in Figure 6a; (**c**) whole SEM morphology of the worn coating; (**d**) local amplified SEM morphology of the worn coating in area B in Figure 6c. 1, 2, 3, 4, 5, 6, 7, 8, 9, 10 and 11 in figure represent the location points of the EDS test, respectively.

4. Discussion

4.1. Formation and Flow Characteristics of the Laser Cladding Molten Pool with Double-Layer Preset Powder Layer

As a relatively economical and new surface modification technology, laser cladding can significantly improve the surface properties of substrate materials such as wear resistance, corrosion resistance and oxidation resistance. The laser cladding process is a non-equilibrium, instantaneous and non-uniform physicochemical metallurgical process. The heat conduction and melt flow in the molten pool determine the microstructure, grain morphology and defect types of the cladding layer, thus affecting the properties of the materials and the service life of the parts.

In the case of laser cladding of the double-layer preset powder layer (Figure 1), under the action of the laser beam, the double-layer preset powder layer absorbs the laser energy and melts quickly to form a molten pool. When the laser beam leaves the molten pool, it solidifies quickly. There are three heat transfer modes in this preset laser cladding: heat radiation from the laser to the powder layer, heat convection between the protective gas and powder layer, and heat conduction between the powder layers [41]. These three heat transfer modes lead to complex physical phenomena, such as melting, solidification and phase transformation during laser cladding. Figure 8 shows a schematic diagram of the flow behavior of the cross-section molten pool during laser cladding with a double-layer preset powder layer.

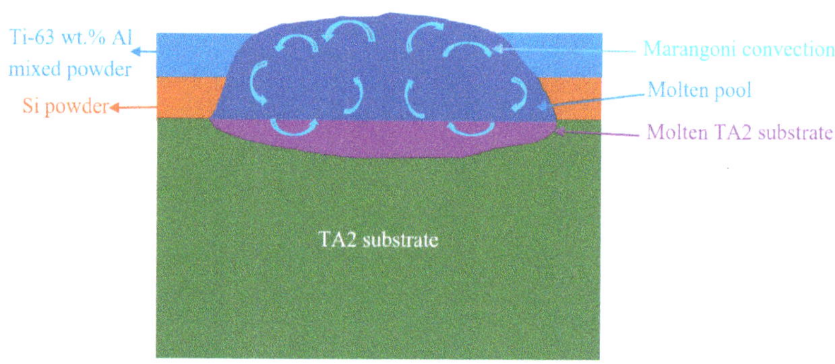

Figure 8. Schematic diagram of flow behavior of cross-section molten pool during laser cladding with double-layer preset powder layer.

During the laser action, the double-layer preset powder layer and part of the substrate melt rapidly due to the absorption of laser energy, forming a large and uneven temperature gradient, resulting in a tension gradient on the surface of the molten pool. Generally, a symmetrical surface tension gradient is generated on the left and right sides of the symmetrical line at the center of the molten pool. Due to the influence of surface tension, the high-temperature melt in the center of the molten pool flows to the surface of the molten pool. At the same time, under the comprehensive action of buoyancy, gravity and other factors, the high-temperature melt flows to both sides along the surface near the center point of the interface. The closer the area is to the edge, the more intense the melt flow and the higher the velocity. Due to the continuity of movement, the low-temperature melt reaching the edge of the molten pool flows along the solid–liquid boundary towards the bottom of the molten pool, and then converges and rises in the middle area of the molten pool, forming the left and right circulating flows of the molten pool (see blue arrows in Figure 8), respectively.

In particular, the change in the surface temperature of the laser molten pool causes a change in the surface tension gradient of the fluid and induces the flow of the surface fluid from the region of low surface tension to the region of high surface tension. This Marangoni flow is the main driving force for mass transmission in the molten pool [42]:

$$\gamma = \gamma_0 + \sigma(T - T_0)$$

where: γ_0 is the surface tension at the reference temperature T_0; σ is the surface tension coefficient.

Moreover, in the process of laser cladding, in addition to the influence of laser cladding process parameters on the temperature field and velocity field of the molten pool, the surface tension coefficient of the material also has an impact on the temperature field and velocity field of the molten pool [43]. When the surface tension coefficient of the material is less than 0, the melt in the molten pool flows from the laser center to the edge of the molten pool. When the material surface tension coefficient is greater than 0, the melt in the molten pool flows from the edge of the molten pool to the laser center, but the peak velocity of both fluid flows appears at the edge of the molten pool surface. The surface tension coefficients of Ti, Si and Al powders used in this paper are −0.16 mN/m, −0.25 mN/m and −0.25 mN/m [44], respectively, which are all negative values, so the flow of the molten pool belongs to the former.

In particular, in the stable zone of the laser cladding layer, the molten pool tends to be stable as the heat input reaches a dynamic equilibrium with the heat lost due to heat transfer. Moreover, under the action of Marangoni flow, the molten double-layer preset powder layer and part of the substrate are mixed and stirred in the molten pool to promote the homogenization of the alloying elements in the molten pool.

In conclusion, under the action of Marangoni flow, Al, Si and Ti alloying elements in the molten pool can be evenly mixed to obtain the effect of laser cladding of a single preset powder layer, and then obtain the laser cladding layer microstructure with basically a uniform composition to ensure the performance and service life of the coating.

4.2. Formation Mechanism of Different Phases in the Coating

In the laser cladding double-layer preset powder layer, once the molten pool is formed, the melted Ti-63 wt.% Al powder layer, the melted Si powder layer and the partially melted TA2 matrix mix with each other under the action of Marongoni flow stirring force. According to the isothermal section diagram of the Ti-Al-Si ternary system at 1000 °C [45], Ti-Al intermetallic compounds and Ti-Si ceramic reinforced phases can be formed by different in situ reactions of Ti, Al and Si powders in different proportions. In addition, according to the Ti-Al binary phase diagram [46], there are mainly $TiAl_3$, $TiAl_2$, TiAl and Ti_3Al among Ti-Al intermetallic compounds. The melting point of Ti_3Al is 1600 °C, which is higher than that of $TiAl_2$ (1425 °C), TiAl (1465 °C) and $TiAl_3$ (1340 °C). According to the Ti-Si binary phase diagram [46], $TiSi_2$, TiSi, Ti_5Si_4, Ti_5Si_3, Ti_3Si and other compounds can be generated when Ti reacts with Si. The melting point of Ti_5Si_3 is 2130 °C, which is the highest melting point of Ti-Si compounds. In addition, according to the calculation results of Zhu [47], the enthalpy changes in ΔH and Gibbs free energy ΔG of Ti_5Si_3 are the lowest among Ti-Si compounds, so Ti_5Si_3 is the easiest to obtain by solidification first when the composition is appropriate. Si has a low solubility in the Ti-Al phase, and Ti_5Si_3 has a high melting point, so Ti_5Si_3 is the easiest to form in the laser cladding process [48]. Therefore, in this paper, a preset powder scheme was designed according to the location range of the Ti_5Si_3 + Ti_3Al phase generated in the literature [45], which was finally verified; that is, only the Ti_5Si_3 and Ti_3Al phases were generated in the coating.

4.3. Wear Mechanism of the Coating and TA2 Matrix

As we all know, friction and wear is an extremely complex process. Compared with lubricating friction, dry friction without lubricating medium results in a more intense friction process, more severe wear, and shorter effective working life of friction pairs. The wear amount is a basic parameter that reflects the wear resistance of the material. The smaller the wear amount per unit time, the better the wear resistance. Wear resistance refers to the ability of the material itself to resist wear damage, and it is usually expressed by the amount of wear. There is a negative correlation between the wear amount and wear resistance; that is, the smaller the wear amount, the better the wear resistance.

The mathematical model of material wear can quantitatively describe the wear process and calculate the normal working life of the friction pairs to better grasp the essence of the wear problem, reduce and control the wear degree, and improve the reliability of the friction pair. In 1953, Archard [49], a professor at the University of Leicester, proposed a computational model for adhesive wear. Archard's wear calculation formula is as follows:

$$Q = k \times (NL)/H$$

where, Q is the amount of adhesive wear on the contact surface; K is the coefficient of adhesive wear, representing the probability of an abrasive particle generated by mutual friction between a pair of micro-convex bodies; N is the normal pressure; H is the hardness of the soft surface materials; L is the relative sliding distance.

Archard's adhesive wear theory describes the general rules of material wear. The wear amount is proportional to the displacement and positive pressure and inversely proportional to the hardness of the softer materials. It connects the wear with the material properties, which makes it convenient to calculate the wear amount. It can be seen that hardness is an important index to evaluate material properties. In general, the higher the hardness, the better the wear resistance of the material.

Specifically in this paper, on the one hand, it can be seen from Figure 5 that the microhardness of the laser cladding coating is higher than that of the TA2 matrix. Therefore, under the same conditions, the wear resistance of the laser cladding coating is better than that of the TA2 matrix (Figure 6).

On the other hand, the wear resistance of materials depends not only on the hardness of the materials but also on the hardness of the abrasives. The microhardness of Ti_5Si_3, Ti_3Al, TA2 and SiC are 1500HV [28], $530HV_{0.2}$ [35], $200HV_{0.1}$ (Figure 5) and 3700HV [28], respectively. Therefore, when the applied load is 45N, it is inevitable that the wear mechanism of the TA2/SiC friction pairs is different from the coating (Ti_5Si_3 + Ti_3Al)/SiC friction pairs. That is, although the two groups of friction pairs cause oxidative wear due to the heat generated by dry friction and wear (as evidenced by the appearance of the O element in Table 3) and the hardness difference of the TA2/SiC friction pairs is larger, TA2 is prone to plastic deformation and adhesive wear, whereas the hardness difference of the coating (Ti_5Si_3 + Ti_3Al)/SiC friction pairs is relatively small. Ti_5Si_3 (Ti_3Al) is prone to brittle fracture and abrasive wear; however because the hardness of Ti_5Si_3 (Ti_3Al) is lower than that of SiC, adhesive wear also occurs. In conclusion, the TA2/SiC friction pairs mainly exhibit adhesive wear and oxidative wear, whereas the coating(Ti_5Si_3 + Ti_3Al)/SiC friction pairs mainly exhibit abrasive wear, adhesive wear and oxidative wear.

5. Conclusions

(1) The wear-resistant Ti_5Si_3/Ti_3Al composite coatings were successfully prepared on TA2 titanium alloy by laser cladding using the double layer preset method of Ti-63 wt.% Al mixed powder layer/Si powder layer.

(2) The coating is mainly composed of coarse primary Ti_5Si_3 phase and fine Ti_5Si_3/Ti_3Al eutectic structure, with more Ti_5Si_3 on the top layer and in the middle of the coating and more Ti_5Si_3/Ti_3Al eutectic on the bottom.

(3) The mass wear rate of the laser cladding coating is 1/5.79 of that of the TA2 matrix. The wear mechanisms of the coatings are mainly abrasive wear, adhesive wear and oxidative wear, whereas the wear mechanisms of the TA2 matrix are adhesive wear and oxidative wear.

(4) The effect of laser cladding of the double preset powder layer is similar to that of the single preset powder layer because of the strong stirring effect of the Marongoni flow in the molten pool.

Author Contributions: Investigation—K.H. and W.H.; writing—original draft preparation, W.H.; writing—review and editing, K.H. All authors have read and agreed to the published version of the manuscript.

Funding: This work was supported by the Open Project Program (No. 2022-KF-19) of the State Key Laboratory of Advanced Technology for Materials Synthesis and Processing, Wuhan University of Technology.

Data Availability Statement: The data of this study are available from the corresponding author upon reasonable request.

Acknowledgments: The authors are grateful to the Analytical and Testing Center of Huazhong University of Science and Technology.

Conflicts of Interest: The authors declare no conflict of interest.

References

1. Budzynski, P.; Kaminski, M.; Surowiec, Z.; Turek, M.; Wiertel, M. Effect of carbon ion implantation and xenon ion irradiation on the tribological properties of titanium and Ti6Al4V Alloy. *Acta Phy. Pol. A* **2022**, *142*, 713–722. [CrossRef]
2. Zhao, L.D.; Lugscheider, E. Reactive plasma spraying of TiAl6V4 alloy. *Wear* **2002**, *253*, 1214–1218. [CrossRef]
3. Srinivasan, R.; Kamaraj, M.; Rajeev, D.; Ravi, S.; Senthilkumar, N. Plasma spray coating of aluminum-silicon-MWCNT blends on titanium grade 5 alloy substrate for enhanced wear and corrosion resistance. *Silicon* **2022**, *14*, 8629–8641. [CrossRef]
4. Koshuro, V.; Fomina, M.; Zakharevich, A.; Fomin, A. Superhard Ta-O-N coatings produced on titanium using induction physical vapor deposition. *Ceram. Int.* **2022**, *48*, 19467–19483. [CrossRef]
5. Hussein, M.A.; Adesina, A.Y.; Kumar, A.M.; Sorour, A.A.; Ankah, N.; Al-Aqeeli, N. Mechanical, in-vitro corrosion, and tribological characteristics of TiN coating produced by cathodic arc physical vapor deposition on Ti20Nb13Zr alloy for biomedical applications. *Thin Solid Film.* **2020**, *709*, 138183. [CrossRef]
6. Chen, X.W.; Ren, P.; Zhang, D.F.; Hu, J.; Wu, C.; Liao, D.D. Corrosion and wear properties of h-BN-modified TC4 titanium alloy micro-arc oxide coatings. *Surf. Innov.* **2022**, *11*, 49–59. [CrossRef]
7. Majumdar, J.D.; Weisheit, A.; Mordike, B.L.; Manna, I. Laser surface alloying of Ti with Si, Al and Si+Al for an improved oxidation resistance. *Mater. Sci. Eng. A* **1999**, *266*, 123–134. [CrossRef]
8. Wang, Z.; Li, J.N.; Zhang, Z.; Li, J.M.; Su, M.L. Microstructure and wear resistance of Cu/La_2O_3 modified laser clad composites. *Laser. Eng.* **2019**, *44*, 33–39.
9. Ali, S.R.A.S.; Hussein, A.H.A.; Nofal, A.; Elnaby, S.I.H.; Elgazzar, H. A contribution to laser cladding of Ti-6Al-4V titanium alloy. *Metall. Res. Technol.* **2019**, *116*, 634.
10. Tian, Y.; Zhang, Z.; Li, J.N.; Huo, Y.S.; Su, M.L. Surface modification of TA7 titanium alloy with laser clad La_2O_3 amorphous coatings. *Laser. Eng.* **2019**, *44*, 331–339.
11. Li, X.D.; Liu, S.S.; Wang, J.W.; Yu, M.X.; Tang, H.B. Effect of different ZrN addition on microstructure and wear properties of titanium based coatings by laser cladding technique. *Coatings* **2019**, *9*, 261. [CrossRef]
12. Sui, X.M.; Lu, J.; Zhang, X.; Sun, L.; Zhang, W.P. Microstructure and properties of TiC-reinforced Ti_2Ni/Ti_5Si_3 eutectic-based laser cladding composite coating. *J. Therm. Spray Technol.* **2020**, *29*, 1838–1846. [CrossRef]
13. Chen, C.N.; Su, M. Study on microstructure and abrasive resistance of laser cladding TiN surface alloyed TC9. *J. Beijing Univ. Aeronaut. Astronaut.* **1998**, *24*, 253–255.
14. Xiang, K.; Chen, L.Y.; Chai, L.J.; Guo, N.; Wang, H. Microstructural characteristics and properties of $CoCrFeNiNb_x$ high-entropy alloy coatings on pure titanium substrate by pulsed laser cladding. *Appl. Surf. Sci.* **2020**, *517*, 146214. [CrossRef]
15. Wu, H.; Liang, L.X.; Lan, X.D.; Yin, Y.; Song, M.; Li, R.D.; Liu, Y.; Yang, H.O.; Liu, L.; Cai, A.H.; et al. Tribological and biological behaviors of laser cladded Ti-based metallic glass composite coatings. *Appl. Surf. Sci.* **2020**, *507*, 145104. [CrossRef]
16. Chai, L.J.; Wang, C.; Xiang, K.; Wang, Y.Y.; Wang, T.; Ma, Y.L. Phase constitution, microstructure and properties of pulsed laser-clad ternary CrNiTi medium-entropy alloy coating on pure titanium. *Surf. Coat. Technol.* **2020**, *402*, 126503. [CrossRef]
17. Ke, J.; Liu, X.B.; Liang, J.; Liang, L.; Luo, Y.S. Microstructure and fretting wear of laser cladding self-lubricating anti-wear composite coatings on TA2 alloy after aging treatment. *Opt. Laser Technol.* **2019**, *119*, 105599. [CrossRef]
18. Guo, C.; Ma, M.L.; Chen, F.; Wei, B.L. Microstructure and space tribological properties of NiCrBSi/Ag composite coating prepared by laser cladding. *Surf. Technol.* **2019**, *48*, 177–184.
19. Guo, C.; Zhou, J.S.; Zhao, J.R.; Wang, L.Q.; Yu, Y.J.; Chen, J.M.; Zhou, H.D. Improvement of the oxidation and wear resistance of pure Ti by laser-cladding Ti_3Al coating at elevated temperature. *Tribol. Lett.* **2011**, *42*, 151–159. [CrossRef]
20. Zhai, Y.J.; Liu, X.B.; Qiao, S.J.; Wang, M.D.; Lu, X.L.; Wang, Y.G.; Chen, Y.; Ying, L.X. Characteristics of laser clad α-Ti/TiC+(Ti,W)C_{1-x}/Ti_2SC+TiS composite coatings on TA2 titanium alloy. *Opt. Laser Technol.* **2017**, *89*, 97–107. [CrossRef]
21. Liu, Y.N.; Yang, L.J.; Yang, X.J.; Zhang, T.G.; Sun, R.L. Optimization of microstructure and properties of composite coatings by laser cladding on titanium alloy. *Ceram. Int.* **2021**, *47*, 2230–2243. [CrossRef]
22. An, Q.; Qi, W.J.; Zuo, X.G. Microstructure and wear resistance of in-situ TiC reinforced Ti-based coating by laser cladding on TA15 titanium alloy surface. *J. Mater. Eng.* **2022**, *50*, 139–146.

23. Ye, Z.Y.; Li, J.N.; Liu, L.Q.; Ma, F.K.; Zhao, B.; Wang, X.L. Microstructure and wear performance enhancement of carbon nanotubes reinforced composite coatings fabricated by laser cladding on titanium alloy. *Opt. Laser Technol.* **2021**, *139*, 106957. [CrossRef]
24. Maliutina, I.N.; Si-Mohand, H.; Sijobert, J.; Lazurenkoa, D.V.; Bataev, I.A. Structure and oxidation behavior of γ-TiAl coating produced by laser cladding on titanium alloy. *Surf. Coat. Technol.* **2017**, *319*, 136–144. [CrossRef]
25. Weng, F.; Yu, H.J.; Liu, J.L.; Chen, C.Z.; Dai, J.J.; Zhao, Z.H. Microstructure and wear property of the Ti_5Si_3/TiC reinforced Co-based coatings fabricated by laser cladding on Ti-6Al-4V. *Opt. Laser Technol.* **2017**, *92*, 156–162. [CrossRef]
26. Jiang, P.; Zhang, J.J.; Yu, L.G.; Wang, H.M. Wear-resistant Ti_5Si_3/TiC composite coatings made by laser surface alloying. *Rare Metal Mater. Eng.* **2000**, *29*, 269–272.
27. Guo, C.; Zhou, J.S.; Chen, J.M.; Zhao, J.R.; Yu, Y.J.; Zhou, H.D. Improvement of the oxidation and wear resistance of pure Ti by laser cladding at elevated temperature. *Surf. Coat. Technol.* **2010**, *205*, 2142–2151. [CrossRef]
28. Zhang, H.X. Investigation on Microstructure and Wear Resistance of Ti-Al-Si Ceramic Composite Coatings on Titanium Alloys by Laser Cladding. Ph.D. Thesis, Shandong University, Jinan, China, 2016.
29. Vojtěch, D.; Novák, M.; Novák, P.; Lejček, P.; Kopeček, J. Unidirectional crystallization and high-temperature oxidation of in situ Ti_3(Al,Si)-Ti_5(Si,Al)$_3$ composite. *Mater. Sci. Eng. A* **2008**, *489*, 1–10. [CrossRef]
30. Wu, J.S.; Qiu, G.H.; Zhang, L.T. The β-Ti(Al,Si)+Ti_5(Si,Al)$_3$ eutectic line in the Ti-Al-Si ternary system. *Mater. Sci. Technol.* **1994**, *2*, 1–7. (In Chinese)
31. Wu, J.S.; Beaven, P.A.; Wagner, R. The Ti_3(Al,Si)+Ti_5(Si,Al)$_3$ eutectic reaction in the Ti-Al-Si system. *Scr. Metall. Mater.* **1990**, *24*, 207–212. [CrossRef]
32. Wu, J.S.; Qiu, G.H.; Zhang, L.T. The β-Ti(Al,Si)+Ti_5(Si,Al)$_3$ eutectic line in the Ti-Al-Si ternary system. *Scr. Metall. Mater.* **1994**, *30*, 213–218. [CrossRef]
33. Liu, J.L.; Zou, Z.R.; Su, B.R. *High Energy Beam Heat Treatment*; China Machine Press: Beijing, China, 1997; pp. 40–46.
34. Jackson, M.J.; Robinson, G.M.; Gill, M.D.H.; Neill, W.O. The effect of nozzle design on laser micro-machining of M2 tool steels. *J. Mater. Process. Technol.* **2005**, *160*, 198–212. [CrossRef]
35. Guo, C.; Chen, J.M.; Yao, R.G.; Zhou, J.S.; Zhang, S.T. Microstructure and tribological properties of Ti_3Al intermetallic compound coating prepared by laser cladding. *Tribol. Lett.* **2013**, *33*, 14–21.
36. Wang, Y.; Xiong, J.; Yan, J.; Fan, H.Y.; Wang, J. Oxidation resistance and corrosion behavior of hot-dip aluminized coatings on commercial-purity titanium. *Surf. Coat. Technol.* **2011**, *206*, 1277–1282. [CrossRef]
37. Shi, G.N.; Wu, S.H.; Lu, X.L.; Liu, X.B. Microstructure and property of laser clad Ti_2SC/TiS self-lubricating anti-wear composite coating on TA2 titanium alloy. *Trans. Mater. Heat Treat.* **2016**, *37*, 198–202.
38. Okamoto, H. O-Ti (Oxygen-Titanium). *J. Phase Equilibria Diffus.* **2011**, *32*, 473–474. [CrossRef]
39. Yu, P.C.; Liu, X.B.; Liu, X.L.; Zhu, G.X.; Chen, Y.; Wu, S.H. Study on tribology and high-temperature oxidation resistance of laser cladding composite coatings on Ti6Al4V alloy. *Chin. J. Lasers* **2015**, *42*, 81–88.
40. Yu, J.J.; Zhang, X.Y.; Ma, L.L.; Li, H.; Zou, L.J.; Zhang, W.P. Microstructures and properties of laser cladding Ti/Ni+Si_3N_4/ZrO_2 composite coatings on TA15 titanium alloys. *Surf. Technol.* **2016**, *45*, 105–109.
41. Cui, Z.Y.; Fang, C.; Zhang, W.L. Effects of Marangoni flow on the thermal behavior and melt flow behavior in laser cladding. *Appl. Laser* **2018**, *38*, 409–416.
42. Lee, Y.S.; Nordin, M.; Babu, S.S.; Farson, D.F. Influence of fluid convection on weld pool formation in laser cladding. *Weld. J.* **2014**, *93*, 292–300.
43. Zhao, M.J.; Li, B.X.; Wu, T.; Zhao, L.Z.; Zeng, J.B.; Jiao, H.T.; Li, J.; Song, L.J. Study on the effect mechanism of laser cladding velocity field based on thermal-fluid coupling model. *J. East China Jiaotong Univ.* **2022**, *39*, 74–83.
44. Mills, K.C.; Su, Y.C. Review of surface tension data for metallic elements and alloys: Part 1—Pure metals. *Int. Mater. Rev.* **2006**, *51*, 329–351. [CrossRef]
45. Raghavan, V. Al-Si-Ti(Aluminum-Silicon-Titanium). *J. Phase Equilibria Diffus.* **2009**, *30*, 82–83. [CrossRef]
46. ASM International. *ASM Metals Handbook Volume 3-Alloy Phase Diagrams*; The Materials Information Company: Houston, TX, USA, 1992; pp. 327, 1420.
47. Zhu, Y.; Yang, Y.Q.; Ma, Z.J.; Chen, Y. Thermodynamic studies on interfacial reactions in SiC-fibre-reinforced Ti-matrix composites. *Rare Met. Mater. Eng.* **2002**, *31*, 279–282.
48. Vojtěch, D.; Novák, M.; Novák, P.; Kopeček, J.; Guhlová, P. Properties of (Ti,Nb)Al–(Ti,Nb)Si eutectic composite. *Compos. Struct.* **2010**, *92*, 1440–1448. [CrossRef]
49. Archard, J.F. Contact and rubbing of flat surfaces. *J. Appl. Phys.* **1953**, *24*, 981–988. [CrossRef]

lubricants

MDPI

Article

Wear Resistance of In Situ NbC-Reinforced Laser Cladding Ni45 Coatings

Yingpeng Liu [1], Kaiming Wang [2], Hanguang Fu [1,*], Bin Zong [1] and Jiguang Zhang [1]

[1] Key Laboratory of Advanced Functional Materials, Ministry of Education, Department of Materials Science and Engineering, Beijing University of Technology, Beijing 100124, China; liuyp@emails.bjut.edu.cn (Y.L.); zongbin@bjut.edu.cn (B.Z.); zhangjiguang@bjut.edu.cn (J.Z.)

[2] College of Automobile and Mechanical Engineering, Changsha University of Science and Technology, Changsha 410114, China; kmwang@csust.edu.cn

* Correspondence: hgfu@bjut.edu.cn; Tel.: +86-10-67396093

Abstract: In situ NbC-reinforced laser cladding Ni45 coatings have the advantages of high bond strengths, low dilution rates, small heat-affected zones and good wear resistance and have broad application prospects in the field of surface strengthening and repair of workpieces such as automotive molds and engine turbines. Previous studies have mostly used pure niobium powder for in situ synthesis to prepare Ni-based NbC coatings with a high production cost. In this paper, NbC was successfully synthesized in situ in Ni45 powder using inexpensive FeNb65 and Cr3C2. The prepared coating has a uniform microstructure and excellent wear resistance, and the reinforced phases are mainly NbC and Cr23C6. Coating 4# with 25 wt.% FeNb65 + Cr3C2 has the highest microhardness of 776.3HV0.2, about 1.45 times that of the Ni45 coating, and its wear resistance is 36.36 min/mg, about 60.6 times that of the Cr12MoV steel base material and about 23.76 times that of the Ni45 coating.

Keywords: laser cladding; in situ NbC-reinforced nickel-based coating; microstructure; microhardness; wear resistance

1. Introduction

Laser cladding has the advantages of a low dilution rate and a fast cooling rate [1–4]. It may be used to strengthen material surfaces or to repair the surface of damaged materials and has very broad application prospects [5–7]. Metal-based composite coatings are high-quality coatings prepared by mixing self-soluble alloy powders with hard-phase powders by laser cladding [8–10]. They have the advantages of high bond strengths, small heat-affected zones, high microhardnesses, and good wear resistance. Nickel-based NbC composite coatings are widely used in the surface strengthening and repair of automotive molds, engine turbines, and combustion turbine blades due to their high microhardness and good wear resistance [11–15].

According to the formation method of hard ceramic phases within the coatings, coating preparation methods can be categorized into the direct addition method and the in situ synthesis method. The direct addition method is convenient and fast, but the hard ceramic phase is less compatible with the substrate. The in situ synthesis method is the main method of coating preparation, as the hard ceramic phase is uniformly distributed and well combined with the substrate. Several scholars have tried to add a NbC hard ceramic phase to coatings by in situ synthesis. Initially, graphite and pure niobium powder were used to synthesize NbC in situ [16,17]. However, due to the large difference in density between graphite and metal powder, graphite tends to float on the surface of the mixed powder when the powder is mixed, resulting in large differences in the organization of different regions of the coating. In addition, graphite has a low ignition point and can be easily burned during laser cladding, thus seriously affecting the stability of the composition in the coating. To solve this problem, the teams of Chang Baohua at Tsinghua University and

Citation: Liu, Y.; Wang, K.; Fu, H.; Zong, B.; Zhang, J. Wear Resistance of In Situ NbC-Reinforced Laser Cladding Ni45 Coatings. *Lubricants* **2023**, *11*, 316. https://doi.org/10.3390/lubricants11080316

Received: 27 June 2023
Revised: 19 July 2023
Accepted: 25 July 2023
Published: 26 July 2023

Tianbiao Yu at Northeastern University tried to replace graphite with B4C to provide a sufficient amount of carbon for the formation of hard ceramic phases in the coatings [18,19]. Fu Hanguang's team at Beijing University of Technology tried to replace graphite with Cr3C2 in an in situ synthesis reaction [20]. While solving the problem of graphite burnout, the added Cr elements can also form hard ceramic phases such as Cr23C6 and Cr7C3 to further increase the microhardness of the coating. In previous studies of NiCrBSi-NbC composite coatings, the Nb element was mostly provided in the coating by adding pure niobium powder. However, due to the high price of pure niobium powder and its high melting point of 2468 °C, it is not feasible to apply it in practical production. In this paper, we try to use FeNb65 (melting point 1570–1650 °C), which is inexpensive and has a low melting point, to replace pure niobium powder. A NiCrBSi-NbC coating with excellent wear resistance and a low price was prepared by in situ synthesis of NbC with Cr3C2. The microstructure and wear resistance were also tested and analyzed, expecting to provide a reference for the application of in situ NbC-reinforced laser melting of nickel-based composite coatings.

2. Experiment

2.1. Materials

The laser cladding experiment was carried out on Cr12MoV die steel with dimensions of 12 mm × 60 mm × 100 mm. Before the test, the upper surface of the Cr12MoV steel substrate was polished with an angle grinder to remove rust. Subsequently, cotton dipped in acetone was used to wipe off the grease on the surface of the substrate as well as any residual debris from the grinding process. The laser cladding material was Ni45 powder produced by Beikuang New Material Technology Co., Ltd., Beijing, China., which is a standard spherical powder in the particle size range of 45~105 μm. The specific composition of Cr12MoV steel and Ni45 powder is shown in Table 1. The materials used for the in situ synthesis of NbC were 45~105 μm FeNb65 and Cr3C2, both of which were purchased from Xing Rongyuan Technology Co., Ltd., Beijing, China. The theoretical mass fraction of Nb in FeNb65 is 65 wt.%, and the actual mass share is 64.76 wt.%. The atomic ratio of Cr to C in Cr3C2 is 3:2, and the mass fractions of Cr and C are 86.7% and 13.3%. The SEM morphology photographs of FeNb65 and Cr3C2 are shown in Figure 1. Initially, FeNb65 and Cr3C2 were mixed according to a 1:1 molar ratio of Nb to C. Subsequently, a mixed powder of FeNb65 and the Cr3C2 and Ni45 powder was placed into a ball mill and mixed well according to different ratios. The prepared powders were put into a drying oven at a constant temperature of 80 °C for 2 h and set aside. The coating number and composition design are shown in Table 2.

Table 1. Composition of Cr12MoV and Ni45 (wt.%).

	Ni	Cr	B	Si	Fe	C	Mn	P	V	Mo	S	Cu
Cr12MoV	0.12	12.30	-	0.27	Bal.	1.44	0.21	0.02	0.27	0.55	0.01	0.13
Ni45	Bal.	10.02	1.89	3.05	5.30	0.33	-	-	-	-	-	-

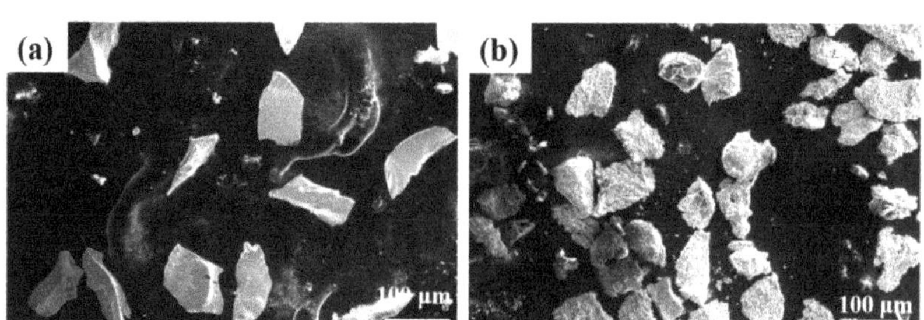

Figure 1. Morphology of Cr3C2 (**a**) and FeNb65 (**b**) powders selected for the in situ synthesis of NbC.

Table 2. Coating numbers and composition (wt.%).

	Ni45	FeNb65 + Cr_3C_2
1#	100	0
2#	85	15
3#	80	20
4#	75	25

2.2. Laser Parameters

A YLS-6000-S2 fiber laser made by IPG of Oxford, MA, USA. was used for the test, with a maximum output power of 6000 W. The laser system generates a continuous laser with a wavelength of 1070 nm and transmits it through the fiber to the cladding head, where it acts on the substrate in a square spot of 5 mm × 5 mm. The melting head was driven by a six-axis linkage robot from ABB. In addition, an intra-optical coaxial powder feeder was connected to the cladding head to transport the powder. High-purity argon gas (99.99% purity) at 15 L/min was fed into the laser cladding process to avoid oxidation of the melt pool. A schematic diagram of laser cladding is shown in Figure 2. The process parameters of laser cladding were as follows: laser power, 2100 w; powder feeding rate, 15 g/min; scanning speed, 4 mm/s; and lap rate, 32%.

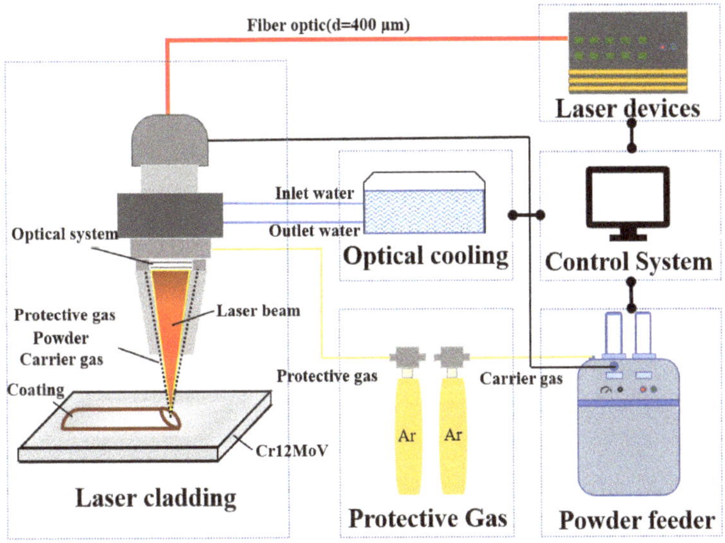

Figure 2. A diagram of the laser cladding experimental setup.

2.3. Microstructure Observation

Metallographic specimens were cut from a single coating in the direction perpendicular to the coating and mechanically ground in a grinding and polishing machine with 240 mesh, 400 mesh, 1000 mesh, 1500 mesh, 2000 mesh and 3000 mesh SiC sandpaper in that order. It was subsequently polished with a diamond polishing paste with a particle size of W2.5. The finished specimens were etched with aqua regia (HNO_3:HCl = 1:3) for 22 s. The microstructure of the coatings was observed using a scanning electron microscope (ZEISS Gemini SEM 300 in Germany) and elemental analysis was performed with an additional energy spectrum probe (EDS). A 19 mm × 12 mm × 12 mm specimen was cut from the coating, and the phase composition of the coating was analyzed using a Cu-Kα radiation X-ray diffractometer (BRUKER D8 Advance in Germany) after mechanical grinding with 240 and 400 mesh SiC sandpaper in turn. The specific parameters were as follows: tube voltage, 40 kV; tube current, 40 mA; scanning speed, 8°/min; scanning range, 20~90° coupled continuous scanning; and step size, 0.02°.

2.4. Microhardness and Wear Resistance

To ensure the accuracy of the microhardness test, the specimens were mechanically ground on a grinding and polishing machine with SiC sandpaper of 240, 400, 1000, 1500, 2000 and 3000 mesh in that order. The prepared specimens were measured using a microhardness tester (MICRO-MET-5103 in Germany) with an indenter load of 1.96 N and a loading time of 10 s. Five locations were selected at 250 µm intervals starting from the upper part of the coating and the average value was taken as the microhardness at that location after removing the best value.

To ensure the testing accuracy of coating wear resistance, specimens of 19 mm × 12 mm × 12 mm were mechanically ground using 240 mesh and 400 mesh SiC sandpaper in turn. The prepared specimens were subjected to a friction wear test using a high-speed ring block friction wear tester (Jnstart MRH-3W in China). According to GB/T 3960-1983, the test load was 196 N, the wear time was 2 h, the speed as 200 r/min, the hardness of the grinding ring was 60 HRC and the material was GCr15 steel. The specimens were ultrasonically cleaned with anhydrous ethanol before and after the frictional wear test, and the weight was recorded by weighing them several times on a balance with an accuracy of 0.1 mg. The wear weight loss of the specimens was determined and the wear resistance of the coating was calculated. The surface morphology and roughness of the worn specimens were observed with a laser confocal microscope (OLS40-CB in Japan) and a scanning electron microscope.

3. Results and Discussion

3.1. Microstructure Analysis

The coatings with different compositions were analyzed by an X-ray diffractometer and the results are shown in Figure 3. By comparing the standard diffraction cards and using Jade 6 analysis software, it can be seen that the possible phases of coating 1# are γ-Ni, Cr23C6, Cr7C3 and CrB. Coatings 2#, 3# and 4# are probably γ-Ni, Cr23C6, CrB and NbC. With the increase in NbC content, the intensity of the three NbC diffraction peaks at 35°, 41° and 58° increases significantly. The results show that it is feasible to synthesize a NbC hard ceramic phase using ferric niobium and Cr3C2 for nickel-based coatings.

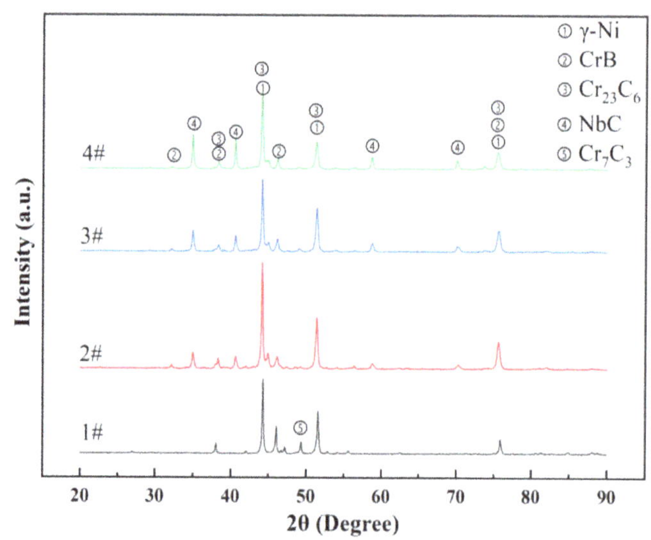

Figure 3. XRD patterns of nickel-based composite coatings with different compositions.

The microstructure photos in Figure 4 show that a large number of bright white phases are distributed inside coatings 2#, 3# and 4#, which is the typical morphology of the NbC phase. This result confirms that the in situ synthesis reaction of Cr3C2 and FeNb65 to generate NbC hard ceramic phases is feasible. Moreover, due to the similar density of NbC and Ni matrices, there is no obvious concentration gradient in each phase in the coating and NbC is uniformly distributed in the coating. In addition, the presence of long gray stripes, black block phases and gray irregularly shaped phases in the coating can be observed in the four sets of images.

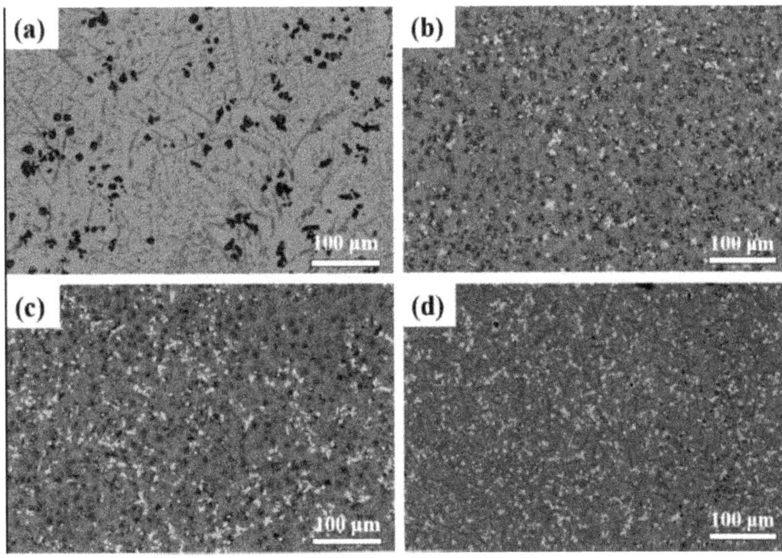

Figure 4. Microstructures of the coating at low magnification under SEM: (**a**) 1#; (**b**) 2#; (**c**) 3#; (**d**) 4#.

To further analyze the morphology of the phases within the coating, the microstructure morphology of the coating was observed at a higher magnification. The results are shown in Figure 5. Coatings 2# and 3# have irregular polygonal blocks of the NbC hard ceramic phase, while in coating 4#, the morphology of NbC is mostly regular quadrilateral and cross shaped. This is caused by the difference in the Nb content in the melt pool. When the amount of Nb in the melt pool is comparatively low, the nucleation and growth of NbC is limited. As the NbC content increases, the octahedral structure of NbC grows along the [100] direction, forming polygonal or cross-shaped structures with regular edges [9]. In addition, long black phases cross-arranged in clusters were observed in coatings 2# and 3#. In contrast, in specimen 4#, these black clustered phases disappeared and were replaced by black block phases. The long gray stripe phase became thicker and its color was clearly visible in BSE mode, presumably due to the solid solution of some elements of the cluster phase within the long gray stripe phase. The composition of the various phases needs to be further analyzed.

To further determine the types of each phase and the distribution of elements, the mapping of coatings 2#, 3# and 4# was performed using EDS in combination with SEM. As shown in Figures 6 and 7, the matrices of coatings 2# and 3# are mainly Ni as well as a solid solution of Fe and Cr, which is known to be γ-Ni solid solution from XRD analysis. This is due to the rapid diffusion of Fe, Cr and Si from the melt pool into the high temperature austenite phase and solid solution strengthening under the irradiation of the laser beam. The main elements of the gray elongated phase are Cr and C, which are known as Cr23C6 from XRD analysis. The bright white bulk phase is mainly composed of Nb and C, which can be identified as NbC in combination with XRD analysis. In addition, a B solid solution can also be seen in the NbC phase, forming a Nb(C, B) solid solution [21]. The main elements of the black cluster-like phase are Cr and B. Combined with the XRD analysis, it is identified as a CrB phase. As shown in Figure 8, the black cluster-like phase largely disappeared in the coating 4#. It is presumed that with the increase in the amount of added FeNb65, a large number of Nb(C, B) hard ceramic phases are formed, which occupy the majority of B. Therefore, with the increase in NbC content, the CrB phase in the coating gradually disappears and the content of Cr23C6 gradually increases.

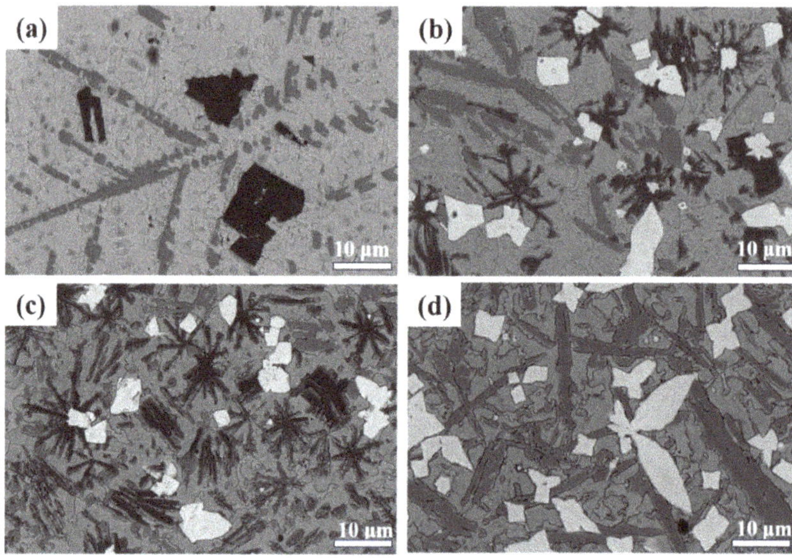

Figure 5. Microstructures of the coating at high power under SEM: (**a**) 1#; (**b**) 2#; (**c**) 3#; (**d**) 4#.

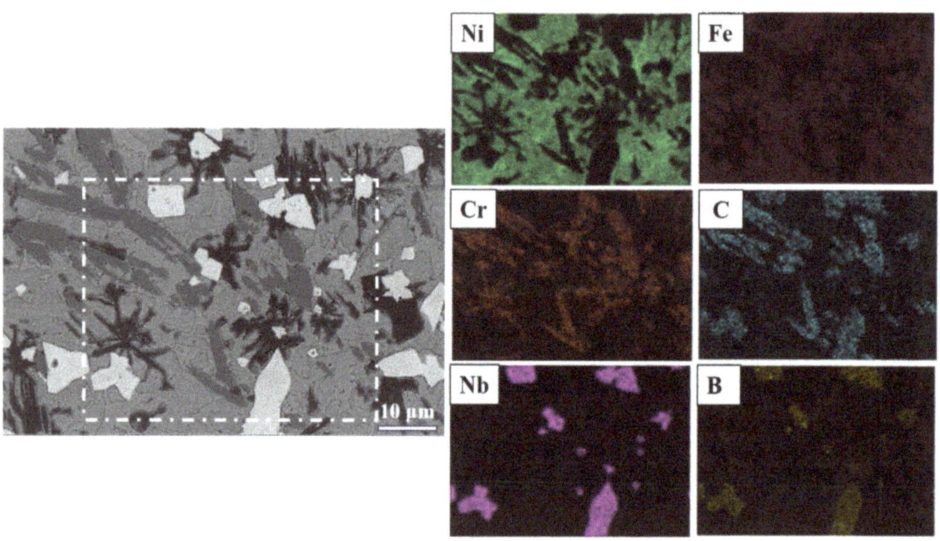

Figure 6. SEM—element distribution mapping of coating 2#.

Figure 7. SEM—element distribution mapping of coating 3#.

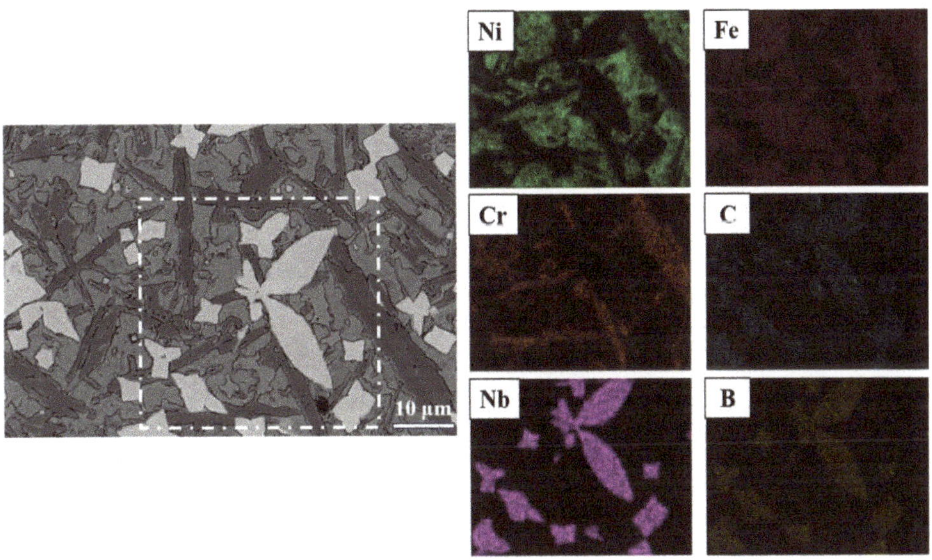

Figure 8. SEM—element distribution mapping of coating 4#.

3.2. Microhardness Analysis

Microhardness is an important indicator of the mechanical properties of coatings, and the microhardness of the coating was tested at different depths along the direction perpendicular to the coating. The microhardness of the four groups of coatings at different depths and the average microhardness of the surface layer (3250 µm) are shown in Figure 9a,b, respectively. As shown in Figure 9a, the microhardness curves of the coatings show the same trend, with a relatively uniform microhardness in the superficial region. The heat-affected zone has a higher microhardness than the substrate due to it experiencing faster thermal cycling. As shown in Figure 9b, the microhardness of coating 1# is the lowest at about 535.3 HV0.2. The microhardnesses of coatings 2#, 3# and 4# increase sequentially, and the microhardness of coating 4# is the highest at 776.33 HV0.2, 45% higher compared to coating 1#. The enhancement of the coating hardness is mainly due to two reasons: One is the in situ synthesis of NbC with a hardness up to 2458.2 HV through ferro-niobium and $Cr3C2$, which effectively improves the microhardness of the coating. The second is the formation of $Cr23C6$ in the coating by the excess addition of $Cr3C2$, which also leads to a great improvement in the microhardness of the coating [22,23]. The relationship between the microhardness of the coating and the wear resistance is generally positively correlated, and the wear resistance of the coating can be effectively improved by enhancing the microhardness of the coating.

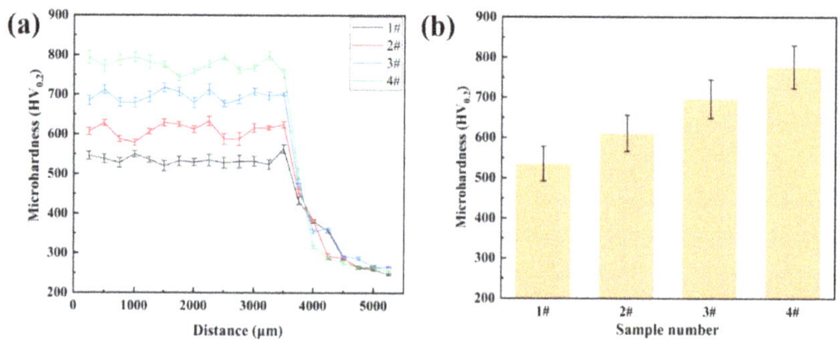

Figure 9. Microhardness of the coatings: (**a**) microhardness at different depths; (**b**) average microhardness.

3.3. *Wear Resistance Analysis*

To study the wear resistance of the coating, the specimens were analyzed for wear resistance using a high-speed ring block friction wear tester. The wear loss and wear resistance of the coating after 2 h at 196 N are shown in Figure 10a. The wear loss of the base material Cr12MoV steel was as high as 201.5 mg, while the wear losses of the remaining four groups of coatings were 78.8 mg, 10.8 mg, 4.8 mg and 3.3 mg, respectively. The wear resistance of the coatings was characterized by the inverse of the average wear weight loss per unit time, and the wear resistances of the substrate and the four groups of coatings were 0.6 min/mg, 1.53 min/mg, 11.11 min/mg, 25 min/mg and 36.36 min/mg, respectively. The wear resistance of the Ni45 coating 1# increased by 155% compared to the substrate. The wear resistances of coatings 2#, 3# and 4# further improved due to the generation of high-hardness NbC and $Cr23C6$ phases. Among them, coating 4# showed the best wear resistance, which was increased by 606% compared to the base material. Figure 10b illustrates the friction coefficients of the four coatings during frictional wear. The friction coefficients of the coatings with the addition of a hard ceramic phase all decrease, but the friction coefficient of coating #2 fluctuates more, which may be due to the exfoliation of the hard phase with a larger particle size during the wear process. The improvement in the surface wear resistance of Cr12MoV steel is more significant compared with previous strengthening methods. Additionally, there are a wealth of very promising applications due to the reduced economic cost of surface strengthening of the material [24–26].

In order to analyze the wear morphology of the coating, the wear surface of the coating was observed using a laser confocal microscope. Three-dimensional images were constructed according to the wear depth, and the wear morphology and wear parameters of the coating surface are shown in Figure 11 and Table 3, respectively. Combining Figure 11 with Table 3, we can easily see that the width and depth of the wear marks and the cross-sectional area of coating 1# are the highest among the four groups of coatings. As the content of NbC increases, the width and depth of the abrasion marks and the cross-sectional area decrease significantly. Compared with coating 1#, the width and depth of wear marks and cross-sectional area of coating 4# decreased by 47.1%, 44.7% and 21.1%, respectively, and the wear resistance of the coating improved remarkably.

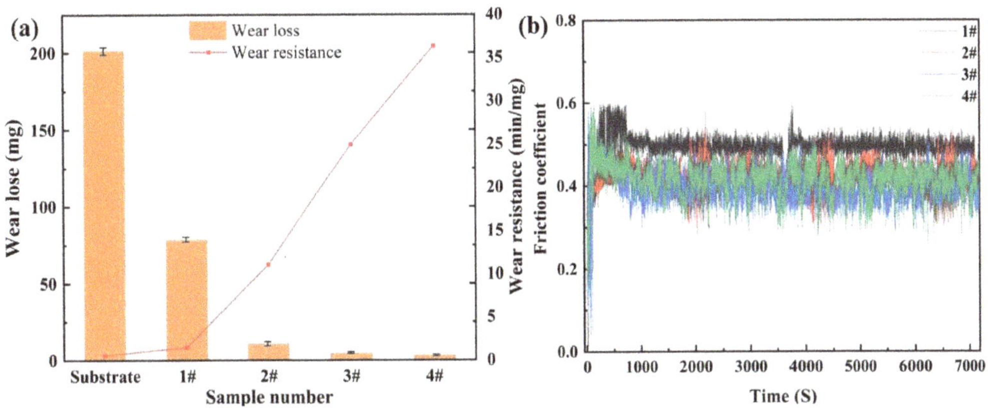

Figure 10. Wear characteristics of coatings: (**a**) wear weight loss and wear resistance; (**b**) friction coefficients.

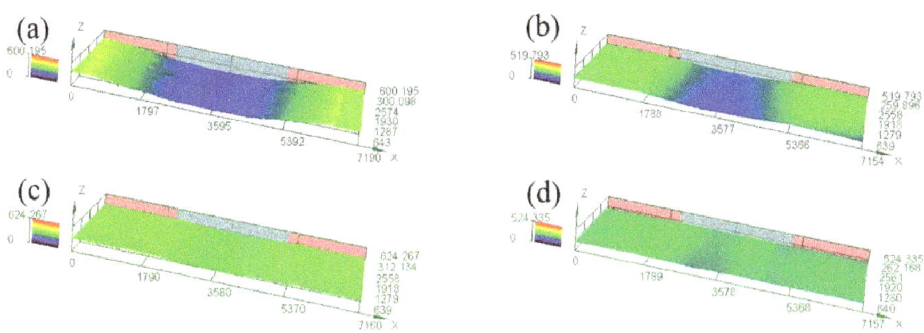

Figure 11. Three-dimensional diagram of the wear pattern on the surface of coatings (**a**) 1#; (**b**) 2#; (**c**) 3#; (**d**) 4#.

Table 3. Width, depth and cross-sectional area of the wear profile of coating surfaces.

	1#	2#	3#	4#
Width (μm)	6489.0	4033.1	3061.8	3059.3
Depth (μm)	26.5	22.6	21.4	11.9
Area (μm^2)	86,122.0	45,636.6	32,808.3	18,129.1

Figure 12 shows the wear morphology under a scanning electron microscope after frictional wear tests for the different compositions of coatings. From Figure 12a, it can be seen that there are plow grooves, a large amount of debris and a compaction layer on the surface of coating 1#, and the whole surface of the coating exhibits a wavy undulation. This is primarily because the hard ceramic phase of the coating flakes off during the wear of the coating and friction. During the relative movement of coating and grinding, ring scratches of different depths are left on the surface of the coating. Subsequently, some of the debris collects near the spalling pits and forms a compacted layer under a load of 196 N and tangential friction. In addition, because the Ni45 coating is relatively soft, the hard ceramic phase was more deeply pressed, forming wave-type undulations on the coating surface. Taken together, the wear mechanism of coating 1#'s surface is severe abrasive wear and intense adhesive wear. From Figure 12b, it can be seen that uniform plow grooves and debris also exist on the wear surface of coating 2#. The presence of hard ceramic phases can be observed in the grooves, and with the protection of these hard ceramic phases, the coating avoids further wear. Figure 12c,d shows that the wear mechanisms of coatings

3# and 4# are similar, with the coating surface uniformly distributed with grooves and a small amount of debris. It can be observed that more hard ceramic phases are formed inside the coating, which enhances the hardness of the coating and reduces the depth of the grooves. The wear mechanism of the coatings of these two samples is mainly abrasive wear. The width of the grooves on the surface of the four groups of coatings was measured, and the average widths were 26.0 μm, 18.0 μm, 15.0 μm and 14.5 μm, respectively. This measurement confirms that the formation of a hard ceramic phase improves the wear resistance of the coatings.

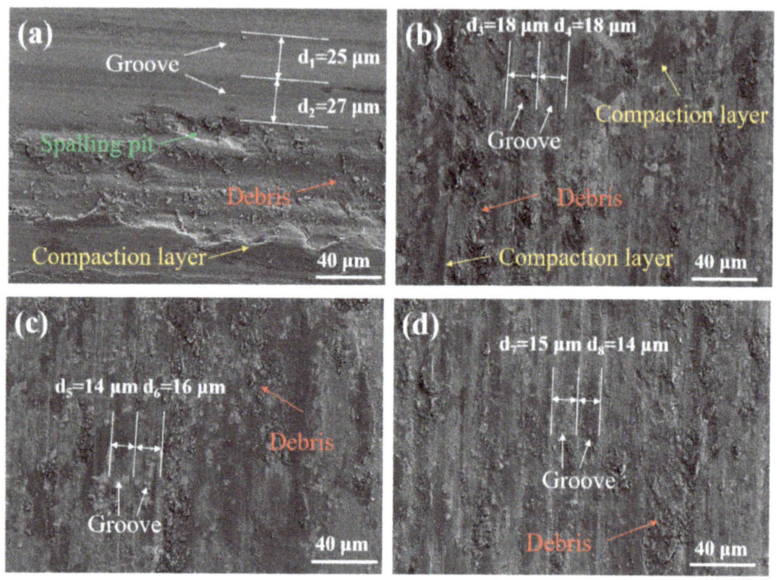

Figure 12. Morphology of the worn surface of coating (**a**) 1#; (**b**) 2#; (**c**) 3#; (**d**) 4#.

The wear weight loss of coatings 1# and 2# is high, and the exfoliated hard ceramic phase with the abraded debris is compacted under the extrusion of the grinding ring. It has been speculated that these compacted layers have some protective effect on the coating, thus improving the wear resistance of the coating [27–30]. To analyze the effect of these compacted layers on the wear resistance of the coating, the friction wear test was adjusted. The specific operation was to fix a ball of cotton in the inner cavity of the friction and wear tester, so that it was in close contact with the grinding ring. Thus, abrasive debris is removed from the surface of the grinding ring over time during the wear process to avoid the formation of a compacted layer. Before and after the test, the specimens were weighed using a balance with an accuracy of 0.1 mg, and the wear morphology of the coating was observed by scanning electron microscopy. By measuring the wear weight loss of the coating after modification of the high-speed ring block friction wear tester and observing the wear morphology of the coating, the effect of the compacted layer on the coating surface was studied. A schematic diagram of the modification of the experimental setup is shown in Figure 13. The specimen was rerun on the modified machine for a wear test with a test load of 196 N and a wear time of 2 h.

The wear morphology of the coating surface after the modification of the frictional wear experimental machine is shown in Figure 14. It can be seen that the addition of cotton on the grinding ring could effectively remove the debris on the surface of the coating, and the depth of the grooves on each coating surface was significantly reduced.

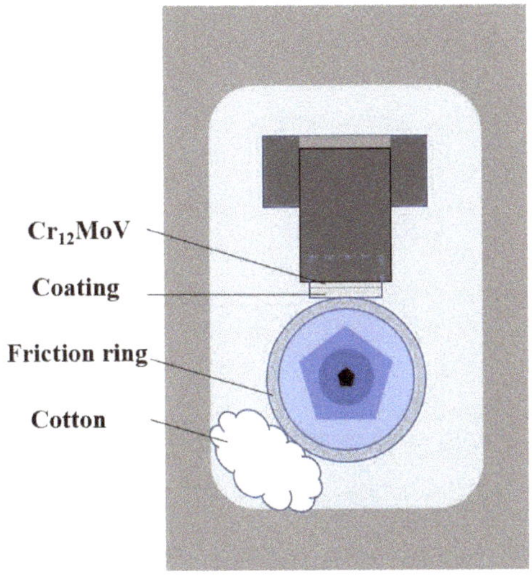

Figure 13. Schematic diagram of the adjusted friction and wear experimental setup.

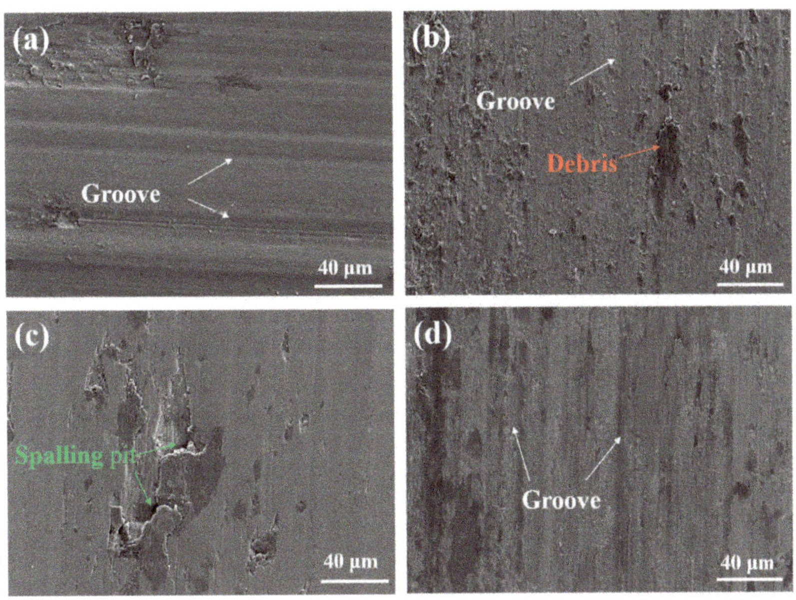

Figure 14. The wear morphology of the coating surface after adding cotton: (**a**) 1#; (**b**) 2#; (**c**) 3#; (**d**) 4#.

A comparison of the wear weight loss of the coatings before and after modification of the high-speed friction and wear tester is shown in the Figure 15. For coatings #1 and #2, which have a higher wear weight loss, the weight of wear increases. This is due to the removal of abrasive debris from the grinding ring, reducing the area of the compacted layer on the coating surface and thus reducing the protection of the coating surface. The decrease in the wear weight loss for coatings 3# and 4#, where abrasive wear dominates, is due to the removal of larger hard ceramic phase particles, which flake from the coating surface to avoid more wear on the coating surface from large hard ceramic phase particles.

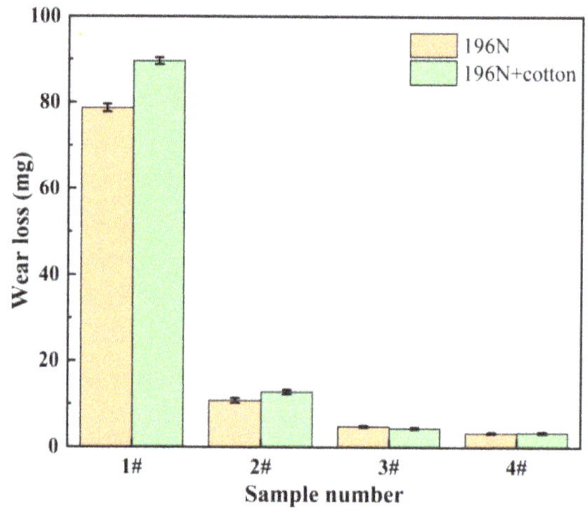

Figure 15. Comparison of wear weight loss of coatings before and after modification of the friction and wear tester.

4. Conclusions

In this study, laser-clad NbC-reinforced nickel-based composite coatings were successfully prepared using a mixture of Ni45, FeNb65 and Cr_3C_2 powders. The changes in microstructure, microhardness and wear resistance of the coatings were investigated in detail. The results show the significant potential of the coatings to be applied in the field of surface strengthening and repair. The findings of the study are summarized as follows:

1. NbC was synthesized in situ from FeNb65 and Cr3C2, and elements such as Fe, Cr and Si diffused into the high temperature austenite phase to form a γ-Ni solid solution phase.
2. With the increase in NbC content, the microhardness of the coating increases. The microhardness of coating 4# reached 776.3 HV0.2, a 45% increase in microhardness compared to coating 1#. This is attributed to the uniform distribution of the NbC phase and $Cr_{23}C_6$ in the coating.
3. The wear resistance of the coatings increased significantly with the increase in NbC content. Coating 4# with 25 wt.% FeNb65 + Cr_3C_2 showed the best wear resistance of 36.36 min/mg, an improvement of 606.0% in abrasion resistance compared to the base material and 227.6% compared to coating 1#.
4. The compacted layer formed by abrasive extrusion protects the coating surface when subjected to wear, reducing the wear of the coating.

Author Contributions: Conceptualization, H.F. and Y.L.; methodology, K.W.; resources, H.F.; data curation, Y.L.; writing—original draft preparation, Y.L.; writing—review and editing, H.F. and Y.L.; visualization, B.Z. and J.Z.; funding acquisition, H.F. All authors have read and agreed to the published version of the manuscript.

Funding: This research was funded by R&D Program of Beijing Municipal Education Commission grant number KZ202210005004.

Data Availability Statement: All data is contained within the article.

Conflicts of Interest: The authors declare no conflict of interest.

References

1. Jiang, L.; Cui, X.; Jin, G.; Tian, Z.; Wen, X.; Tian, H.; Liu, E. Design and characterization of a novel $Cu_{2.3}Al_{1.3}Ni_{1.7}SnCr_{0.3}$ multi-principal element alloy coating on magnesium alloy by laser cladding. *J. Mater. Sci. Technol.* **2023**, *152*, 220–236. [CrossRef]
2. Zhu, Z.-X.; Liu, X.-B.; Liu, Y.-F.; Zhang, S.-Y.; Meng, Y.; Zhou, H.-B. Effects of Cu/Si on the microstructure and tribological properties of FeCoCrNi high entropy alloy coating by laser cladding. *Wear* **2023**, *512*, 204533. [CrossRef]
3. Wu, S.; Liu, Z.; Gong, Y.; Liang, X.; Wu, Y.; Zhao, X. Analysis of the sequentially coupled thermal–mechanical and cladding geometry of a Ni60A-25%WC laser cladding composite coating. *Opt. Laser Technol.* **2023**, *167*, 109595. [CrossRef]
4. Budde, L.; Biester, K.; Coors, T.; Faqiri, M.Y.; Lammers, M.; Hermsdorf, J.; Hassel, T.; Pape, F.; Overmeyer, L. Influence of shielding gas coverage during laser hot-wire cladding with high carbon steel. *Int. J. Adv. Manuf. Technol.* **2023**, *127*, 3195–3207. [CrossRef]
5. Bhatnagar, S.; Mullick, S. A study on the influence of reinforcement particle size in laser cladding of TiC/Inconel 625 metal matrix composite. *Opt. Laser Technol.* **2023**, *161*, 109115. [CrossRef]
6. Thompson, S.M.; Bian, L.K.; Shamsaei, N.; Yadollahi, A. An overview of Direct Laser Deposition for additive manufac-turing; Part I: Transport phenomena, modeling and diagnostics. *Addit. Manuf.* **2015**, *8*, 36–62.
7. Szost, B.A.; Terzi, S.; Martina, F.; Boisselier, D.; Prytuliak, A.; Pirling, T.; Hofmann, M.; Jarvis, D.J. A comparative study of additive manufacturing techniques: Residual stress and microstructural analysis of CLAD and WAAM printed Ti–6Al–4V components. *Mater. Des.* **2016**, *89*, 559–567. [CrossRef]
8. Sun, D.; Zhu, L.; Cai, Y.; Yan, Y.; Ge, F.; Shan, M.; Tian, Y.; Han, J.; Jiang, Z. Tribology comparison of laser-cladded CrMnFeCoNi coatings reinforced by three types of ceramic (TiC/NbC/B_4C). *Surf. Coat. Technol.* **2022**, *450*, 129013. [CrossRef]
9. Paul, C.P.; Gandhi, B.K.; Bhargava, P.; Dwivedi, D.K.; Kukreja, L.M. Cobalt-Free Laser Cladding on AISI Type 316L Stainless Steel for Improved Cavitation and Slurry Erosion Wear Behavior. *J. Mater. Eng. Perform.* **2014**, *23*, 4463–4471. [CrossRef]
10. Apolinário, L.H.R.; Torres, E.A.; Araújo, H.R.; Vicente, A.d.A.; Santos, T.F.d.A. Effect of laser cladding parameters in NbC reinforced 316L austenitic stainless steel composite depositions on a mild steel. *Int. J. Adv. Manuf. Technol.* **2022**, *122*, 3095–3113. [CrossRef]
11. Jie, G.; Qingchao, M.; Yan, S.; Kangning, W.; Qiang, S.; Canming, W. Effect of Nb content on microstructure and corrosion resistance of Inconel 625 coating formed by laser cladding. *Surf. Coat. Technol.* **2023**, *458*, 129311. [CrossRef]
12. Wang, J.; Zhang, Q.; Shen, W.; Liang, Z.; Chang, C.; Yang, L.; Li, J.; Huang, F. Failure Analysis of a Chromium Plating Layer on a Piston Rod Surface and the Study of Ni-Based Composite Coating with Nb Addition by Laser Cladding. *Metals* **2022**, *12*, 1194. [CrossRef]
13. Pizzatto, A.; Teixeira, M.F.; Rabelo, A.; Falcade, T.; Scheid, A. Microstructure and Wear Behavior of NbC-Reinforced Ni-Based Alloy Composite Coatings by Laser Cladding. *Mater. Res. Ibero Am. J. Mater.* **2021**, *24*, e20200447. [CrossRef]
14. Lian, G.F.; Yue, K.; Zeng, J.Y.; Feng, M.Y.; Lan, R.Q.; Kong, L.H. Microstructures and Properties of NbC-Reinforced Ni-Based Coatings Synthesized In Situ by Ultrasonic Vibration-Assisted Laser Cladding. *Materials* **2023**, *16*, 1704. [CrossRef] [PubMed]
15. Shi, B.; Huang, S.; Zhu, P.; Xu, C.; Zhang, T. Microstructure and Wear Behavior of In-Situ NbC Reinforced Composite Coatings. *Materials* **2020**, *13*, 3459. [CrossRef] [PubMed]
16. Li, Q.; Lei, Y.; Fu, H. Laser cladding in-situ NbC particle reinforced Fe-based composite coatings with rare earth oxide addition. *Surf. Coat. Technol.* **2014**, *239*, 102–107. [CrossRef]
17. Dong, G.; Yan, B.; Deng, Q.; Yu, T. Microstructure and wear resistance of in situ NbC particles reinforced Ni-based alloy composite coating by laser cladding. *J. Wuhan Univ. Technol. Sci. Ed.* **2012**, *27*, 231–237. [CrossRef]
18. Wang, K.; Du, D.; Liu, G.; Chang, B.; Ju, J.; Sun, S.; Fu, H. Microstructure and property of laser clad Fe-based composite layer containing Nb and B4C powders. *J. Alloys Compd.* **2019**, *802*, 373–384. [CrossRef]
19. Chen, L.; Yu, T.; Xu, P.; Zhang, B. In-situ NbC reinforced Fe-based coating by laser cladding: Simulation and experiment. *Surf. Coat. Technol.* **2021**, *412*, 127027. [CrossRef]
20. Sun, S.; Fu, H.; Ping, X.; Lin, J.; Lei, Y.; Wu, W.; Zhou, J. Reinforcing behavior and microstructure evolution of NbC in laser cladded Ni45 coating. *Appl. Surf. Sci.* **2018**, *455*, 160–170. [CrossRef]
21. Chen, Z.; Yan, H.; Zhang, P.; Yu, Z.; Lu, Q.; Guo, J. Microstructural evolution and wear behaviors of laser-clad Stellite 6/NbC/h-BN self-lubricating coatings. *Surf. Coat. Technol.* **2019**, *372*, 218–228. [CrossRef]
22. Jellad, A.; Labdi, S.; Benameur, T. On the hardness and the inherent ductility of chromium carbide nanostructured coatings prepared by RF sputtering. *J. Alloys Compd.* **2009**, *483*, 464–467. [CrossRef]
23. Yu, T.; Tang, H. Microstructure and high-temperature wear behavior of laser clad TaC-reinforced Ni-Al-Cr coating. *Appl. Surf. Sci.* **2022**, *592*, 153263. [CrossRef]
24. Huang, L.; Zhou, J.; Xu, J.; Huo, K.; He, W.; Meng, X.; Huang, S. Microstructure and wear resistance of electromagnetic field assisted multi-layer laser clad Fe901 coating. *Surf. Coat. Technol.* **2020**, *395*, 125876. [CrossRef]
25. Yan, H.; Chen, Z.F.; Zhao, J.; Zhang, P.L.; Yu, Z.S.; Lu, Q.H. Enhancing tribological properties of WS_2/NbC/Co-based self-lubricating coating via laser texturing and laser cladding two-step process. *J. Mater. Res. Technol.* **2020**, *9*, 9907–9919. [CrossRef]
26. Gao, Y.; Tong, Y.; Guohui, L.; Lu, P.; Zhang, D. Microstructure and Mechanical Properties of Ni-Based Alloy Composite Coating on Cr12MoV by Laser Cladding. *Coatings* **2022**, *12*, 1632. [CrossRef]
27. Wu, J.Q.; Li, Z.; Luo, Y.; Li, Y.; Gao, Z.L.; Zhao, Y.B.; Wu, C.W.; Liao, Y.X.; Jin, M. The effects of double ceramic particles (B_4C-SiC) on the performance, microstructure, and friction-wear mechanisms of copper-based PM. *Tribol. Int.* **2022**, *175*, 107865. [CrossRef]

28. Yin, H.G.; Li, J.H.; Zhao, L.W.; Yao, F. Simulation of in-situ preheating of ni-based multi-layer and multi-pass coatings on H_{13} steel. *J. Mech. Sci. Technol.* **2022**, *36*, 4671–4680. [CrossRef]
29. Zhu, Y.; Song, X.; Liu, M.; Jia, W.; Yin, Z.; Wang, Q.; Zhang, Y.; Jiang, D.; Song, Q.; Cui, H. Improved wear resistance of TiAl composite coatings through NbC@$(Ti,Nb)_2AlC$ core-shell structure and nanoscale precipitation strengthening. *Vacuum* **2023**, *213*, 112091. [CrossRef]
30. Liu, X.; Bi, J.; Meng, Z.; Li, R.; Li, Y.; Zhang, T. Tribological behaviors of high-hardness Co-based amorphous coatings fabricated by laser cladding. *Tribol. Int.* **2021**, *162*, 107142. [CrossRef]

lubricants

Article

Significant Improvement in Wear Resistance of CoCrFeNi High-Entropy Alloy via Boron Doping

Haitao Zhang [1], Junwei Miao [1], Chenglin Wang [2], Tingju Li [1], Longjiang Zou [1] and Yiping Lu [1,3,*]

1 Engineering Research Center of High Entropy Alloy Materials (Liaoning Province), School of Materials Science and Engineering, Dalian University of Technology, Dalian 116024, China; zhanghaitao2021@mail.dlut.edu.cn (H.Z.); jwmiao@mail.dlut.edu.cn (J.M.); tjuli@dlut.edu.cn (T.L.); zoulong@dlut.edu.cn (L.Z.)
2 Key Laboratory for Light-Weight Materials, Nanjing Technology University, Nanjing 210009, China; clwang1990@163.com
3 Key Laboratory of Solidification Control and Digital Preparation Technology (Liaoning Province), School of Materials Science and Engineering, Dalian University of Technology, Dalian 116024, China
* Correspondence: luyiping@dlut.edu.cn

Abstract: CoCrFeNi high-entropy alloy (HEA) exhibits excellent mechanical properties but relatively poor wear resistance. In particular, when the load reaches a certain level and the deformation mechanism of the CoCrFeNi HEA changes, the formation of shear bands leads to a significant increase in wear rate. Although numerous studies have been conducted on alloying strategies to improve the wear resistance of alloys, there is still limited research on the influence of deformation mechanism adjustment on wear resistance. Therefore, in order to fill this research gap, this study aims to use boron doping to regulate the deformation mechanism and successfully improve the wear resistance of CoCrFeNi HEA by 35 times. By observing the subsurface microstructure, the mechanism behind the significant improvement in wear resistance was further revealed. The results indicate that the reduction of shear bands and the formation of nanostructured mixed layers significantly improve wear resistance. The proposed strategy of boron doping to change the deformation mechanism and improve wear resistance is expected to provide new enlightenment for the development of wear-resistant HEAs.

Keywords: boron doping; deformation mechanism; high-entropy alloys; nanostructured mixing layer; wear resistance

Citation: Zhang, H.; Miao, J.; Wang, C.; Li, T.; Zou, L.; Lu, Y. Significant Improvement in Wear Resistance of CoCrFeNi High-Entropy Alloy via Boron Doping. *Lubricants* **2023**, *11*, 386. https://doi.org/10.3390/lubricants11090386

Received: 10 August 2023
Revised: 1 September 2023
Accepted: 5 September 2023
Published: 9 September 2023

1. Introduction

High-entropy alloys (HEAs) with multiple principal elements have unique properties that traditional alloys cannot achieve [1,2]. For example, face-centered cubic (FCC) HEAs maintain high ductility and fracture toughness at cryogenic temperatures [3]. However, these single-phase FCC HEAs generally have low strength and poor wear resistance [4–8]. As the load increases, the wear of the material becomes increasingly severe [9,10]. Therefore, improving the wear resistance of materials is highly essential. Many studies have tuned the composition and microstructure of FCC HEAs to overcome these shortcomings and improve wear resistance. Wu et al. found that the addition of Al could promote the formation of the body-centered cubic (BCC) phase and increase hardness, thereby greatly improving the wear resistance of the Al_xCoCrCuFeNi HEA [11]. Joseph et al. added Al to the CoCrFeNi HEA, and the formation of alumina and BCC phases also significantly improved wear resistance [12]. Many other studies found that the preparation of self-lubricating CoCrFeNi HEA could also considerably improve wear resistance. For instance, excellent wear resistance was obtained by adding solid lubricant graphite (MoS, Ag, and Cu) to the CoCrFeNi HEA [13–15].

In addition, preparing coatings is also an effective way to enhance the wear resistance of HEAs, and boriding is one of the most popular surface treatment methods for improving the surface performance of HEAs. Boriding is also defined as boronizing. Boronizing is a thermochemical surface treatment method that involves the transfer of active boron released from boron-producing substances onto the alloy surface through thermal diffusion between 800 and 1050 °C, with a holding time ranging from 30 min to 10 h [16]. During the thermal diffusion process, the active boron atoms diffuse into the interstitial positions in the lattice of the alloy, resulting in the formation of a metal boride layer on the workpiece [17]. Hou et al. prepared a boronized layer on the surface of Al0.25CoCrFeNi HEA using the solid-boronizing method. The Vickers hardness of the boronized sample surface is close to 1136 HV, which is 6.0 times higher than that in the unboronized condition. The wear resistance of the boronized Al0.25CoCrFeNi HEA was improved by 12 times compared to the unboronized alloy [18]. Wu et al. improved the tribological properties of single-phase FCC Al0.1CoCrFeNi HEAs by the pack-boronizing method. The surface-borided layer mainly consists of (Co, Fe)B, CrB, and NiB hard phases, which increase the surface hardness and reduce the wear rate [19].

Generally, the wear resistance of FCC HEAs is improved by alloying, which brings limitations to the design of wear-resistant HEAs. The deformation mechanism inevitably changes during friction and wear [20,21]. The effect of deformation on wear resistance has always been ignored. As a result, methods of adjusting the deformation mechanism to improve wear resistance, which may significantly affect FCC HEAs, have rarely been explored. According to a previous study, doping boron could effectively adjust the deformation mechanism of CoCrFeNi HEA [22]. Based on this, the HEAs doped with boron can provide solutions for industrial applications where the equipment deteriorates due to abrasive wear [23,24]. Therefore, exploring the deformation mechanism of materials with boron doping in the wear process may help establish the relationship between the deformation mechanism and wear resistance, thus overcoming the limitation of the wear resistance of FCC HEAs. Surface wear during sliding friction is a complex phenomenon involving many wear mechanisms [25,26]. The current research on friction and wear of HEAs mainly focuses on the wear mechanism from the perspective of surface topography analysis [27]. In fact, the basic factor determining the wear mechanism is the deformation and fragmentation of the subsurface layer, which change the structure and properties of this layer, thus determining the final wear resistance [28,29]. Consequently, the wear mechanism that affects wear resistance in the wear process remains to be further studied, especially under high loads.

The wear performance of HEAs is closely related to their microstructure, and most of them are based on composition control to improve wear resistance. However, a method to improve the wear resistance by adjusting the deformation mechanism of the alloy has not been proposed. In this study, boron doping was used to change the deformation mechanism of CoCrFeNi HEA for the first time to achieve high wear resistance. The mechanism that causes high wear resistance was revealed. The research results might provide a theoretical basis for the design and development of high-wear-resistant HEAs in the future.

2. Experimental Procedures

CoCrFeNi and CoCrFeNiB0.3 HEAs were prepared using the arc melting method in a vacuum arc melting furnace equipped with a water-cooled copper crucible. Before melting, the raw materials were placed in a water-cooled copper crucible from low to high melting point, and then the furnace chamber was closed. The furnace chamber was vacuumed to below 3×10^{-3} Pa, then reverse charged with protective high-purity argon to 0.05 Pa before starting the melting process. During smelting, titanium ingots need to be melted first to absorb residual oxygen in the furnace chamber, and then the alloy needs to be melted. In order to ensure the melting of all raw materials and obtain an alloy with uniform composition, all alloy samples need to be repeatedly melted 5 times, and the sample needs to be flipped once for each melting. Each smelting time is 2 min, and the smelting current is about 300 A. Electron probe microanalysis (JXA-8530F PLUS) and TEM (JSM-F200) were used to observe the microstructures of CoCrFeNi and CoCrFeNiB0.3 HEAs. The samples were cut into two parallel sides by the electric spark cutting method, and then they were polished with 80 #, 240 #, 400 #, 800 #, 1000 #, 1500 #, and 2000 # sandpapers, and the surface was cleaned with alcohol to ensure the smoothness and cleanliness of the surface. The hardness test was performed using the Vickers hardness tester model (MH-60) with loads of 1 kg and a duration of 15 s. Seven measurements for each ingot were performed to calculate the average data and ensure accuracy. The disc surfaces were ground with 4000-mesh grit papers before the wear tests.

The discs (10 mm diameter and 3 mm thick) taken from the ingot were used for wear tests. Ball-on-disc wear tests were performed using a WTM-2E controlled atmosphere at room temperature with loads of 2 N, 5 N, and 8 N for 1 h, and a sliding velocity of 0.188 m/s. The friction ball is a Si_3N_4 ceramic ball with a diameter of 6 mm. To ensure the accuracy of the wear tests, at least three samples were tested in the same condition. After the tests, the wear rate of all disc samples was measured by the volume loss method before ultrasonic cleaning with ethanol. The surface profile of the wear trajectory was obtained using a QLS4000 confocal laser scanning microscope to quantify the volume loss and calculate the wear rate W using the following equation:

$$W = \frac{V}{L \times P}$$

where V is the wear volume loss (mm^3), L is the sliding wear distance (m), and P is the applied normal load (N). Three independent wear tests were carried out for each test condition, and the average wear rate and standard deviation were obtained.

Thin foils of CoCrFeNi and CoCrFeNiB0.3 HEAs were prepared using a focused ion beam system (Helios G4 UX) for TEM observations. Firstly, the bulk sample was polished, which can be slightly corroded. It was placed under an electron microscope for observation to find the ideal position. By cutting the sample with an ion beam, a sheet of the same size was obtained and then welded to a copper column. Finally, the standard FIB program was used to further polish the sheet thickness to the nanometer level, usually less than 100 nm, and then a low-voltage electron beam was used to clean the surface of the sheet, and finally the sample was obtained. The sampling position was on the wear scar, and the directions normal to the sliding surface, along the sliding direction, and perpendicular to the sliding direction in the sliding plane were defined as ND and SD, respectively. The ND–SD cross-section under the sliding surface of the alloys was analyzed using TEM, high-resolution TEM (HRTEM), and high-angle annular dark-field scanning transmission electron microscopy (HAADF-STEM).

3. Results and Discussion

3.1. Microstructure and Wear Properties

Figure 1a,b shows the microstructure of CoCrFeNiB0.3 HEA. According to previous studies, the microstructure of CoCrFeNi HEA was only a single FCC phase. After adding boron, a large number of precipitates appeared in the CoCrFeNiB0.3 HEA [30]. As shown in the corresponding selected area electron diffraction (SAED) pattern in the inset, the matrix was still in the FCC phase, and the precipitate was Cr_2B [31]. Figure 1c shows the Vickers hardness (HV) of CoCrFeNi and CoCrFeNiB0.3 HEAs. The HV of CoCrFeNiB0.3 HEA was significantly higher than that of CoCrFeNi HEA, mainly because of the formation of Cr_2B precipitates. Figure 1d,e shows the 3D profiles of the worn surfaces of CoCrFeNi and CoCrFeNiB0.3 HEAs under different loads. The wear scar depth increased with the increase in loads. Under the same load, the wear scar depth of CoCrFeNi HEA was significantly larger than that of CoCrFeNiB0.3 HEA. The results indicated that Cr_2B was formed after adding boron, the hardness of the alloy increased, and the resistance of the matrix against deformation improved, resulting in higher wear resistance [23]. Figure 1f shows the wear rates of CoCrFeNi and CoCrFeNiB0.3 HEAs under different loads. For the CoCrFeNi HEA, the wear rate under 2 N, 5 N, and 8 N loads was 2.6×10^{-5}, 2.9×10^{-5}, and 3.57×10^{-4} mm^3/(N·m), respectively. On the contrary, for the CoCrFeNiB0.3 HEA, the wear rate under 2 N, 5 N, and 8 N loads was 8.3×10^{-6}, 8.6×10^{-5}, and 8.9×10^{-5} mm^3/(N·m), respectively. Generally, the wear rate increased with the load. The wear rate of CoCrFeNiB0.3 HEA was lower than that of CoCrFeNi HEA, indicating that the wear resistance of CoCrFeNiB0.3 HEA was better. The wear rate of CoCrFeNi HEA was two to three times that of CoCrFeNiB0.3 HEA under 2 N and 5 N loads, while the wear rate of CoCrFeNi HEA was 35 times that of CoCrFeNiB0.3 HEA under the 8 N load. Obviously, the increase in wear resistance under low-load conditions could be attributed to the higher hardness. However, under higher loads, the significantly improved wear resistance of CoCrFeNiB0.3 HEA indicated an impact on some other factors besides the increase in hardness. Therefore, wear scars and subsurface structures were observed to further investigate the underlying mechanism.

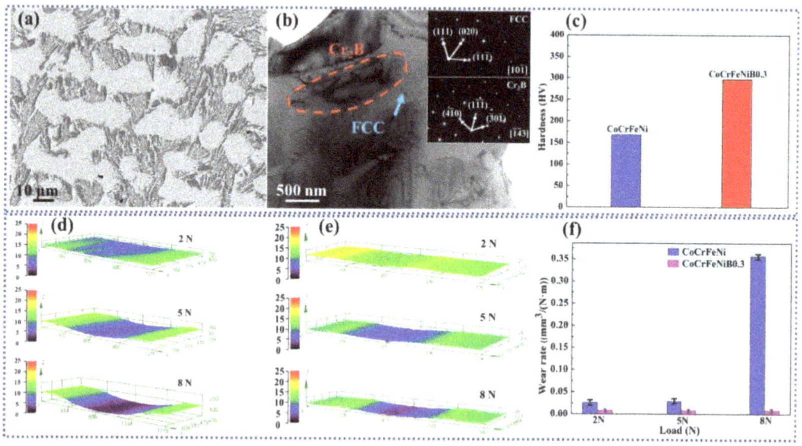

Figure 1. (**a**) Backscattered electron images of CoCrFeNiB0.3 HEA; (**b**) bright-field images of CoCrFeNiB0.3 HEA with SAED pattern; (**c**) Vickers hardness of CoCrFeNi and CoCrFeNiB0.3 HEAs; (**d**) 3D profiles of the worn surfaces of CoCrFeNi HEA under different loads; (**e**) 3D profiles of the worn surfaces of CoCrFeNiB0.3 HEA under different loads; (**f**) wear rates of CoCrFeNi and CoCrFeNiB0.3 HEAs under different loads.

3.2. Worn Morphology

Figure 2 shows the wear scars of CoCrFeNi HEA under different loads. As shown in Figure 2a,c,e, the width of the wear scar increased with the load. The wear behavior was typical of abrasive wear [13]. A large number of grooves appeared on the worn surfaces, which were caused by the hard, rough surface sliding across the softer surfaces, as shown in Figure 2b,d,f. Moreover, the number and depth of grooves increased with the increase in load, indicating an increase in the wear rate of the alloy.

Figure 3 shows the wear scar of CoCrFeNiB0.3 HEA under different loads. As shown in Figure 3a,c,e, the width of the wear scar increased with the increase in load [32]. Delamination features appeared on the worn surfaces of the alloy, demonstrating that the addition of boron transformed the wear behavior of CoCrFeNi HEA from abrasive wear to cracking and delamination. The width of the wear scar of CoCrFeNiB0.3 HEA decreased clearly, and the worn surfaces were smoother compared with CoCrFeNi HEA, indicating that the wear resistance of CoCrFeNiB0.3 HEA was better than that of CoCrFeNi HEA. Under the 2 N load, almost no plastic deformation occurred, and small grooves appeared on the worn surface. Although some grooves were observed on worn surfaces under 5 N and 8 N loads, they were not obvious.

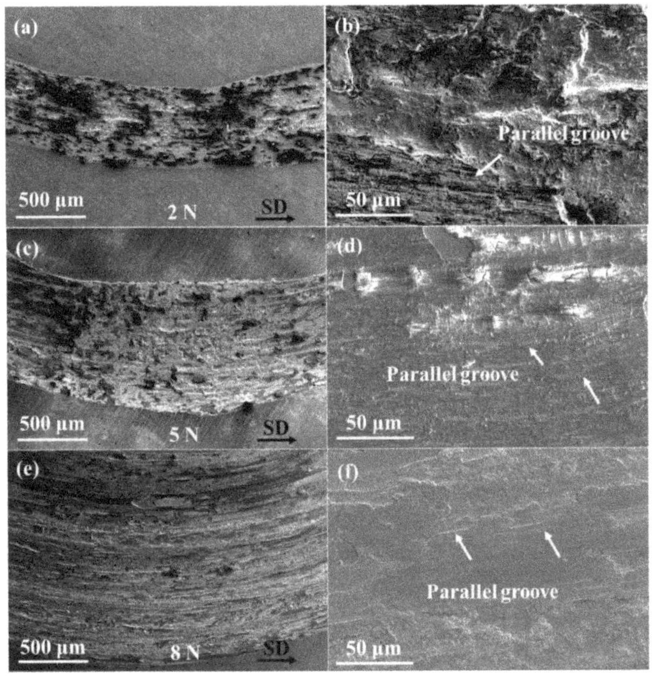

Figure 2. Wear scars of CoCrFeNi HEA with the corresponding enlarged image under different loads: (**a**) and (**b**) 2 N; (**c**) and (**d**) 5 N; (**e**) and (**f**) 8 N.

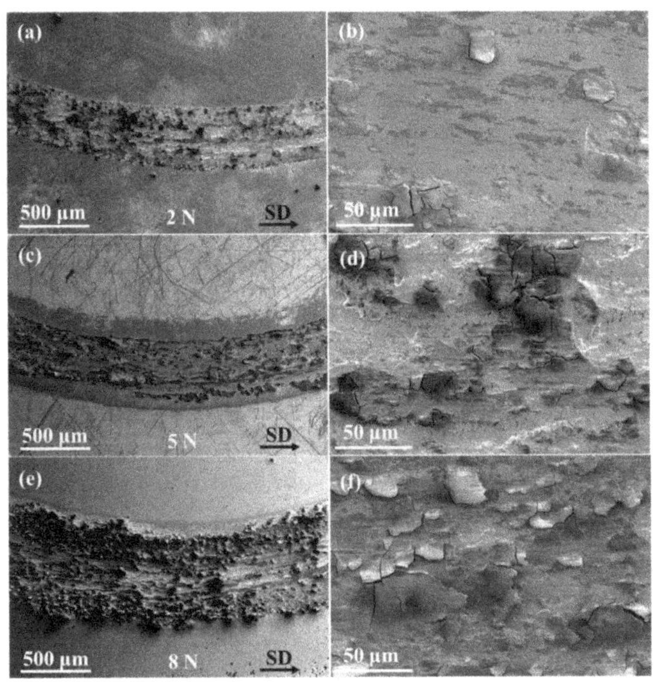

Figure 3. Wear scars of CoCrFeNiB0.3 HEA with the corresponding enlarged image under different loads: (**a**,**b**) 2 N; (**c**,**d**) 5 N; (**e**,**f**) 8 N.

3.3. Worn Subsurface Structure

The worn subsurface structures of the alloys were further analyzed to reveal the origin of the large difference in wear rates between CoCrFeNi and CoCrFeNiB0.3 HEAs under the 8 N load. Figure 4 shows the sliding wear-induced ND–SD cross-sectional subsurface structure of the CoCrFeNi HEA under the 8 N load. The subsurface layer consisted of oxide, equiaxed nanocrystalline, nanolaminated, and deformation layers, which was consistent with previous studies [14]. Additionally, previous research results showed that the formation of shear bands was the main reason for the sharp increase in the wear rate under higher loads [33]. A large number of shear bands were also formed in the nanocrystalline layer of CoCrFeNi HEA, as shown in Figure 4b, which should be mainly caused by severe damage from deformation twins under higher loads. This indicated that localized deformation occurred in the form of shear bands during stress concentration. Nucleation and growth of cracks often occurred in the shear bands, which directly led to high wear rates.

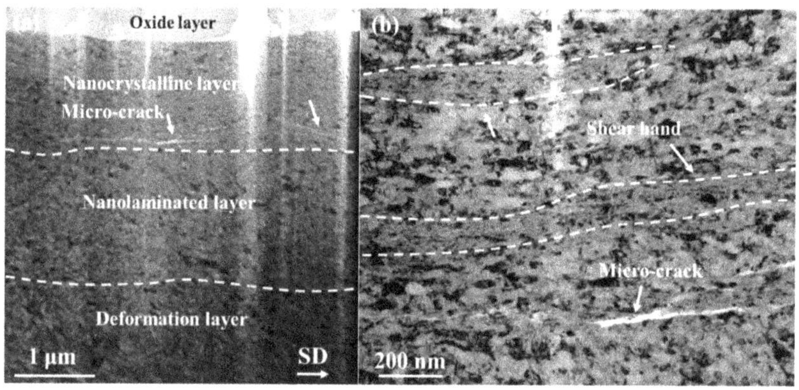

Figure 4. Sliding wear-induced ND–SD cross-sectional subsurface structure of the CoCrFeNi HEA under the 8 N load: (**a**) bright-field image and (**b**) enlarged image of the nanocrystalline layer.

Figure 5 shows the sliding wear-induced ND–SD cross-sectional subsurface structure of the CoCrFeNiB0.3 HEA under the 2 N load. The subsurface layer consisting of oxide, equiaxed nanocrystalline, nanolaminated, and deformation layers was formed in the CoCrFeNiB0.3 HEA. The average thickness of the nanocrystalline layer was 0.5 μm. As shown in Figure 5b and the inset, the grain size was 50 nm. The nanocrystalline layer was followed by a nanolaminated layer with a thickness of 1 μm and a grain size of ~50 to 200 nm, which was elongated along the sliding direction. The remaining part was the deformation layer, where the deformation twins were formed, as shown in Figure 5d.

Figure 6 shows the sliding wear-induced ND–SD cross-sectional subsurface structure of the CoCrFeNiB0.3 HEA under the 5 N load. The composition of the subsurface structure under the 5 N load was the same as that under the 2 N load. The increase in load had an obvious influence on the thickness of the subsurface structure. Beneath the oxide layer was the nanocrystalline layer with an average thickness of 0.5 μm. As shown in Figure 6b and the inset, the grain size was ~50 nm. The nanocrystalline layer was followed by a nanolaminated layer, with a thickness of 1 μm and a grain size of ~200 to 300 nm, which was elongated along the sliding direction. The remaining part was the deformation layer with the deformation twins, as shown in Figure 6d.

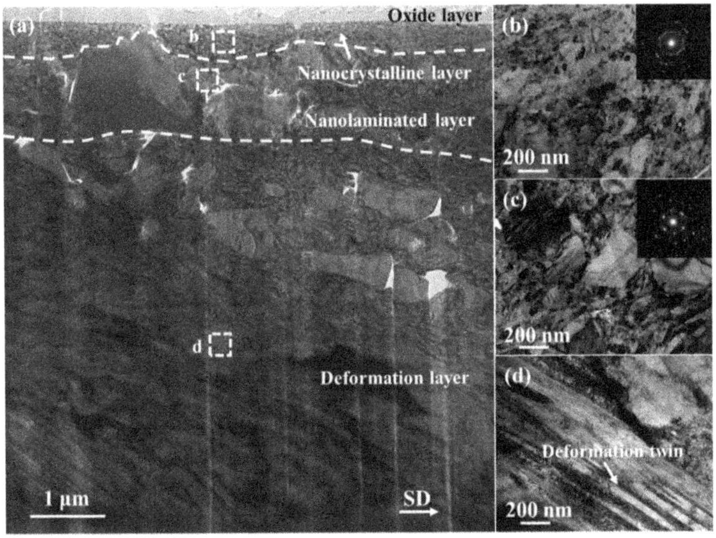

Figure 5. Sliding wear-induced ND–SD cross-sectional subsurface structure of the CoCrFeNiB0.3 HEA under the 2 N load: (**a**) bright-field TEM (BFTEM) image and (**b**–**d**) high-magnification images of selected regions in (**a**).

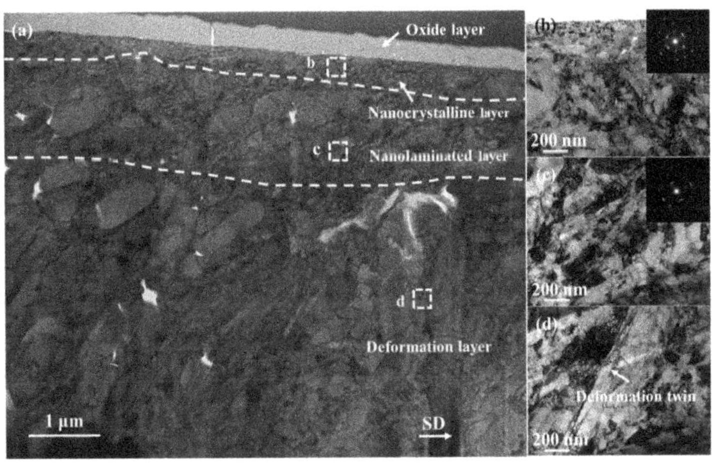

Figure 6. Sliding wear-induced ND–SD cross-sectional subsurface structure of the CoCrFeNiB0.3 HEA under the 5 N load. (**a**) BFTEM image and (**b**–**d**) high-magnification images of selected regions in (**a**).

Figure 7 shows the sliding wear-induced ND–SD cross-sectional subsurface structure of the CoCrFeNiB0.3 HEA under the 8 N load. The subsurface structure was different from those under 2 N and 5 N loads. Meanwhile, the subsurface structure of CoCrFeNiB0.3 HEA was obviously different from that of CoCrFeNi HEA under the 8 N load (Figure 3). The topmost region of the subsurface structure was observed to be streamlined. The grains in CoCrFeNiB0.3 HEA were finer, about 10–20 nm, compared with those in CoCrFeNi HEA, as shown in Figure 7b, indicating that boron doping could contribute to grain refinement. However, the microstructure should be a mixed region composed of boride and matrix, referred to as the nanostructured mixing layer [34]. Next, beneath the nanostructured mixing layer was the nanolaminated layer, with a thickness of 2 μm. The grains in the nanolaminated layer were markedly elongated, with an average diameter of 50–100 nm, and the elongated direction was parallel to the sliding direction, as shown in Figure 7c. As boride was a hard and brittle phase and its deformation was inconsistent with the matrix during the deformation process, the stress trended to accumulate at the interface between the precipitates and the matrix, resulting in the fracture of the precipitates [35,36]. The layer beneath was the deformation layer, whose deformation direction was parallel to the sliding direction. The dislocation cells were found in the nanolaminated and deformation layers, and the density was much higher than that of CoCrFeNi HEA. Correspondingly, the number of nanoscale deformation twins was significantly reduced. This was mainly due to the change in the deformation mechanism caused by boron doping. According to previous studies, the formation of borides in FCC HEAs could significantly increase the stacking fault energy (SFE), thereby promoting the movement of dislocations and suppressing the formation of deformation twins [37–39]. Moreover, the dominant motion mode of dislocation changed from plane slip to wave slip.

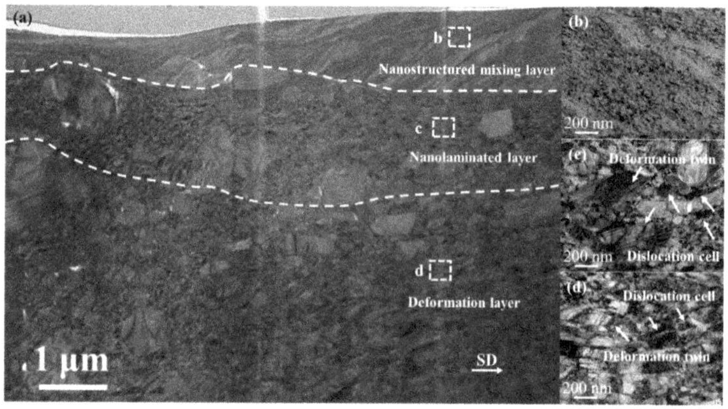

Figure 7. Sliding wear-induced ND–SD cross-sectional subsurface structure of the CoCrFeNiB0.3 HEA under the 8 N load. (**a**) Bright-field image and (**b**–**d**) high-magnification BFTEM image corresponding to the selected region in (**a**).

Figure 8a The gray part in the nanostructured mixing layer was Cr_2B and streamlined along the sliding direction. According to the SAED pattern in Figure 8b, the matrix of the nanostructured mixing layer was still in the FCC phase. During the process of friction and wear, Cr_2B was fragmented and its size significantly reduced. With the crushing of Cr_2B, a crack area appeared in the subsurface layer, and its strength was lower than that of the surrounding materials, which led to a huge shear instability, resulting in a turbulent plastic flow in the subsurface layer [40]. Venkataraman et al. found that for Al alloy, besides the shear instability caused by loads, the reinforcing particles were also a necessary condition for the formation of the nanostructured mixing layer. Therefore, the combination of shear instability and the hard formation of Cr_2B resulted in a nanostructured mixing layer [34,41]. Moreover, the hardness of this nanostructured mixing layer was significantly higher than that of the matrix, which obviously improved the wear resistance of the material [41].

Besides the formation of the nanostructured mixing layer, the change in the deformation mechanism was another important reason for the significant improvement in wear resistance. Compared with the CoCrFeNi HEA, no shear bands were found in the nanostructured mixing layer of the CoCrFeNiB0.3 HEA, indicating that adding boron could effectively suppress the formation of shear bands. The results showed that low SFE strongly promoted the formation of shear bands [42–44]. It was consistent with the experimental results that the density of the shear band increased with the decrease in SFE. The relationship between SFE and the stacking fault (SF) width could be expressed as follows [45]:

$$\gamma = \frac{G\vec{b_1}\cdot\vec{b_2}}{2\pi d}$$

where G is the shear modulus, γ is the SFE, b_1 and b_2 are Burgers vectors of two partial dislocations, and d is the width of the stacking fault (SF). As shown in Figure 9, the SF width of CoCrFeNi and CoCrFeNiB0.3 HEAs was 4.3 and 1.7 nm, respectively. According to the formula, the SF width was inversely proportional to the SFE. Therefore, the SFE of CoCrFeNi HEA was significantly lower than that of CoCrFeNiB0.3 HEA. The addition of boron to the CoCrFeNi HEA led to the formation of boride, which increased the SFE, thereby suppressing the formation of shear bands and significantly improving wear resistance [22]. The aforementioned conclusion was related to the microscopic size, and the same conclusion was also obtained for the macroscopic size, indicating that it was universal. The evolution of microstructure was found to be closely related to the influence of the SFE. More importantly, the influence mechanism of the SFE on the formation of shear bands not only provided an excellent opportunity to deeply understand the deformation mechanism of FCC HEAs but also expanded the application range of interstitial elements. This finding had promising application prospects in preparing HEAs with excellent wear resistance.

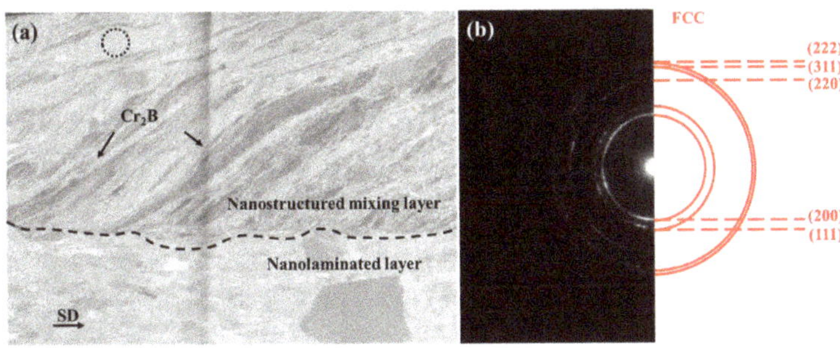

Figure 8. (**a**) HAADF image of the CoCrFeNiB0.3 HEA in the nanostructured mixing layer; (**b**) corresponding SAED pattern of the selected area in (**a**).

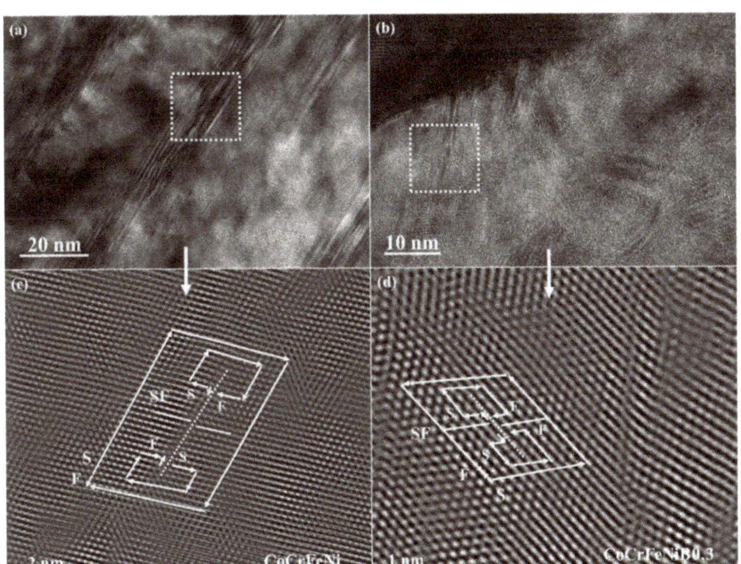

Figure 9. HRTEM image and the corresponding inverse fast Fourier transform (IFFT) image with Burgers circuits surrounding the entire stacking fault (SF): (**a**,**c**) CoCrFeNi HEA; (**b**,**d**) CoCrFeNiB0.3 HEA.

4. Conclusions

In this study, wear experiments using CoCrFeNi and CoCrFeNiB0.3 HEAs were conducted under different loads. The wear resistance of CoCrFeNi HEA was significantly improved by changing the deformation mechanism via boron doping. The mechanism for enhancing wear resistance was systematically investigated using TEM. The results indicated that the formation of a nanostructured mixing layer under higher loads had a positive effect in terms of improving wear resistance. The following conclusions were obtained through the systematic study of the wear properties of CoCrFeNi and CoCrFeNiB0.3 HEAs:

1. The wear formation of CoCrFeNi HEA mixed with boron changed from abrasive wear to delamination wear.
2. The shear instability caused by the fragmentation of precipitation and the joint action of hard particles led to the formation of a nanostructured mixing layer.
3. The formation of a hard phase made only a small contribution to the improvement in wear resistance, whereas the formation of a nanostructured mixing layer and the reduction in the shear band were key to the substantial improvement in the wear resistance of materials.
4. By adding interstitial elements to change the deformation mechanism and improve wear resistance, this study may provide a new strategy for the design of wear-resistant HEAs.

Author Contributions: H.Z. contributed the central idea, analyzed most of the data, and wrote the initial draft of the paper. Y.L. contributed writing—reviewing and editing, visualization, and supervision. The remaining authors contributed to refining the ideas and finalizing this paper. All authors have read and agreed to the published version of the manuscript.

Funding: This work was supported by the National Natural Science Foundation of China (Nos. 51822402 and U20A20278), the National Key Research and Development Program of China (Nos. 2019YFA0209901 and 2018YFA0702901), the fund of the State Key Laboratory of Solidification Processing in NWPU (No. SKLSP201902).

Data Availability Statement: Not applicable.

Conflicts of Interest: The authors declare no conflict of interest.

References

1. Cantor, B.; Chang, I.T.H.; Knight, P.; Vincent, A.J.B. Microstructural development in equiatomic multicomponent alloys. *Mater. Sci. Eng. A* **2004**, *375–377*, 213–218. [CrossRef]
2. Yeh, J.W.; Chen, S.K.; Lin, S.J.; Gan, J.Y.; Chin, T.S.; Shun, T.T.; Tsau, C.H.; Chang, S.Y. Nanostructured high-entropy alloys with multiple principal elements: Novel alloy design concepts and outcomes. *Adv. Eng. Mater.* **2004**, *6*, 299–303. [CrossRef]
3. Gludovatz, B.; Hohenwarter, A.; Catoor, D.; Chang, E.H.; George, E.P.; Ritchie, R.O. A fracture-resistant high-entropy alloy for cryogenic applications. *Science* **2014**, *345*, 1153–1158. [CrossRef]
4. Jiang, H.; Jiang, L.; Qiao, D.; Lu, Y.; Wang, T.; Cao, Z.; Li, T. Effect of Niobium on Microstructure and Properties of the CoCrFeNbxNi High Entropy Alloys. *J. Mater. Sci. Technol.* **2017**, *33*, 712–717. [CrossRef]
5. Liu, Y.; Xie, Y.; Cui, S.; Yi, Y.; Xing, X.; Wang, X.; Li, W. Effect of Mo Element on the Mechanical Properties and Tribological Responses of CoCrFeNiMox High-Entropy Alloys. *Metals* **2021**, *11*, 486. [CrossRef]
6. Zhang, Z.; Zhang, B.; Zhu, S.; Tao, X.; Tian, H.; Wang, Z. Achieving enhanced wear resistance in CoCrNi medium-entropy alloy co-alloyed with multi-elements. Materials Letters. *Mater. Lett.* **2022**, *313*, 131650. [CrossRef]
7. Hao, X.; Zhen, J.; Zhao, X.; Ma, J.; Chen, H.; Guo, S.; Wang, C.; Wang, C. Effect of Sn addition on the tribological behaviors of CoCrFeNi high entropy alloys. *J. Alloys Compd.* **2022**, *909*, 164657. [CrossRef]
8. He, M.; Eizadjou, M.; Chen, H.; Liu, H.; Chang, L.; Ringer, S.P. Microstructure and properties of CoCrFeNi-based multi-principal element alloys containing C and Sc. *J. Mater. Sci.* **2022**, *57*, 9442–9453. [CrossRef]
9. Du, Y.; Pei, X.; Tang, Z.; Zhang, F.; Zhou, Q.; Wang, H.; Liu, W. Mechanical and tribological performance of CoCrNiHfx eutectic medium-entropy alloys. *J. Mater. Sci. Technol.* **2021**, *90*, 194–204. [CrossRef]
10. Lan, L.W.; Wang, X.J.; Guo, R.P.; Yang, H.J.; Qiao, J.W. Effect of environments and normal loads on tribological properties of nitrided Ni45(FeCoCr)40(AlTi)15 high-entropy alloys. *J. Mater. Sci. Technol.* **2020**, *42*, 85–96. [CrossRef]
11. Wu, J.; Lin, S.; Yeh, J.; Chen, S.; Huang, Y.; Chen, H. Adhesive wear behavior of AlxCoCrCuFeNi high-entropy alloys as a function of aluminum content. *Wear* **2006**, *261*, 513–519. [CrossRef]
12. Joseph, J.; Haghdadi, N.; Shamlaye, K.; Hodgson, P.; Barnett, M.; Fabijanic, D. The sliding wear behaviour of CoCrFeMnNi and AlxCoCrFeNi high entropy alloys at elevated temperatures. *Wear* **2019**, *428*, 32–44. [CrossRef]

13. Zhang, A.; Han, J.; Su, B.; Li, P.; Meng, J. Microstructure, mechanical properties and tribological performance of CoCrFeNi high entropy alloy matrix self-lubricating composite. *Mater. Des.* **2017**, *114*, 253–263. [CrossRef]
14. Yang, L.; Cheng, Z.; Zhu, W.; Zhao, C.; Ren, F. Significant reduction in friction and wear of a high-entropy alloy via the formation of self-organized nanolayered structure. *J. Mater. Sci. Technol.* **2021**, *73*, 1–8. [CrossRef]
15. Verma, A.; Tarate, P.; Abhyankar, A.C.; Mohape, M.R.; Gowtam, D.S.; Deshmukh, V.P.; Shanmugasundaram, T. High temperature wear in CoCrFeNiCux high entropy alloys: The role of Cu. *Scr. Mater.* **2018**, *161*, 28–31. [CrossRef]
16. Dong, J.; Wu, H.; Chen, Y.; Zhang, Y.; Wu, Y.; Yin, S.; Du, Y.; Hua, K.; Wang, H. Study on self-lubricating properties of AlCo CrFeNi2.1 eutectic high entropy alloy with electrochemical boronizing. *Surf. Coat. Technol.* **2022**, *433*, 128082. [CrossRef]
17. Nakajo, H.; Nishimoto, A. Boronizing of CoCrFeMnNi high-entropy alloys using spark plasma sintering. *J. Manuf. Mater. Process.* **2022**, *6*, 29. [CrossRef]
18. Hou, J.; Zhang, M.; Yang, H.; Qiao, J.; Wu, Y. Surface strengthening in $Al_{0.25}$CoCrFeNi high-entropy alloy by boronizing. *Mater. Lett.* **2019**, *238*, 258–260.
19. Wu, Y.H.; Yang, H.J.; Guo, R.P.; Wang, X.J.; Shi, X.H.; Liaw, P.K.; Qiao, J.W. Tribological behavior of boronized Al0.1CoCrFeNi high-entropy alloys under dry and lubricated conditions. *Wear* **2020**, *460–461*, 203452. [CrossRef]
20. Meng, A.; Liang, F.; Gu, L.; Mao, Q.; Zhang, Y.; Chen, X.; Zhao, Y. An exceptionally wear-resistant CoFeNi2 medium entropy alloy via tribo-induced nanocrystallites with amorphous boundaries. *Appl. Surf. Sci.* **2023**, *614*, 156102. [CrossRef]
21. Miao, J.; Liang, H.; Zhang, A.; He, J.; Meng, J.; Lu, Y. Tribological behavior of an AlCoCrFeNi2.1 eutectic high-entropy alloy sliding against different counterfaces. *Tribol. Int.* **2021**, *153*, 106599. [CrossRef]
22. Zhang, H.; Wang, C.; Shi, S.; Li, T.; Zou, L.; Lu, Y.; Liaw, P.K. Tuning deformation mechanisms of face-centered-cubic high-entropy alloys via boron doping. *J. Alloys Compd.* **2022**, *911*, 165103. [CrossRef]
23. Aguilar-Hurtado, J.Y.; Vargas-Uscategui, A.; Zambrano-Mera, D.; Palma-Hillerns, R. The effect of boron content on the micro structure and mechanical properties of Fe50-XMn30Co10Cr10BX (x = 0, 0.3, 0.6 and 1.7 wt%) multi-component alloys prepared by arc-melting. *Mater. Sci. Eng. A* **2019**, *748*, 244–252. [CrossRef]
24. Aguilar-Hurtado, J.Y.; Vargas-Uscategui, A.; Paredes-Gil, K.; Palma-Hillerns, R.; Tobar, M.J.; Amado, J.M. Boron addition in a non-equiatomic Fe50Mn30Co10Cr10 alloy manufactured by laser cladding: Microstructure and wear abrasive resistance. *Appl. Surf. Sci.* **2020**, *515*, 146084. [CrossRef]
25. Nagarjuna, C.; You, H.; Ahn, S.; Song, J.; Jeong, K.; Madavali, B.; Song, G.; Na, Y.; Won, J.W.; Kim, H.; et al. Worn surface and subsurface layer structure formation behavior on wear mechanism of CoCrFeMnNi high entropy alloy in different sliding conditions. *Appl. Surf. Sci.* **2021**, *549*, 149202. [CrossRef]
26. Cheng, Z.; Yang, L.; Huang, Z.; Wan, T.; Zhu, M.; Ren, F. Achieving low wear in a μ-phase reinforced high-entropy alloy and associated subsurface microstructure evolution. *Wear* **2021**, *474*, 203755. [CrossRef]
27. Zhang, X.; Tong, Y.; Hu, Y.; Liang, X.; Chen, Y.; Wang, K.; Zhang, M.; Xu, J. Microstructure and Performance of Fe50Mn30Cr10Ni10 High-Entropy Alloy Produced by High-Efficiency and Low-Cost Wire Arc Additive Manufacturing. *Lubricants* **2022**, *10*, 344. [CrossRef]
28. Singh, J.B.; Wen, J.; Bellon, P. Nanoscale characterization of the transfer layer formed during dry sliding of Cu-15 wt.% Ni-8 wt.% Sn bronze alloy. *Acta Mater.* **2008**, *56*, 3053–3064. [CrossRef]
29. Hughes, D.A.; Dawson, D.B.; Korellis, J.S.; Weingarten, L.I. Near surface microstructures developing under large sliding loads. *J. Mater. Eng. Perform.* **1994**, *3*, 459–475. [CrossRef]
30. Zhang, L.J.; Yu, P.F.; Fan, J.T.; Zhang, M.D.; Zhang, C.Z.; Cui, H.Z.; Li, G. Investigating the micro and nanomechanical properties of CoCrFeNi-Cx high-entropy alloys containing eutectic carbides. *Mater. Sci. Eng. A* **2020**, *796*, 140065. [CrossRef]
31. Tu, J.; Xu, K.; Liu, Y.; Luo, J.; Zhou, Z.; Ding, L. Characterization of deformation substructure evolution in metastable $Fe_{49}Mn_{30}Co_{10}Cr_{10}B_1$ interstitial high entropy alloy. *Intermetallics* **2022**, *144*, 107508. [CrossRef]
32. Günen, A.; Soylu, B.; Karakaş, Ö. Titanium carbide coating to improve surface characteristic, wear and corrosion resistance of spheroidal graphite cast irons. *Surf. Coat. Technol.* **2022**, *437*, 128280. [CrossRef]
33. Mohseni, H.; Nandwana, P.; Tsoi, A.; Banerjee, R.; Scharf, T.W. In situ nitrided titanium alloys: Microstructural evolution during solidification and wear. *Acta Mater.* **2015**, *83*, 61–74. [CrossRef]
34. Chen, X.; Han, Z.; Lu, K. Wear mechanism transition dominated by subsurface recrystallization structure in Cu-Al alloys. *Wear* **2014**, *320*, 41–50. [CrossRef]
35. Wang, J.; Guo, X.; Qin, J.; Zhang, D.; Lu, W. Microstructure and mechanical properties of investment casted titanium matrix composites with B4C additions. *Mater. Sci. Eng. A* **2015**, *628*, 366–373. [CrossRef]
36. Luan, J.H.; Jiao, Z.B.; Chen, G.; Liu, C.T. Effects of boron additions and solutionizing treatments on microstructures and ductility of forged Ti-6Al-4V alloys. *J. Alloys Compd.* **2015**, *624*, 170–178. [CrossRef]
37. Christian, J.W.; Mahajan, S. Deformation twinning. Progress. *Mater. Sci.* **1995**, *39*, 1–157.
38. Meyers, M.A.; Vöhringer, O.; Lubarda, V.A. The onset of twinning in metals: A constitutive description. *Acta Mater.* **2001**, *49*, 4025–4039. [CrossRef]
39. Han, W.Z.; Zhang, Z.F.; Wu, S.D.; Li, S.X. Combined effects of crystallographic orientation, stacking fault energy and grain size on deformation twinning in fcc crystals. *Philos. Mag.* **2008**, *88*, 3011–3029. [CrossRef]
40. Rosenfield, A.R. A shear instability model of sliding wear. *Wear* **1987**, *116*, 319–328. [CrossRef]

The distribution of alloying elements was inhomogeneous in the coating under the rapid heating and cooling rate [20,21]. The inhomogeneity distribution of alloy composition and microstructure had great influence on the coating hardness and wear resistance [22–26].

The microstructures of the heat-affected zone (HAZ) and properties of the cladding layer could be improved by the post-heat treatment [27]. The phase proportion of FeCoCrNiAlx laser-cladding laser layer could be controlled by using auxiliary heat treatment, resulting in the effective improvement for the mechanical properties [28].

In this paper, the inhomogeneity of T15 coating on 42CrMo steel was studied and the post-heat treatment was put forward to improve the microstructure uniformity and tribological properties of the cladding layer. The microstructure, phase, hardness and tribological properties were analyzed for the original and heat-treated coatings. The wear mechanisms of the coatings were also investigated.

2. Experimental Methods

42CrMo steel was used as the substrate with the dimension of 100 × 100 × 10 mm. Before the laser cladding experiment, the oxide layer and oil stain on the surface should be removed. T15 powder with the particle size of 50~80 μm was selected as the coating matrix powder, as shown in Figure 1. The powder was heated in a vacuum drying box at 90 °C for 2 h. The chemical compositions were obtained by a direct-reading spectrometer for 42CrMo and by ICP-OES for the T15 powder, and the testing results are shown in Table 1.

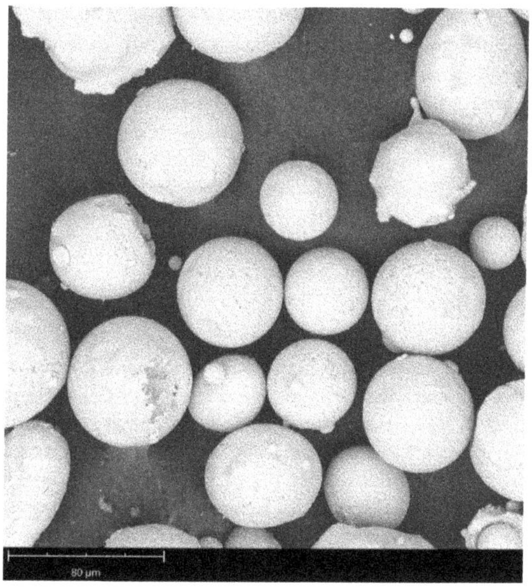

Figure 1. Micromorphology of T15 powder.

Table 1. Chemical composition of 42CrMo and T15 powder (wt. %).

	C	Co	Cr	Mo	Mn	Si	W	V	Fe
42CrMo	0.42	...	0.99	0.19	0.63	0.21	...	...	Bal.
T15	1.6	5.4	4.5	...	0.45	0.48	11.7	4.7	Bal.

The laser cladding experimental equipment was a Disk laser (TruDisk 4002, Trumpf, Ditzingen, Germany) and a KUKA robot system (KR60-3, KUKA, Augsburg, Germany). Argon gas was used as the powder carrier and protective gas to prevent the melt pool from oxidizing. The protective gas flow was 20 L/min [29]. T15 cladding layer thickness was about 1 mm. The process parameters of laser cladding are shown in the Table 2. After cladding, the coatings were cut into small specimens with size 10 × 10 × 10 mm by wire cutting for heat treatment and further analysis.

Table 2. Laser cladding process parameters.

Order	Power (W)	Powder Feeding Voltage (V)	Scanning Speed (mm/s)	Overlap Rate
1	2000	50	7	30%
2	2300	40	6	40%
3	2300	50	8	40%
4	2300	60	9	40%
5	2300	70	6	40%

The cladding samples were heated by muffle furnace (FB, IRM, Lilienthal, Germany). The cladding layer was quenched at 1100 °C, 1190 °C and 1240 °C, respectively, to study its effect on the microstructure uniformity and wear resistance of the cladding layer. The process parameters of heat treatment were shown Figure 2.

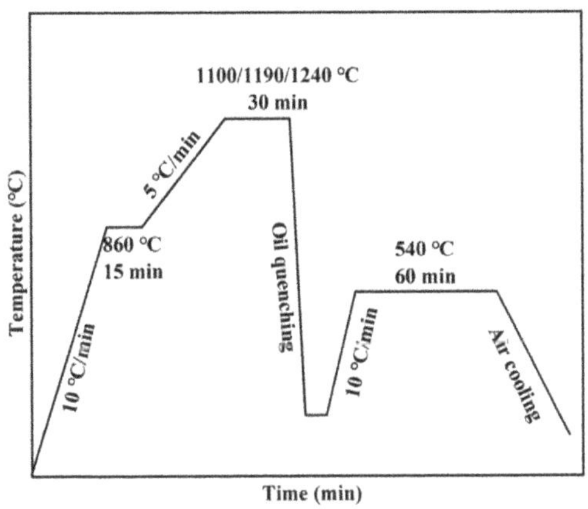

Figure 2. Heat treatment process curve of cladding layer.

The hardness of samples was tested by the Vickers hardness tester (HVS-1000A, Huayin, Laizhou, China). The test load was 300 g and the loading time was 15 s. Twenty points with an interval of 0.1 mm were measured continuously, and the average value of three times was taken as the hardness value of this point.

The dry sliding wear properties at room temperature were tested by a CFT-I friction and wear testing system (CFT-I, Zhongke Kaihua, Lanzhou, China). The test configuration is schematically illustrated in Figure 3. The grinding ball was YG6 with a hardness of 93 HRC and a diameter of 6 mm. The wear mark is located in the middle of the non-overlapping area. The test conditions are shown in Table 3.

Table 3. Details of the wear test conditions.

Load (N)	Speed (r/min)	Test Duration (min)	Sliding Distance (m)	Reciprocating Length (mm)
50	100	120	120	5

The wear volume (V) was measured by the contour scanning instrument. The wear volumes were calculated by using the equation: where V was the volume loss (mm^3),

L is the experimental load (N) and S is total test distance (m). The test configuration is schematically illustrated in Figure 4.

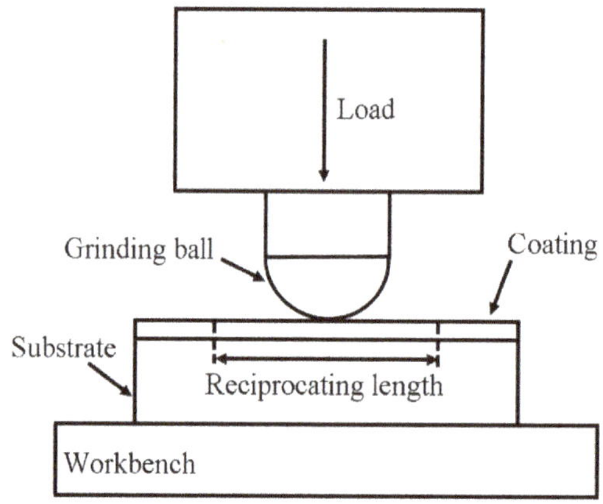

Figure 3. Schematic diagram of the wear test.

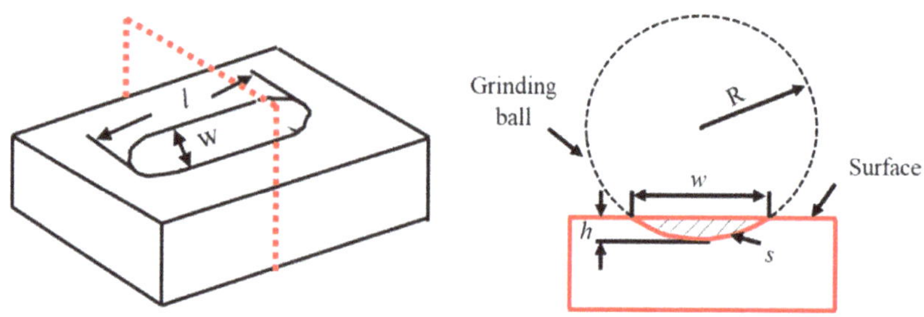

(**a**) (**b**)

Figure 4. Schematic diagram of the wear mark: (**a**) sample wear scar, (**b**) wear mark section. (l is the length of wear scar, w is the width of wear scar, h is the depth of wear scar, s is the cross sectional area of wear scar, R is the Grinding ball radius).

From the calculated wear volume, the wear rates were evaluated using the following Equation (1) [30].

$$W = \frac{V}{L \times S} \tag{1}$$

The microstructures and morphology of the wear trace were characterized by the scanning electron microscope (SEM, ZEISS, Sigma500, Oberkochen, Germany) equipped with energy dispersive spectroscopy (EDS) analysis system. The phase of the coating was identified by X-ray diffraction (XRD, Empyrean, Panaco, Almelo, Holland).

3. Results and Discussion

3.1. Microstructures and Hardness

After grinding, polishing and corrosion, there were obvious "black-and-white" areas on the surface of all cladding layers. The macro morphology of sample 1 was shown in Figure 5a. The "black-and-white" areas were marked as position 1 and 2, respectively. It can be found from Figure 5b that there was a large hardness gradient on the coating surface. Figure 5c,d showed microstructure of the T15 coating at positions 1 and 2. The coating was mainly composed of the equiaxed crystal and alloy carbides. The average size of grain size at position 2 was about 3.8 μm, which was much smaller than at position 1 with about 7 μm.

The network alloy carbides were situated on the grain boundary, because alloy elements were excluded to the grain boundary in the process of forming equiaxed crystals [31]. The number of stress corrosion cracks at position 1 were more than position 2, which were formed at the combined action of large residual tensile stress and acid corrosive agent [32].

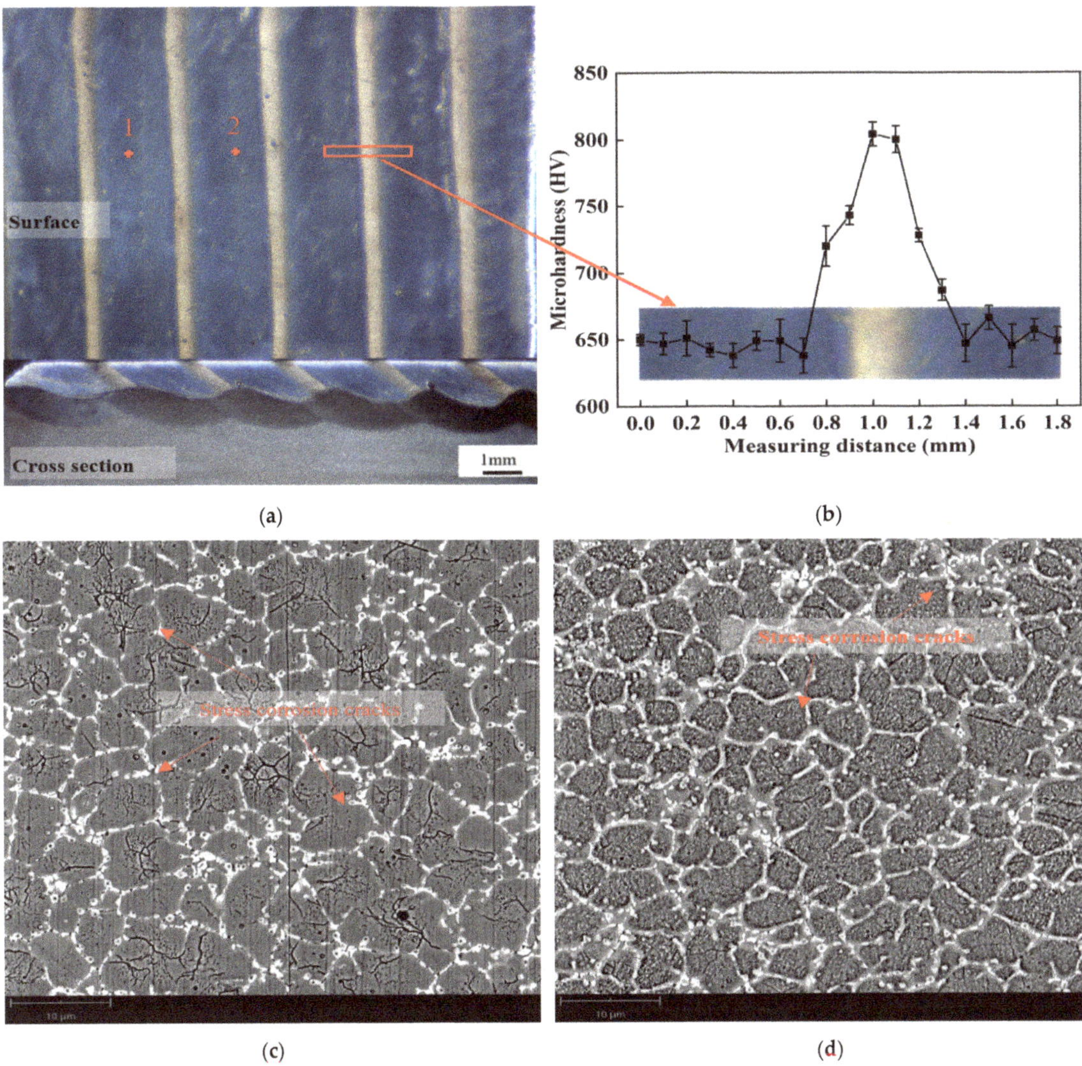

Figure 5. Surface macro morphology, microstructure and hardness of cladding layer: (**a**) macro morphology, (**b**) hardness gradient, (**c**) position 1, (**d**) position 2.

Figure 6 shows the surface scanning results of EDS at positions 1 and 2, respectively. Element segregation existed in both "black-and-white" areas of the cladding layer. A large number of W, V and Cr elements were enriched at the grain boundary [33], because these alloy elements with larger atomic radius were difficult to be a solid soluble with iron element [31]. However, the Co element was companied with the Fe element distributed in the inner grain. The lattice structure, atomic radius and electronic structure of the Co element may be similar to that of the Fe element in the matrix [34].

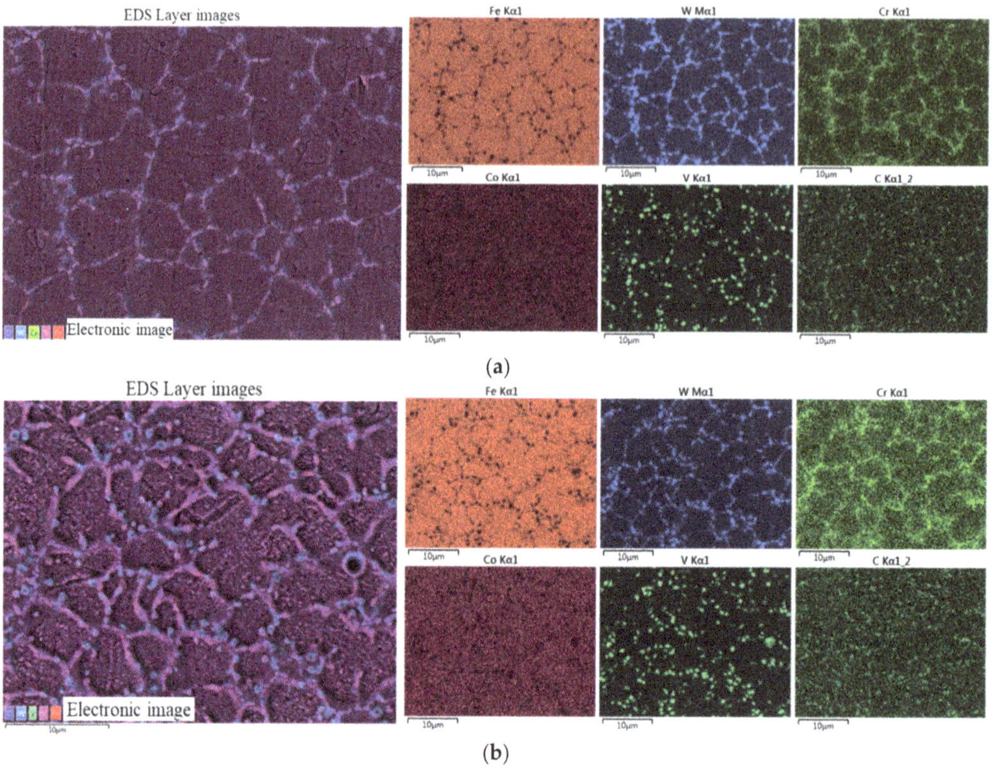

Figure 6. The EDS mapping results of T15 coating at different areas: (**a**) position 1, (**b**) position 2.

3.2. SEM Morphology after Quenching

A multi-channel transverse lap model was constructed, as shown in Figure 7. A heat-affected zone (HAZ) was only formed on the 42CrMo matrix during the first cladding surface. In subsequent cladding, the number of HAZ was changed because of the remelting zone (RZ). One HAZ was located on the arc surface of the prior cladding layer and the other was located on the 42CrMo substrate. Combined with the macro morphology of the coating, it was speculated that the black area was cladding zone (CZ) and the white bright zone was the HAZ of the arc surface of the front cladding layer.

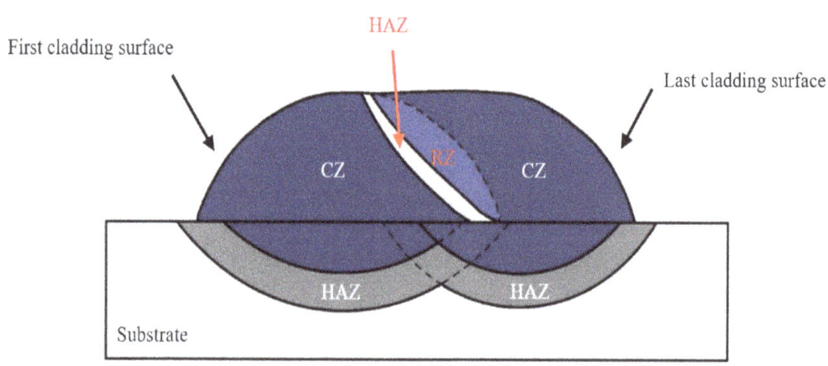

Figure 7. Schematic diagram of multi-channel transverse overlapping.

Figure 8 shows the surface macro morphology of sample 1 after different quenching temperatures. The "black-and-white" area of the coatings surface and cross-section was eliminated after quenching. The structural uniformity of the coating was improved because the alloy carbides and elements were melted into the matrix and uniformly precipitated again at high temperature quenching [35].

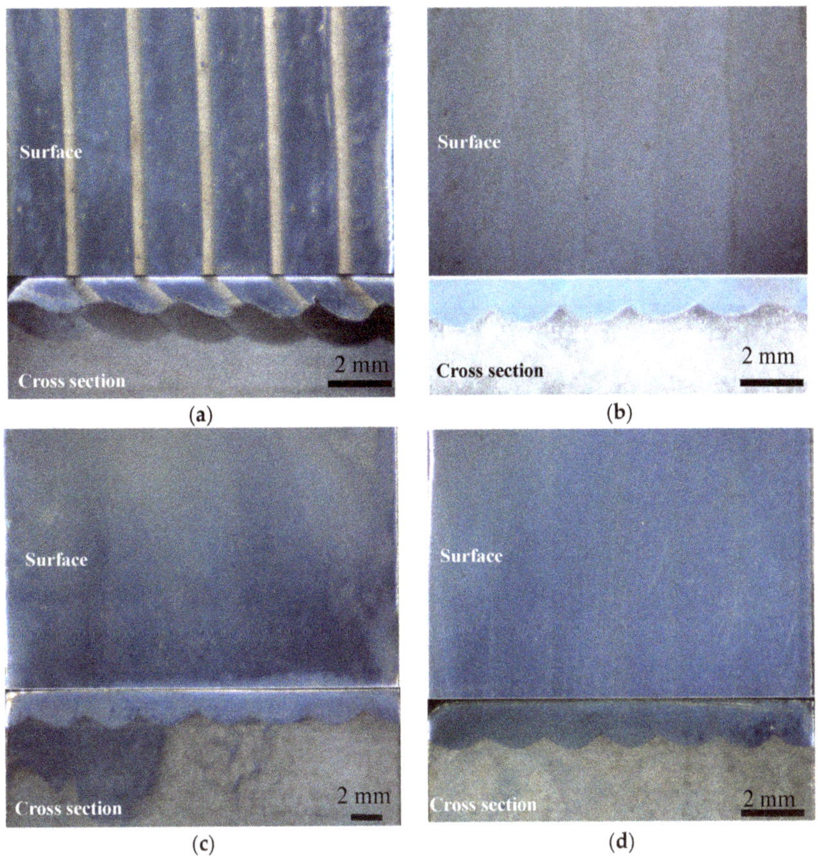

Figure 8. Surface macro morphology of cladding layer at different quenching temperatures: (**a**) original coating, (**b**) 1100 °C, (**c**) 1190 °C, (**d**) 1240 °C.

According to the wear rate of sample 1 at different quenching temperatures (shown in Table 4), it can be clearly seen that the wear resistance of the coating can be further improved after quenching at 1100 °C. Therefore, the following focused on the changes of microstructure, hardness and tribological performance of cladding layers at 1100 °C quenching.

Table 4. Average wear rate of coatings at different heat treatment processes.

Heat Treatment Process	Average Wear Rate ($\times 10^{-6}$ mm^3·N^{-1}·m^{-1})
Original coating	0.73
Quenched (1100 °C)	0.54
Quenched (1190 °C)	1.2
Quenched (1240 °C)	1.34

For the quenched coating (1100 °C), there was no significant difference in microstructure at scanning electron microscope. The surface microstructure of the cladding layer was shown as in Figure 9a. During the process of high temperature quenching, the network alloy carbides were gradually dissolved on the grain boundary of the cladding layer. A large number of fine carbides gradually integrated into the matrix. Some unmelted particles may be VC because of their high melting point [36].

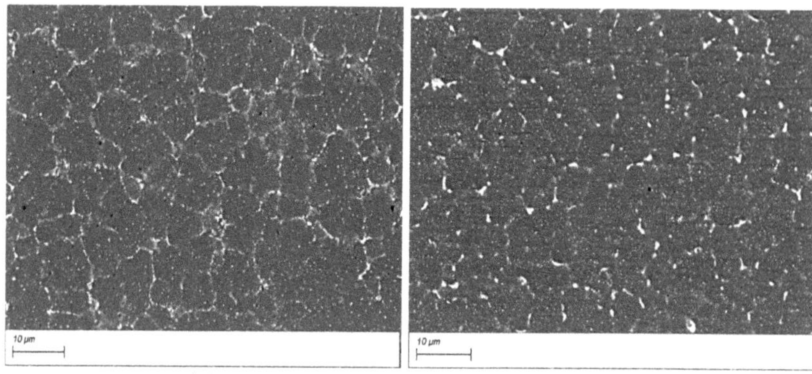

(a) (b)

Figure 9. The microstructure of the cladding layer after heat treatment: (**a**) quenched, (**b**) quenched-tempered.

After quenching, there were a lot of residual stress, retained austenite and brittle phases in the cladding layer. Tempering was immediately carried out to avoid cracks. Figure 9b shows the microstructure of the quenched cladding layer after tempering treatment. A large number of carbides in the grains of the cladding layer were uniformly distributed on the matrix, which were mainly spherical and very small in size.

As shown in Figure 10a,b, the Fe, W, Cr, and Co alloying elements were uniformly distributed except for the V element after post-heat treatment, which was consistent with the above guess of class V carbides.

Figure 11 shows the XRD patterns of the original and heat-treated T15 cladding layer. The phases of original CZ and HAZ could not be measured, respectively, due to the large diameter of diffraction spot. Therefore, the phases of two different regions were not distinguished. The phases compositions of the coatings after heat treatment were similar to those of the original coating, which mainly consisted of martensite, austenite, MC and M6C carbide. The percentage of alloy carbide increased obviously after the quenching and tempering process, for the reason of the secondary precipitation of carbide during heat treatment. Compared with the original cladding layer, the martensite diffraction peak of heat-treated cladding shifted to the right and changed from double-peak to single-peak, which indicated that the martensite structure had been changed by the heat treatment.

As seen from Figure 12, the hardness gradient of the cladding layer could significantly be eliminated after heat treatment. The cladding layer increased to 910 HV after quenching. This is because a large amount of martensite was formed in the cladding layer. After quenching–tempering, the hardness reduced to 750 HV, because of the decomposition of martensite or without good secondary hardening.

3.3. *Wear Rate and Friction Coefficient*

As seen from Figure 13, the cladding layer had experienced running in period and stable period during the wear process. The friction coefficients increased rapidly during the running-in period, which was accompanied with increased wear rate of T15 coatings. The friction coefficient tended to be stable from rising, falling and rising. The friction coefficients of the coatings varied from 0.5 to 0.8.

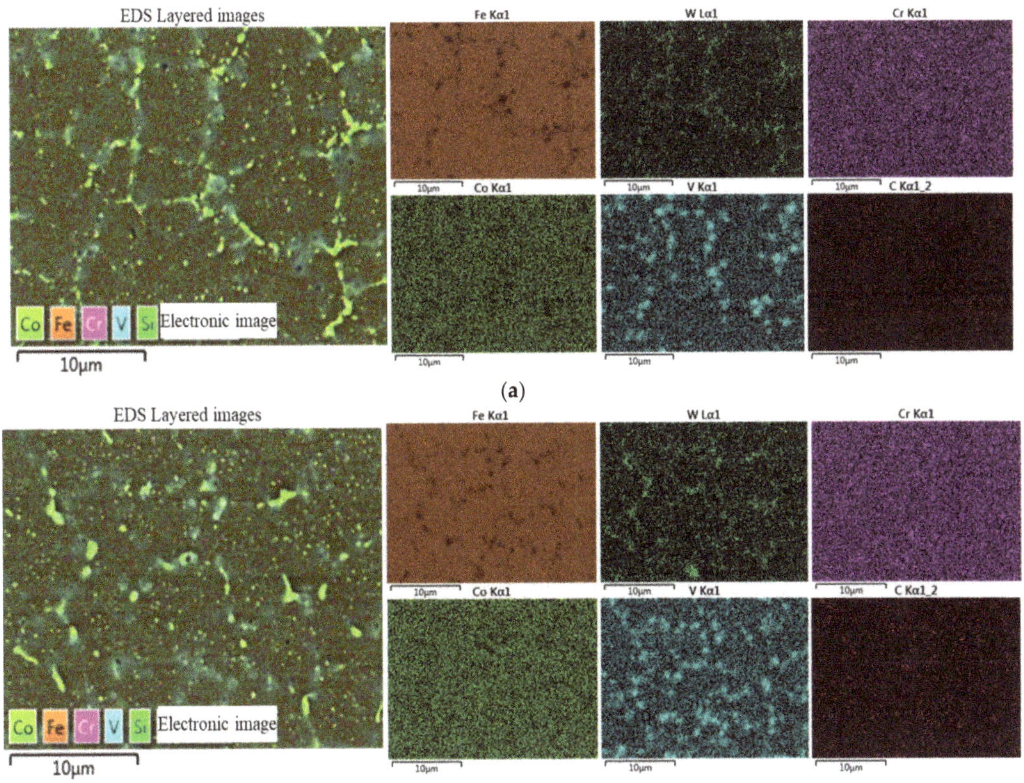

Figure 10. EDS area scanning results of the cladding layer after heat treatment: (**a**) quenched, (**b**) quenched–tempered.

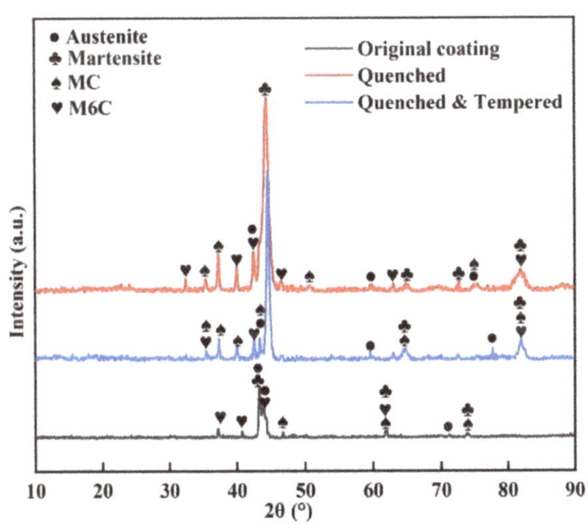

Figure 11. XRD patterns of the T15 coating.

For original CZ, HAZ, quenched and quenched and tempered cladding layers, the average friction coefficients were 0.7, 0.71, 0.76 and 0.68, respectively, as shown in Figure 14.

Post-heat treatment had no significant effect on the friction coefficient, because it was a comprehensive characteristic coefficient affected by many factors. The small friction coefficient of tempered samples did not mean that its wear rate was small [37]. The average wear rate of original CZ (7.3×10^{-7} $mm^3 \times N^{-1} \times m^{-1}$) was nearly twice that of HAZ. After the quenching and quenching–tempering process, the microstructures and hardness gradient of the cladding layer was uniform. The wear rate was 5.6×10^{-7} $mm^3 \times N^{-1} \times m^{-1}$ and 8.5×10^{-7} $mm^3 \times N^{-1} \times m^{-1}$ respectively.

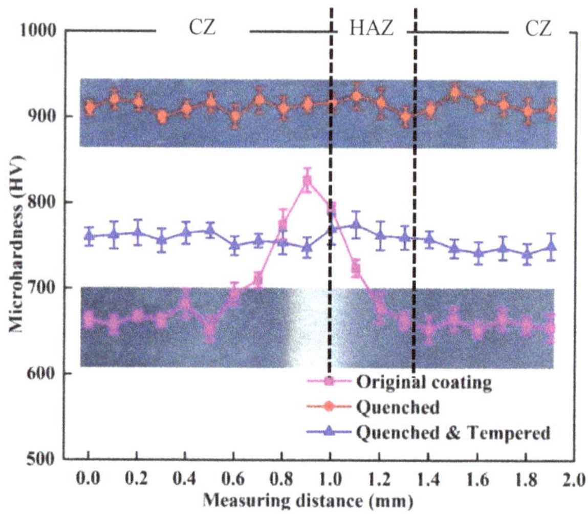

Figure 12. Surface microhardness of cladding layer.

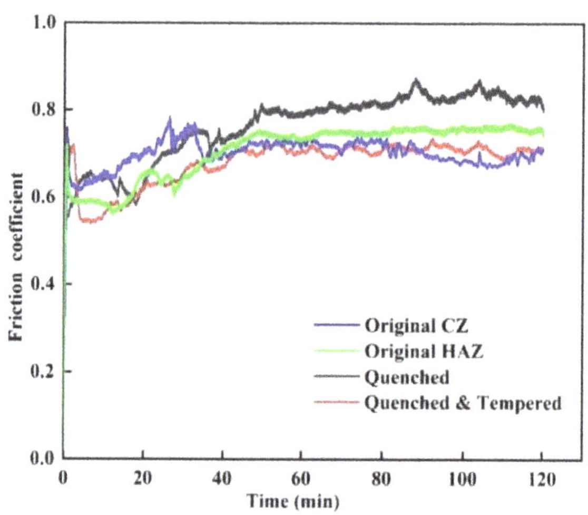

Figure 13. Friction coefficient of cladding layer.

Figure 15 showed the wear trace morphology of different cladding layers. As shown in Figure 15a, there were a large number of fatigue cracks, spalling and furrows on the wear surface of original CZ for large residual stress and brittle phases. It indicated that the main wear mechanism was fatigue failure and abrasive wear [38]. Under the cyclic alternating load, microcracks appeared on surface, and then the residual stress promoted the propagation of microcracks into macrocracks [39]. After quenching and quenching–tempering, the wear mark surface of the sample was relatively smooth. This was due to the hardness improving and the residual stress reducing, avoiding the initiation and rapid propagation of microcracks [40]. In addition, the number of fatigue cracks and the fatigue shedding phenomenon was significantly reduced, as shown in Figure 15b,c. The results showed that the wear mechanism was still mainly slight-abrasive wear and fatigue wear.

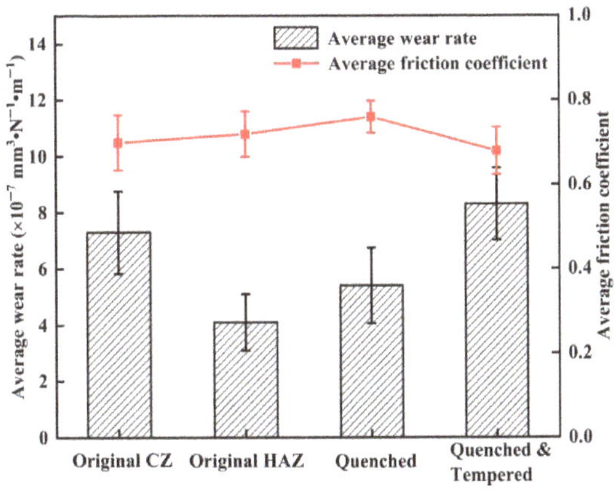

Figure 14. Average wear rate and friction coefficient of cladding layers.

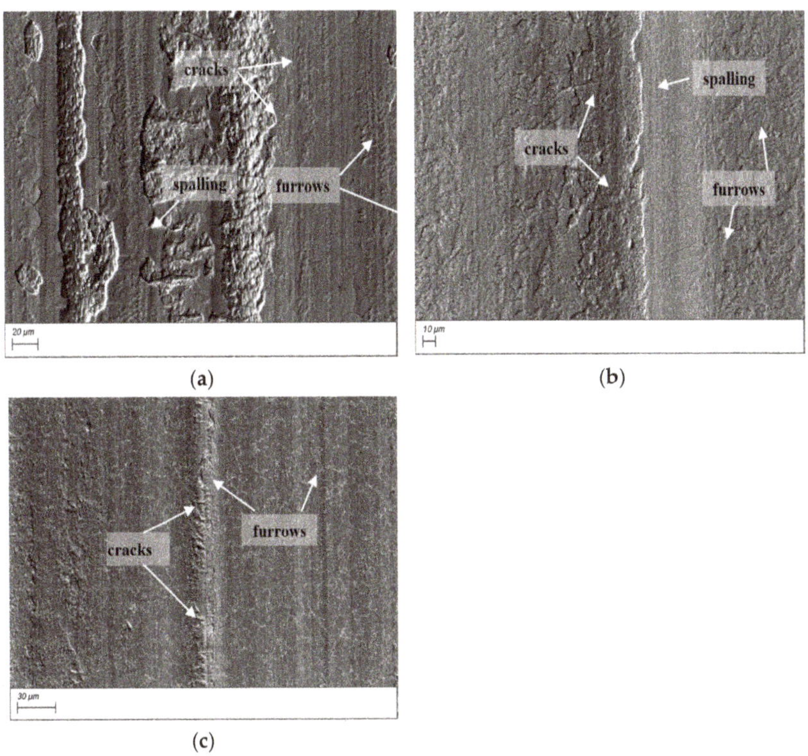

Figure 15. The SEM wear trace morphology of cladding layer: (**a**) original CZ, (**b**) quenched, (**c**) quenched–tempered.

Compared with quenched and tempered samples, there were many fatigue cracks in the wear marks of quenched samples, because there were a lot of brittle martensite and residual stress on the surface. The plasticity and strength of tempered coating was improved for a lot of dispersed carbides. Due to the low surface hardness of the tempered sample, deep furrows were formed under the action of hard wear debris.

As seen from Figure 16, the average friction coefficient of coatings with different parameters hovered around 0.7. For different coatings, the standard deviation of the average wear rate was 0.21 $mm^3 \times N^{-1} \times m^{-1}$. After quenching (1100 °C) and quenching–tempering (1100 °C–540 °C), it was reduced to 0.07 and 0.09 $mm^3 \times N^{-1} \times m^{-1}$, respectively. According to the wear data of the above coating with different parameters, the experimental accuracy was verified by the same post-heat treatment process.

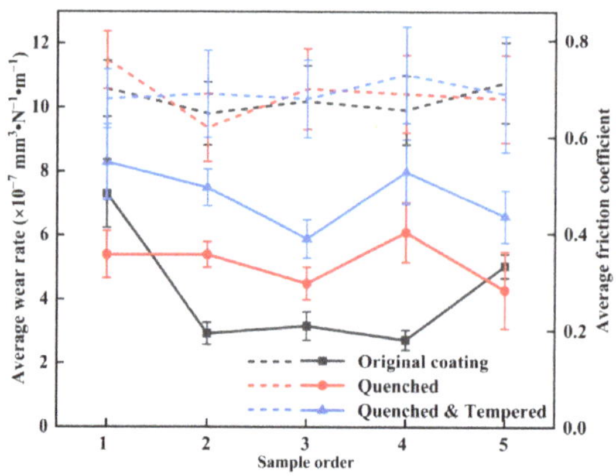

Figure 16. Average wear rate and friction coefficient of coatings with different parameters.

4. Conclusions

There were obvious "black-and-white" areas (cladding zone and heat-affected zone) on the surface of the T15 cladding layer. The average size of equiaxed grains was finer. The microhardness was about 200 HV higher in the CZ than HAZ. The average wear rate of original CZ and HAZ was 7.3×10^{-7} $mm^3 \cdot N^{-1} \cdot m^{-1}$ and 4.3×10^{-7} $mm^3 \cdot N^{-1} \cdot m^{-1}$. The alloy compounds were distributed in the grain boundaries for both regions. The microstructure uniformity and hardness gradient of the surface was eliminated by the post-heat treatment. The alloy elements except for V were dissolved into grains during the quenching process.

The microhardness of cladding layers was 910 HV after quenching and 750 HV after tempering. Post-heat treatment had no significant effect on the friction coefficient. After quenching and quenching–tempering, the wear rate of sample 1 was 5.6×10^{-7} $mm^3 \cdot N^{-1} \cdot m^{-1}$ and 8.5×10^{-7} $mm^3 \cdot N^{-1} \cdot m^{-1}$, respectively. The wear mechanism of the cladding layer was mainly abrasive wear and fatigue wear. The quenching residual stress could be eliminated by the subsequent tempering process.

For the cladding layers with different cladding process parameters, the average friction coefficient of coatings hovered around 0.7. In addition, the standard deviation of wear rate was decreased after post-heat treatment.

Author Contributions: All of the authors contributed to this idea and method. Sample preparation, data acquisition, and analysis were all performed by Y.M. The first draft of the manuscript was written by Y.M. Corresponding author, Y.Z., read the comments on the previous version of the manuscript and approved the revised manuscript. All authors have read and agreed to the published version of the manuscript.

Funding: This work was financially supported by the National Natural Science Foundation of China (No. 51905148 and No. 51875169) and by the Fundamental Research Funds for the Central Universities (No. B200202218).

Data Availability Statement: All data that support the findings of this study are available from the corresponding author upon reasonable request.

Conflicts of Interest: The authors declare no conflict of interest. The funders had no role in the design of the study, in the collection, analyses, or interpretation of data, in the writing of the manuscript, or in the decision to publish the results.

References

1. Zhao, F.S.; Zhang, Z.H.; Shao, M.H.; Bi, Y.J.; Zhao, Y.Q.; Wang, Z.W.; Li, Y.; Li, H.H.; Xu, X.G.; He, Y.Y. Mechanical and Wear Properties of 42CrMo Steel by Plasma Nitriding assisted Hollow Cathode Ion Source. *Mat. Res.* **2021**, *24*, 1–10. [CrossRef]
2. Zhu, L.J.; Liu, Y.H.; Li, Z.W.; Zhou, L.; Li, Y.J.; Xiong, A.H. Microstructure and properties of Cu-Ti-Ni composite coatings on gray cast iron fabricated by laser cladding. *Opt. Laser Technol.* **2020**, *122*, 105879. [CrossRef]
3. Karmakar, D.P.; Muvvala, G.; Nath, A.K. Effect of scan strategy and heat input on the shear strength of laser cladded Stellite 21 layers on AISI H13 tool steel in as-Deposited and heat treated conditions. *Surf. Coat. Technol.* **2020**, *384*, 125331. [CrossRef]
4. Das, A.K. Effect of rare earth oxide additive in coating deposited by laser cladding: A review. *Mater. Today Proc.* **2022**, *52*, 1558–1564. [CrossRef]
5. Liu, J.S.; Shi, Y. Microstructure and wear behavior of laser-Cladded Ni-Based coatings decorated by graphite particles. *Surf. Coat. Technol.* **2021**, *412*, 127044. [CrossRef]
6. Zhang, Z.; Yu, T.; Kovacevic, R. Erosion and Corrosion Resistance of Laser Cladded AISI 420 Stainless Steel Reinforced with VC. *Appl. Surf. Sci.* **2017**, *410*, 225–240. [CrossRef]
7. He, B.; Zhang, L.J.; Zhu, Q.H.; Wang, J.; Yun, X.; Luo, J.S.; Chen, Z.K. Effect of solution treated 316L layer fabricated by laser cladding on wear and corrosive wear resistance. *Opt. Laser Technol.* **2020**, *121*, 105788. [CrossRef]
8. Feng, Y.L.; Yao, C.W.; Shen, C.; Feng, Y.Q.; Feng, K.; Li, Z.G. Influence of in-Situ synthesized carboborites on microstructure evolution and the wear resistance of laser clad Fe-Base composite coatings. *Mater. Charact.* **2020**, *164*, 110326. [CrossRef]
9. Hou, Q.Y. Influence of molybdenum on the microstructure and properties of a FeCrBSi alloy coating deposited by plasma transferred arc hard facing. *Surf. Coat. Technol.* **2013**, *225*, 11–20. [CrossRef]
10. Han, B.; Li, M.Y.; Wang, Y. Microstructure and wear resistance of laser clad $Fe-Cr_3C_2$ composite coating on 35CrMo steel. *Mater. Eng. Perform.* **2013**, *22*, 3749–3754. [CrossRef]
11. Qi, K.; Yang, Y.; Hu, G.F.; Lu, X.; Li, J.D. Thermal expansion control of composite coatings on 42CrMo by laser Cladding. *Surf. Coat. Technol.* **2020**, *397*, 125983. [CrossRef]
12. Saeedi, R.; Razavi, R.S.; Bakhshi, S.R.; Erfanmanesh, M.; Bani, A.A. Optimization and characterization of laser cladding of NiCr and NiCr–TiC composite coatings on AISI 420 stainless steel. *Ceram. Int.* **2021**, *47*, 4097–4110. [CrossRef]
13. Monkova, K.; Monka, P.; Jakubeczyova, D. The Research of the High Speed Steels Produced by Powder and Casting Metallurgy from the View of Tool Cutting Life. *Appl. Mech. Mater.* **2013**, *302*, 269–274. [CrossRef]
14. Alouane, C.; Kasser, A. Consolidation by atmospheric pressure of T15 tool steel powder. *Powder Technol.* **2019**, *352*, 331–339. [CrossRef]
15. Morozow, D.; Siemiatkowski, Z.; Gevorkyan, E.; Ruckiet, M.; Matijošius, J.; Kilikevicius, A.; Caban, J.; Krzysiak, Z. Effect of Yttrium and Rhenium Ion Implantation on the Performance of Nitride Ceramic Cutting Tools. *Materials* **2020**, *13*, 4687. [CrossRef]
16. Zhang, G.Q.; Yuan, H.; Jiao, D.L.; Li, Z.; Zhang, Y.; Liu, Z.W. Microstructure evolution and mechanical properties of T15 high speed steel prepared by twin-Atomiser spray forming and thermo-Mechanical processing. *Met. Mater. Int.* **2012**, *558*, 566–571. [CrossRef]
17. Han, X.; Li, C.; Yang, Y.P.; Gao, X.; Gao, H.X. Experimental research on the influence of ultrasonic vibrations on the laser cladding process of a disc laser. *Surf. Coat. Technol.* **2021**, *406*, 126750. [CrossRef]
18. Nyadongo, S.T.; Olakanmi, E.O.; Pityana, S.L. Experimental and numerical analyses of geometrical and microstructural features of Tribaloy T-800 composite coating deposited via laser cladding-Assisted with pre-Heat (LCAP) process. *J. Manuf. Process.* **2021**, *69*, 84–111. [CrossRef]
19. Li, W.Y.; Yang, X.F.; Xiao, J.P.; Hou, Q.M. Effect of WC mass fraction on the microstructure and friction properties of WC/Ni60 laser cladding layer of brake discs. *Ceram. Int.* **2021**, *47*, 28754–28763. [CrossRef]
20. Jiao, X.Y.; Wang, C.M.; Gong, Z.Q.; Wang, G.M.; Sun, H.F.; Yang, H.R. Effect of Ti on T15M composite coating fabricated by laser cladding technology. *Surf. Coat. Technol.* **2017**, *325*, 643–649. [CrossRef]
21. Szala, M.; Latka, L.; Walczak, M.; Winnicki, M. Comparative Study on the Cavitation Erosion and Sliding Wear of Cold-Sprayed Al/Al_2O_3 and Cu/Al_2O_3 Coatings, and Stainless Steel, Aluminium Alloy, Copper and Brass. *Metals* **2020**, *10*, 856. [CrossRef]
22. Lv, H.; Li, Z.J.; Li, X.D.; Yang, K.; Li, F.; Xie, H.L. Effect of Vanadium Content on the Microstructure and Mechanical Properties of IN718 Alloy by Laser Cladding. *Materials* **2021**, *14*, 2362. [CrossRef] [PubMed]
23. Morozow, D.; Barlak, M.; Werner, Z.; Pisarek, M.; Konarski, P.; Zagorski, J.; Rucki, M.; Chalko, L.; Lagodzinski, M.; Narojczyk, J.; et al. Wear Resistance Improvement of Cemented Tungsten Carbide Deep-Hole Drills after Ion Implantation. *Materials* **2021**, *14*, 239. [CrossRef] [PubMed]
24. Diniz, A.E.; Machado, A.R.; Correa, J.G. Tool wear mechanisms in the machining of steels and stainless steels. *Int. J. Adv. Manuf. Technol.* **2016**, *87*, 3157–3168. [CrossRef]

25. Ben, Q.; Zhang, Y.; Sun, L.; Wang, L.; Wang, Y.; Zhan, X. Wear and Corrosion Resistance of FeCoCrxNiAl High-Entropy Alloy Coatings Fabricated by Laser Cladding on Q345 Welded Joint. *Metals* **2022**, *12*, 1428. [CrossRef]
26. Wang, Y.Y.; Shi, H.; Hao, X.H.; Liu, H.X.; Zhang, X.W. Microstructure and Wear Resistance of Fe60 Laser Cladding Coating Assisted by Steady Magnetic Field–Mechanical Vibration Coupling Field. *Coatings* **2022**, *12*, 751. [CrossRef]
27. Wang, X.Y.; Liu, Z.D.; Li, J.Y.; Chen, L.; Li, B.K. Effect of heat treatment on microstructure, corrosion resistance, and interfacial characteristics of Inconel 625 laser cladding layer. *Optik* **2022**, *270*, 169930. [CrossRef]
28. Cai, Y.C.; Shan, M.D.; Manladan, S.M.; Zhu, L.S.; Gao, F.F.; Sun, D.; Han, J. Effect of high temperature heat treatment on microstructure and properties of FeCoCrNiAl high-Entropy alloy laser cladding layer. *Mater. Charact.* **2022**, *191*, 112137. [CrossRef]
29. Song, M.J.; Wu, L.S.; Liu, J.M.; Hu, Y. Effects of laser cladding on crack resistance improvement for aluminum alloy used in aircraft skin. *Opt. Laser. Technol.* **2021**, *133*, 106531. [CrossRef]
30. Zhang, Y.T.; Ma, Y.L.; Wang, G.; Ji, X.L.; Li, Z.C. Experimental Study on the Wear Properties of 42CrMo Steel with Different Microstructures and T15 Laser Cladding. *J. Mater. Eng. Perform.* **2022**, *31*, 4232–4241.
31. Shan, B.; Chen, J.L.; Chen, S.Y.; Ma, M.Z.; Ni, L.L.; Shang, F.M.; Zhou, L. Laser cladding of Fe-Based corrosion and wear-resistant alloy: Genetic design, microstructure, and properties. *Surf. Coat. Technol.* **2022**, *433*, 128117. [CrossRef]
32. Joseph, G.B.; Valarmathi, T.N.; Rajan, J.A. Effect of welding parameters on mechanical and corrosion behavior of Stellite 6 cladded SA387 Gr 91 with ERNiCr-3 buffer layer. *T. Can. Soc. Mech. Eng.* **2020**, *45*, 37–47. [CrossRef]
33. Craig, P.; Ramandi, H.L.; Chen, H.H.; Vandermaat, D.; Crosky, A.; Hagan, P.; Hebblewhite, B.; Saydam, S. Stress corrosion cracking of rockbolts: An in-situ testing approach. *Constr. Build. Mater.* **2021**, *269*, 121275. [CrossRef]
34. Li, Y.J.; Dong, S.Y.; Yan, S.X.; Li, E.Z.; Liu, X.T.; He, P.; Xu, B.S. Deep pit repairing of nodular cast iron by laser cladding NiCu/Fe-36Ni low-Expansion composite alloy. *Mater. Charact.* **2019**, *151*, 273–279. [CrossRef]
35. Jiao, X.Y.; Wang, J.; Wang, C.M.; Gong, Z.Q.; Pang, X.X.; Xiong, S.M. Effect of laser scanning speed on microstructure and wear properties of T15M cladding coating fabricated by laser cladding technology. *Opt. Lasers Eng.* **2018**, *110*, 163–171. [CrossRef]
36. Gong, F.B.; Shen, J.; Gao, R.H.; Zhang, T.; Xie, X.; Li, Y. Influence of heat treatment on microstructure and mechanical properties of FeCrNi coating produced by laser cladding. *T. Nonferr. Metal. Soc.* **2016**, *26*, 2117–2125. [CrossRef]
37. Zhang, L.; Bai, Y.; He, T.; Fan, W.; Ma, Y.S.; Liu, H.B.; Wang, Y.; Chang, Z.D.; Kang, Y.X. Performance of plasma-Sprayed thermal barrier coating with engineered unmelted nano-Particle contents. *Surf. Coat. Technol.* **2019**, *368*, 67–78. [CrossRef]
38. Ning, K.; Wang, J.M.; Shi, M.J.; Ning, W.G.; Zhang, F. Friction-Increasing Mechanism of Contact Pair Using Different Surface Treatment Processes. *Tribol. Int.* **2022**, *165*, 107337. [CrossRef]
39. Salvati, E. Residual stress as a fracture toughening mechanism: A Phase-Field study on a brittle material. *Theor. Appl. Fract. Mech.* **2021**, *114*, 103021. [CrossRef]
40. Wang, D.S.; Liang, E.J.; Chao, M.J.; Yuan, B. Investigation on the microstructure and cracking susceptibility of laser-Clad V_2O_5 /NiCrBSiC alloy coatings. *Surf. Coat. Technol.* **2008**, *202*, 1371–1378. [CrossRef]

lubricants

MDPI

Article

Dry and Minimum Quantity Lubrication Machining of Additively Manufactured IN718 Produced via Laser Metal Deposition

Ozan Can Ozaner [1], Angshuman Kapil [1], Yuji Sato [2], Yoshihiko Hayashi [2], Keiichiro Ikeda [2], Tetsuo Suga [2], Masahiro Tsukamoto [2], Sener Karabulut [3], Musa Bilgin [3] and Abhay Sharma [1,*]

1 Faculty of Engineering Technology, Department of Materials Engineering, KU Leuven, Campus De Nayer, 2860 Sint-Katelijne Waver, Belgium; ozancan.ozaner@kuleuven.be (O.C.O.); angshuman.kapil@kuleuven.be (A.K.)
2 Joining and Welding Research Institute, Osaka University, Osaka 567-0047, Japan; sato@jwri.osaka-u.ac.jp (Y.S.); yoshihiko-hayashi@ofic.co.jp (Y.H.); keiichirou-ikeda@ofic.co.jp (K.I.); suga@jwri.osaka-u.ac.jp (T.S.); tukamoto@jwri.osaka-u.ac.jp (M.T.)
3 Department of Mechanical Program, Hacettepe University, Ankara 06930, Turkey; senerkarabulut@hacettepe.edu.tr (S.K.); musabilgin@hacettepe.edu.tr (M.B.)
* Correspondence: abhay.sharma@kuleuven.be

Abstract: Inconel 718 (IN718), a Ni-based superalloy, is immensely popular in the aerospace, nuclear, and chemical industries. In these industrial fields, IN718 parts fabricated using conventional and additive manufacturing routes require subsequent machining to meet the dimensional accuracy and surface quality requirements of practical applications. The machining of IN718 has been a prominent research topic for conventionally cast, wrought, and forged parts. However, very little attention has been given to the machinability of IN718 additively manufactured using laser metal deposition (LMD). This lack of research can lead to numerous issues derived from the assumption that the machining behavior corresponds to conventionally fabricated parts. To address this, our study comprehensively assesses the machinability of LMDed IN718 in dry and minimum quantity lubrication (MQL) cutting environments. Our main goal is to understand how LMD process variables and the cutting environment affect cutting forces, tool wear, surface quality, and energy consumption when working with LMDed IN718 walls. To achieve this, we deposited IN718 on SS309L substrates while varying the following LMD process parameters: laser power, powder feed rate, and scanning speed. The results unveil that machining the deposited wall closer to the substrate is significantly more difficult than away from the substrate, owing to the variance in hardness along the build direction. MQL greatly improves machining across all processing parameters regardless of the machining location along the build direction. Laser power is identified as the most influential parameter, along with the recommendation for a specific combination of power feed rate and scanning speed, providing practical guidelines for optimizing the machining process. While MQL positively impacts machinability, hourly energy consumption remains comparable to dry cutting. This work offers practical guidance for improving the machinability of LMDed IN718 walls and the successful adoption of LMD and the additive–subtractive machining chain. The outcomes of this work provide a significant and critical understanding of location-dependent machinability that can help develop targeted approaches to overcome machining difficulties associated with specific areas of the LMDed structure. The finding that MQL significantly improves machining across all processing parameters, particularly in the challenging bottom region, offers practical guidance for selecting optimal cutting conditions. The potential economic benefits of MQL in terms of tool longevity without a substantial increase in energy costs is also highlighted, which has implications for incorporating MQL in several advanced manufacturing processes.

Keywords: Inconel 718; laser metal deposition (LMD); additive manufacturing; machinability; minimum quantity lubrication (MQL); tool wear; cutting forces; energy consumption

Citation: Ozaner, O.C.; Kapil, A.; Sato, Y.; Hayashi, Y.; Ikeda, K.; Suga, T.; Tsukamoto, M.; Karabulut, S.; Bilgin, M.; Sharma, A. Dry and Minimum Quantity Lubrication Machining of Additively Manufactured IN718 Produced via Laser Metal Deposition. *Lubricants* **2023**, *11*, 523. https://doi.org/10.3390/lubricants11120523

Received: 4 November 2023
Revised: 30 November 2023
Accepted: 6 December 2023
Published: 10 December 2023

1. Introduction

The utilization of laser metal deposition (LMD), a variant of the directed energy deposition (DED)-based additive manufacturing (AM), holds widespread potential for the manufacture, remanufacture, and repair of critical high-value-added components [1], and the fabrication of wear- and corrosion-resistant coatings [2]. Owing to the flexibilities and benefits like high power density, production and geometric freedom, stability, controllability, low thermal input (i.e., small heat-affected zone), strong metallurgical bonding, less post-deposition deformation, ability to process difficult-to-machine materials (e.g., Ti alloys, Ni-based superalloys), and low buy-to-fly ratio (~1.5:1) [3], LMD has anchored its position as a competitive and highly versatile AM technology [4]. The LMD process, due to its highly localized deposition nature, allows for the addition of the right material at the right place for components with different degrees of geometrical complexity [5]. The process can produce parts with almost 100% density and reliable metallurgical properties that meet the requirement for direct usage, thus enabling the preceding use of costly forming dies and tooling [6]. Despite the multiple technological advantages, the surface quality and dimensional accuracy of the parts fabricated via LMD do not meet the geometrical and mechanical requirements of practical applications [7], thereby necessitating a subsequent post-processing or finishing step. AM using LMD typically produces surfaces with an average roughness of $\geq$10 μm [7], whereas most mechanical systems in critical applications require much smoother surfaces [8]. Although laser-based surface finishing approaches like laser re-melting [7] and laser polishing [9] exist, the achievable improvement in surface roughness more often than not satisfies the stringent requirements of international standards. Thus, part fabrication using LMD almost always requires a successive end-machining process.

LMD-based AM is being rapidly adapted for fabricating and repairing high-performance alloys [10]. Among several of these alloys, Ni-based superalloy Inconel 718 (IN718), ubiquitous in industrial gas turbines, jet engine components (turbine blades, disks, shafts, stators, and casing), nuclear power plant components, supporting structures, and pressure vessels in aerospace and oil and gas sectors [11], has received extensive attention. While the AM of IN718 parts with the LMD technique is ever-increasing, challenges pertaining to its machinability remain unresolved, leading to the significant usage of resources and costs in terms of tool wear, material waste, and lead time [12]. Thus, even though the LMD process is often deemed vastly productive, the productivity of LMD for IN718 remains a critical factor that merits further investigation.

IN718, despite its exceptional properties, is among the most difficult-to-machine materials [13]. Due to the high concentration of alloying elements, IN718 resists plastic deformation during machining, leading to rapid strain hardening [8]. Under such situations, the machining of IN718 often leads to the formation of undesirable residual stresses and dimensional inaccuracy. In addition, the poor thermal conductivity of IN718 results in a rapid temperature rise of the machined surface, leading to degraded surface quality and occasional burns [14]. Machinability of IN718 fabricated by conventional and DED AM routes has been evaluated and compared in limited studies in recent years. Bagherzadeh et al. [15] reported that IN718 deposited via selective laser melting and DED parts is easy to machine compared to the wrought alloy, especially for machining with lubrication. However, in the case of dry machining, a larger cutting force was reported for laser-clad IN718 compared to the wrought parts, owing to differences in grain geometries [16]. The machinability of DED IN718 parts improves with high-temperature machining (400 °C) of the fabricated parts, leading to higher surface quality and significantly lower tool wear than room-temperature machining [17]. The machining aspects of IN718/SS316L functionally graded material (FGM) fabricated by laser DED AM were recently evaluated by Li et al. [18] and Zhang et al. [19]. In both studies, the machining of the sections in the FGM containing more IN718 than SS316L led to higher cutting forces and the generation of more cutting heat, owing to the higher hardness and strength of the IN718. New techniques like cryogenic cooling with liquid nitrogen have been effective in significantly reducing the cutting forces and tool wear during the machining of LMDed IN718 [20].

Although significant technical knowledge is available to tackle the challenges associated with machining conventionally cast, wrought, and forged IN718 parts, there is a dearth of scientific research and practical know-how of the intricacies of machining LMDed IN718. The absence or lack thereof of such studies can lead to multifold problems derived from the basic assumption that the machining behavior corresponds to that of conventionally fabricated parts. Development of a fundamental understanding of how AM affects the machining operation is rather critical, as the properties (mechanical, chemical, etc.) of AM parts [21], and in particular, those fabricated by LMD [22], are strikingly different from conventionally manufactured parts. Considering the increased push for LMD integration in industries, it is crucial to highlight that machining an LMDed difficult-to-machine material like IN718 implies additional challenges, reflected in the final surface quality and dimensional accuracy, and thus, it is vital to understand the LMD process effect on the machinability aspects of IN718. Pereira et al. [23] evaluated the cutting forces and surface roughness for LMDed IN718 preforms. The research highlighted considerably improved surface finish with down milling and significantly higher cutting forces while machining in the zone near the substrate (i.e., build platform). The reason for differences along the build direction was attributed to the LMD process-generated rigidity and residual stress; however, it was not scientifically explored. A proper understanding of this behavior is significant as it can guide tooling and machining parameter selection. The effect of post-deposition heat treatment on the machining of LMD parts has been another area of recent focus. Studies conducted by both Careri et al. [24] and Calleja et al. [25] revealed that the machining heat-treated LMDed IN718 parts led to higher surface roughness values and significant tool wear, implying the employment of heat treatments downstream of the machining operations. The study of the chip morphology for the machining of LMDed IN718 has also been a subject of interest. Unlike the continuous chip formation in the forged [8] and cast [26] parts, machining the LMDed IN718 surface leads to irregular lamellar chip formation. These differences arise from the anisotropy in material properties in AM, particularly with the growth of coarse grains along the build-up direction. The irregular chip formation was reflected in the steep fluctuations in measured cutting forces. Zhou et al. [27] optimized the laser parameters for LDED of alloy GH4169 (an alloy equivalent of IN718 developed in China); however, the objective of the work was to evaluate the effect of interrupted machining passes in between deposition passes on the phase formation, element segregation, and performance of the GH4169 alloy.

Central to the seamless implementation of the LMD AM in industries is the minimization of the post-deposition machining requirements. The present investigation is motivated by the critical review of the literature, wherein it is evident that although preliminary work focusing on the machining of LMDed IN718 is available, the previous studies do not elucidate the fundamentals governing the interplay between the process and the ensuing machinability of the fabricated part for variation in LMD processing parameters. Moreover, existing studies on the machining of LMDed IN718 do not address the influence of the cutting environment on the machinability aspects. In this work, the term "cutting environment" pertains to the cooling and lubrication applied in the cutting zone, encompassing both dry machining, where no cutting fluids are used, and minimum quantity lubrication machining (MQL), where the controlled application of cutting fluid is applied. In the MQL approach, a very small amount of cutting fluid is mixed with a carrier gas such as air, CO_2, N_2, etc. The cutting fluid makes a layer on the cutting surface and favorably impacts the cooling and lubrication phenomenon, as explained later in Section 2. The MQL is proven to be useful in the machining of difficult-to-cut materials [14].

The purpose of this study is to address the critical research gaps with practical implications by incorporating a comprehensive assessment of the machinability of IN718 over a wide-ranging set of LMD processing parameters and considering two different types of machining environments, i.e., dry and minimum quantity lubrication (MQL). Recent work by Gupta et al. [28] has proven the efficacy of machining (turning) 2205 duplex steel under MQL. The implementation of the MQL, a sustainable machining approach, led to significant improvement in surface quality and lowering of the tool wear and energy consumption. Our main goal is to understand how LMD process variables and the cutting environment affect cutting forces, surface quality, surface temperatures, tool wear, and energy consumption when working with LMDed IN718 walls. For the mentioned goal, IN718 is deposited on cold-rolled SS309L substrates with varying LMD process parameters, including laser power, powder feed rate, and scanning speed, and the effect of variation in these parameters on the machinability is examined. The machinability and its variation (if any) along the build direction are correlated with hardness distribution, the most significant LMD process parameters(s) are identified, the effect of process parameters on machining under both dry and MQL environments is critically analyzed, followed by the identification of suitable process parameter-cutting environment combination for the ease of machining. The rest of this paper is organized as follows: Section 2 presents the details about the materials and experimental procedure; Section 3 provides a detailed discussion of the obtained results; and finally, Section 4 draws conclusions from this study and provides future research directions.

2. Materials and Methods

The experimental workflow in this study is divided into two stages, i.e., material deposition and post-deposition machining. A flowchart of the experimental plan is provided in Figure 1, with elaborate details provided in Sections 2.1 and 2.2.

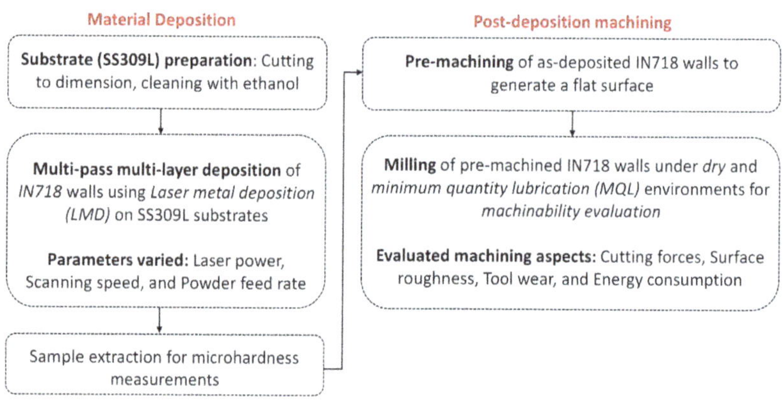

Figure 1. Flowchart depicting the experimental workflow.

2.1. Material Deposition

IN718 powder with an average diameter of 30 μm is deposited on SS309L substrates with a disk laser with a wavelength of 1030 nm and a beam diameter of 2 mm. The choice of stainless steel (SS) as the substrate material arises from the fact that the fabrication of sound metallurgical IN718-SS AM joints has been repeatedly demonstrated in the literature [29]. In addition, the use of steel as the substrate instead of IN718 acts as a cost reduction means in the AM chain and is thus of industrial interest. During the deposition, combinations of laser power, scanning speed, and powder feed rate are used at five levels, as presented in Table 1. Multi-pass, multi-layer deposition is carried out to reach the desired part thickness per the deposition scheme in Figure 2a. A total of 6 passes with a hatch distance of 1 mm are employed for each layer. The total number of layers varied depending on the individual layer height, directly related to the operating process parameters. The substrate's initial temperature was equivalent to the standard room temperature during the deposition. The final dimensions of the wall geometry of IN178 and the substrate dimensions are depicted in Figure 2b. For post-deposition, samples approximately 10 mm wide from the left and right edges of the main material are cut for hardness mapping, as depicted in Figure 2c. Standard metallographic procedures are employed for grinding, polishing, and etching the samples. A low-force Vickers hardness test (HV0.2) based on EN ISO 6507-1 standard with a 0.2 kgf load and dwell time of 15 s is performed on the deposited parts using a ZwickRoell-make DuraScan Microhardness tester. Detailed information on the hardness test is provided in Section 3.

Table 1. LMD process parameters.

Sample	Laser Power (W)	Powder Feed Rate (g/min)	Scanning Speed (m/min)	Number of Layers *
1	600	10	0.6	56
2	700	10	0.6	52
3	800	10	0.6	52
4	900	10	0.6	52
5	1000	10	0.6	40
6	800	4	0.6	80
7	800	7	0.6	56
8	800	10	0.6	52
9	800	13	0.6	32
10	800	16	0.6	32
11	800	10	0.2	14
12	800	10	0.4	25
13	800	10	0.6	52
14	800	10	0.8	68
15	800	10	1.0	94

* The number of layers represents the layers achieved to attain the fixed sample height, as shown in Figure 2b.

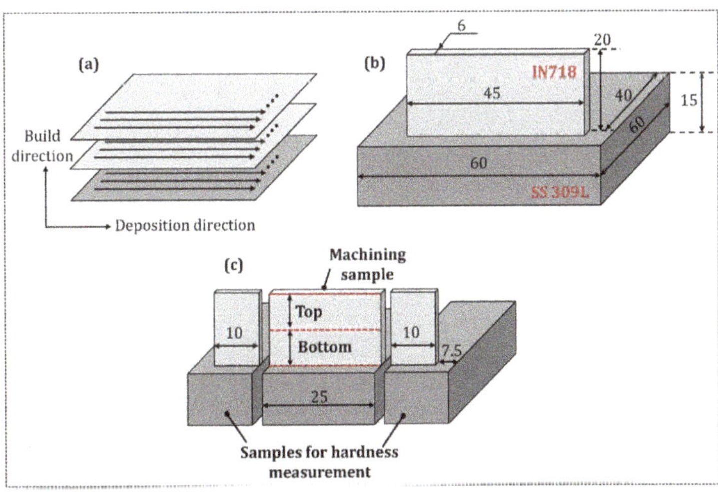

Figure 2. (**a**) Deposition scheme for LMD; (**b**) geometry and dimensions of the SS 309L substrate and deposited IN718 wall; and (**c**) the location and dimensions of samples for machining and hardness measurement (Schematics not drawn to scale).

2.2. Post-Deposition Machining

The experimental setup utilized for the machinability evaluation is presented in Figure 3. A 3-axis Frontier MCV-866 CNC vertical milling machine with a maximum spindle speed of 10,000 rpm is used. While machining, the cutting forces are recorded with a Kistler 9272 dynamometer at a sampling frequency of 20 kHz. The collected force data provide the radial/axial forces and the resultant forces. To accurately identify chatter and peak states in cutting forces, the machining tests are performed using a single insert at an entering angle of 90°. PVD-coated AOMT123608PEER-M VP15TF cutting inserts with a nose radius of 0.8 mm are used for the machinability evaluation experiments. The insert is mounted on a standard BBT 40 tool holder with a diameter of 16 mm, and each machining test is carried out using a fresh edge. The optimized milling parameters provided in Table 2 are utilized for the machining study under dry and MQL cutting environments. The MQL technique presents an environmentally friendly alternative for machining operations, involving the supply of a minimal amount of oil to the cutting zone in the form of an emulsion with a carrier gas, typically air. This oil + air emulsion spreads across the tool–chip interface, enhancing heat transfer by influencing the heat transfer coefficient. Additionally, it significantly reduces friction at the interface in a discernible manner, allowing for the friction coefficient to be expressed in terms of the machining parameters [30]. Consequently, MQL has proven to be particularly advantageous in the machining of difficult-to-machine materials such as titanium and nickel alloys.

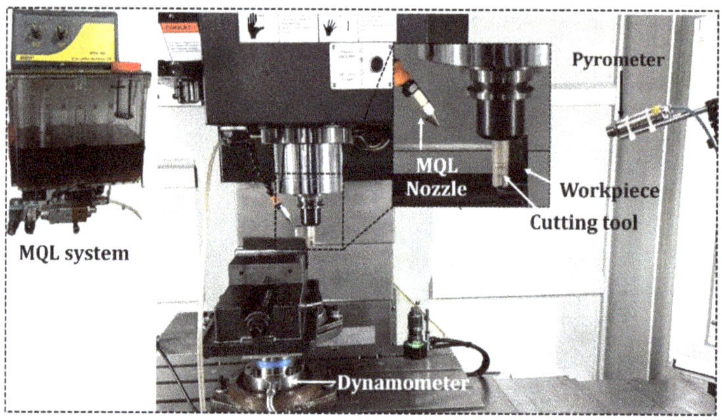

Figure 3. Experimental setup for machinability evaluation experiments.

Table 2. Machining parameters.

Machining Parameter	Value
Cutting speed, V_c (m/min)	50
Feed per tooth, f_z (mm/tooth)	0.1
Depth of cut, a_p (mm)	0.8
Cutting width, a_e (mm)	10

The post-deposition machining parameters are determined based on two factors. Firstly, several pilot experiments are conducted to reach suitable values, and secondly, catalog data provided by the cutting tool manufacturer are utilized to narrow down the parameter selection process. During machining, the collected force data provide the active forces (radial F_y/axial F_x) and the passive forces (F_z). Since the depth of the cut value is constant, the effect of the passive force is negligible [31,32]; therefore, the resultant force was calculated with radial and axial forces. This study compares the maximum resulting forces during cutting in all cutting conditions, as shown in the representative force-time plot in Figure 4a. During the machining process, the temperature of the machined surfaces is measured using an MI3 pyrometer and a Testo 872 thermal camera. Before the actual experiments, the pyrometer and the thermal camera were calibrated. The deposited IN718 wall is heated using an open-torch flame, and the temperatures during cooling are recorded using thermocouples connected to a data acquisition unit. Simultaneously, the temperature measurements are also conducted using the pyrometer and the thermal camera, and their emissivity factors are adjusted until the temperatures measured by the pyrometer and the thermal camera match the temperature read from the thermocouple. The temperature of the cutting zones is measured while the positions of the pyrometer and the thermal camera are fixed. The temperature data analysis was performed using the IR Soft software. After each cutting operation, as seen in Figure 4b, the maximum temperature of the cutting zone during machining was detected using IR Soft. For the MQL, a biodegradable cutting oil called Viscol Viscut is employed. This cutting oil has a kinematic viscosity of 37–40 cSt at 40 °C and a density of 0.88 g/cm^3 at 15 °C, with a flash point exceeding 240 °C. The oil is used with a Werte STN40 lubrication system to facilitate lubrication and air cooling at the machining interface. It is directly applied to the cutting area to ensure effective lubrication. The MQL nozzle is precisely positioned 15 mm from the cutting point, maintaining a 20° angle relative to the machine's X–Y plane.

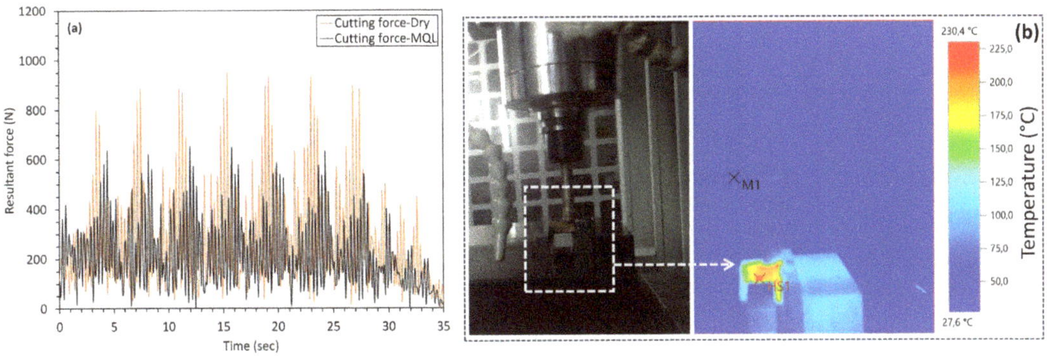

Figure 4. (**a**) Representative resultant force–time plot for machining of sample deposited at the laser power, powder feed rate, and scanning speed of 800 Watt, 4 g/min and 0.6 m/min, respectively, (**b**) Representation of the maximum temperature measurement during machining. (Note: The commas in the (**b**) are used for decimal signs).

The flow rate of the oil is set at 10 mL/min. The milling parameters are kept constant to observe the effect of the cutting environment on the machined cutting forces, tool wear, surface quality, and energy consumption in the machining of LMD-fabricated IN718 workpieces. The energy consumption of the machine is recorded using a Fluke 435 power quality and energy analyzer. The milling of IN718 walls is conducted in two regions, i.e., the top and bottom, as represented in Figure 2c. The top region is machined first, followed by machining of the bottom region using the same tool. There is an interruption between the machining of the top and bottom regions. The distinction between the top and bottom regions is critical, as the bottom region is related to the initial deposition layers, whereas the top region, where the final layers are deposited, may be subject to very different thermo-mechanical process conditions, geometric distortions, and resulting mechanical properties. The front and back surfaces of the IN718 wall are machined under dry and MQL environments, respectively. Post machining, the roughness of the machined surface is evaluated using a contact-based Mitutoyo surftest SJ-210 tester. Surface roughness measurements of machined surfaces are conducted in alignment with both the machining and deposition directions at three locations on each part. Optical microscopy was used to analyze the cutting insert wear.

3. Results and Discussion

3.1. Surface Appearance

The LMDed IN718 wall has an irregular surface resulting from the presence of unmelted powder particles that stick to the side surfaces of the walls, as seen in Figure 5. The degree of unmelted powder particles sticking to the surface is directly related to the size of the molten pool, which in turn is influenced by the laser power. Due to the shape of the heat source, i.e., the laser beam, the energy density in the middle of the molten pool is greater than that near the edge. The high-energy-density region corresponds to higher temperatures, whereas toward the edge, there is a gradual decrease in temperature due to depleting energy density. As a consequence, there is a higher probability that the incoming powder particles remain unmelted at the edges and get stuck to the deposited surface. With a smaller melt pool, the number of powder particles that remain unmelted is expected to be greater. The laser power dictates the melt pool size, i.e., larger melt pools at higher laser power. The interplay of laser power and the ensuing melt pool size decides the quality and appearance of the as-deposited LMDed wall surfaces, verified by the results presented in Figure 5a–c.

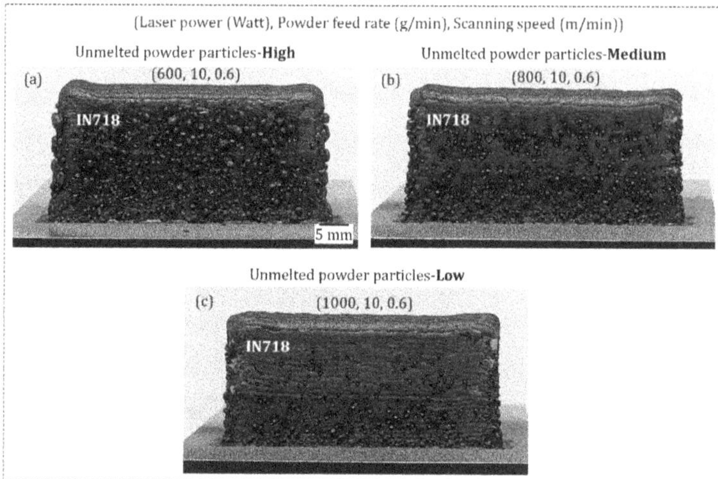

Figure 5. As-deposited IN718 walls at different laser powers.

As a consequence of the unmelted powder particles sticking to the surface of the deposited part, the targeted part thickness of 6 mm of the IN718 wall is not achieved. Evaluating the machinability with highly irregular surfaces, as in this case, is a precursor to erroneous results. In particular, variable forces are obtained since the depth of the cut value cannot be fixed due to the high surface roughness. For this reason, prior to the machinability evaluation, the LMDed wall surfaces are pre-machined to remove the unmelted powder particles and generate a flat surface, as per the procedure described in Appendix A.

3.2. Build Direction Hardness Variation

It is well established in the literature that in LMD, the laser energy density primarily affects the thermal history and determines deposition features like microstructural morphology and size. During the deposition of the initial layers in LMD, the G/R ratio (where G is the temperature gradient and R is the solidification growth rate) is fairly large, whereas along the build direction, the ratio has a gradual decrease [1]. This variation is reflected in the anisotropy of the mechanical properties, in particular, the hardness distribution. A heterogeneous hardness distribution along the build direction leads to non-uniformity in the mechanical properties, which can specifically have a significant effect on the machinability of the deposited part.

To examine the variation in hardness (if any) in the LMDed IN718 walls, hardness mapping was conducted both perpendicular (Figure 6) and parallel (Figure 7) to the build direction of the parts up to the top of the part. The hardness measurements conducted parallel to the build direction also evaluated the effect of the laser power. Figure 6 depicts the through-thickness hardness values (i.e., perpendicular to the build direction) of the LMDed IN718 wall deposited at a laser power of 1000 Watt, powder feed rate of 10 g/min, and scanning speed of 0.6 m/min, as a representative case. For each region, indentation is conducted in two lines, each containing around 50 to 54 indents, depending on the sample thickness at the location. The measurements conducted in three different regions of the sample, i.e., top, middle, and bottom, provide sufficient evidence that the hardness in the top region, i.e., IN718 deposited at a considerable distance from the substrate, had the lowest values. In contrast, the bottom and middle regions had much higher hardness. This observation was consistent for all the investigated samples. The average hardness values recorded in the top, middle, and bottom regions are 256.32 ± 10.06 HV, 294.38 ± 9.11 HV, and 285.78 ± 14.53 HV, respectively. The results presented in Figure 6 suggest that although there is considerable variation in hardness along the build direction, the through-thickness hardness remains within a narrow band (variation ≤ 5%), irrespective of the location along the build direction.

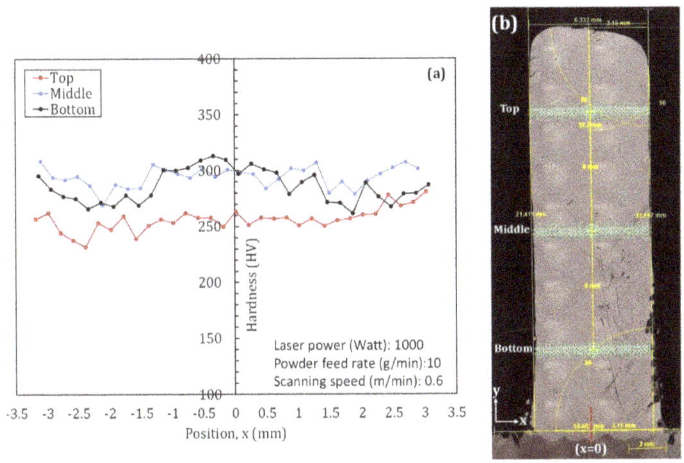

Figure 6. (**a**) Hardness measurement perpendicular to the build direction in different regions of the LMDed IN718 wall, and (**b**) the actual sample depicting the location of the indents.

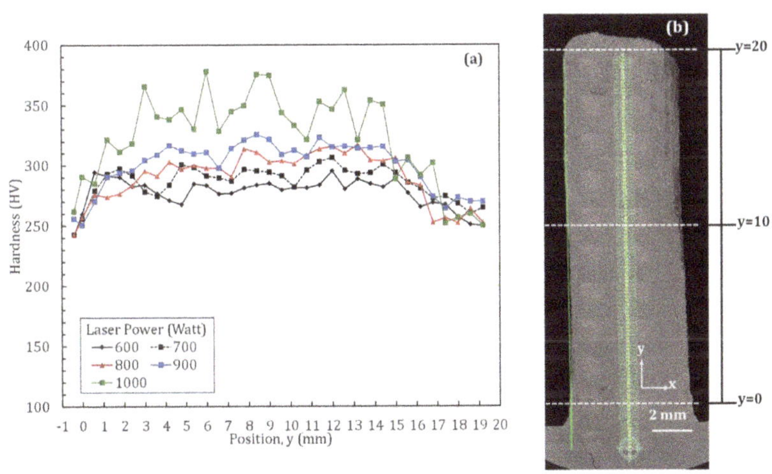

Figure 7. (**a**) Effect of laser power on hardness along the build direction for LMDed IN718 wall, and (**b**) the actual sample depicting the location of the indents.

The hardness mapping presented in Figure 6 highlights the hardness heterogeneity along the build direction. The higher hardness of the deposited IN718 wall near the bottom and middle regions compared to the top region is similar to previously obtained results in the works of Tian et al. [33], Stevens et al. [34], and Li et al. [35]. Similar to previous investigations [33–35], an increment in the hardness in the build direction is observed in the first few layers, followed by a plateau and subsequently a decrease in the hardness in the upper layers, i.e., away from the substrate, the hardness drops (Figure 7). This hardness pattern is however distinct and is contrary to the observed hardness patterns in steels [36]. For steels, like this study, the initial layers exhibit higher hardness due to rapid cooling, followed by a decrease in hardness due to delayed cooling and annealing by the successive layers. The upper layers of AMed steels demonstrate an increase in hardness as the upper layers do not undergo annealing cycles.

The different regions along the build direction experience different thermal cycles, and these slight changes in thermal cycles influence microstructure heterogeneity. The observed hardness heterogeneity in this work can be attributed to the differences in the precipitation of the strengthening phase. Due to the high cooling rate of the LMD process, the γ'' strengthening phase in IN718 is not expected to form initially during rapid solidification but rather develop due to the subsequent heating cycles during multi-layer deposition. The lower hardness of the top region may be due to a decrease in the γ'' strengthening phase. Moreover, the build part experiences discrepancy in elemental segregation. Near the middle region, where the highest average hardness is recorded, the cooling rate is lower. This provides an opportunity for more Nb segregation, causing enhanced γ'' precipitation during multiple thermal gyrations. The closer to the top, the lower heat accumulation and less thermal cycling limit the segregation.

In LMD, the laser power dictates the thermal history, heat accumulation, and ensuing changes in microstructure and mechanical properties. For a constant scanning speed, the heat input varies with the laser power according to the relation in Equation (1):

$$\text{Heat input (HI)} = P/V \tag{1}$$

where P is the laser power (Watt) and V is the scanning speed (m/min). The melt pool widens for higher laser powers, resulting in a lower thermal gradient and, consequently, a slower cooling rate than the parts fabricated with a lower laser power [2]. At lower laser powers, the heat accumulation in the deposited material is very low, which allows for a faster cooling rate and the formation of smaller grain sizes. Thus, based on the literature [37] and general convention, in LMD, a higher hardness is expected for parts fabricated at lower laser powers. However, as depicted in Figure 7, the LMDed IN718 walls in this work displayed higher hardness at higher laser powers, in contrast to results in the literature. This observation is crucial as it leads to the understanding of the primary factor that controls the hardness-laser power relationship. Based on the results presented in Figure 7, it can be conclusively stated that precipitation phenomena dominate over the grain size effect in controlling the influence of laser power on hardness in LMD of IN718. For the lowest laser power of 600 W, a higher cooling rate and lower heat accumulation limit the formation of the γ'' strengthening phase. On the other hand, at a higher laser power of 1000 W, slower cooling and significant heat accumulation promote the formation of the γ'' strengthening phase. It is important to note that a further increase in laser power can lead to significantly high heat input and heat accumulation, leading to Nb dissolving into the matrix and, consequently lower hardness, as observed in the literature on a very high laser power of 1700 W [37].

A sharp increase in hardness values (~250 to 290 HV) is observed at a distance of approximately 2.5 mm from the interface (y = 0) for all the deposited walls (Figure 7). Subsequently, an overall increasing trend in hardness, albeit uneven, is observed up to around 15 mm. The variation in the hardness values between 2.5 mm and 15 mm is approximately 5.98%, 7.74%, 8.34%, 9.58%, and 14.57% for samples fabricated at laser powers of 600, 700, 800, 900, and 1000 Watts, respectively. Notably, the hardness values in the region near the substrate (0–10 mm), i.e., the bottom region, are generally higher than those in the top region (10–20 mm) of the IN718 wall, as demonstrated previously in Figure 6.

3.3. Machinability Evaluation

3.3.1. Effect of Build Direction Anisotropy and Cutting Environment

To unveil the effect of build direction property anisotropy on machinability, the cutting forces, machined surface roughness, and machining-generated surface temperatures are evaluated. The hardness results presented in Figure 7 suggest that the top region of the AM IN718 is softer, whereas both the middle and bottom regions have almost similar hardness. Following this observation, the machinability evaluation was conducted for the top and bottom regions. To further evaluate the effect of the cutting environment, machining is conducted under dry and MQL environments. Box–whisker plots presented in Figure 8a–c depict the variation in resultant cutting forces, surface roughness, and maximum temperature, respectively, over the entire range of experimental conditions. Several critical observations can be derived from Figure 8. Firstly, there is a considerable impact of machining location on the cutting forces and surface quality. The machining of the bottom region generates significantly higher cutting forces, surface roughness, and temperatures than when the top region is machined. This result correlates with the hardness results, as the bottom region exhibits much higher hardness than the top region. Secondly, while the force, roughness, and temperature values have a much larger variation for the bottom region across different samples, the variation is minimal in the top region. Thirdly, the employment of MQL considerably eases the machining of the IN718 wall, particularly in the bottom region. This is reflected in Figure 8a–c, with lower recorded cutting forces, surface roughness, and temperatures when machining is conducted under an MQL environment. The employment of MQL had the most positive influence on the surface temperature.

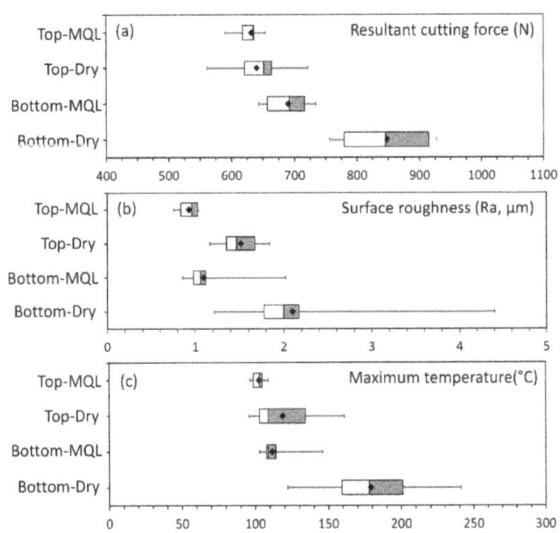

Figure 8. Comparison of machining aspects in the top and bottom regions under dry and MQL cutting environments (**a**) Resultant cutting force, (**b**) surface roughness, and (**c**) maximum temperature.

Dry cutting of the bottom region produced the highest cutting forces, surface roughness, and temperatures across all processing conditions. Notably, in dry cutting, the machined surface roughness in the bottom region exhibited an Ra higher than that of the machined surface in the top region due to the increased cutting forces. For dry cutting, the mean surface roughness value in the bottom region is 1.4 times that of the top region, whereas the maximum surface roughness in the bottom region is 2.4 times that of the top region. Application of the MQL cutting environment reduces the cutting forces up to 20% in the bottom region compared to the dry machining. This is because the applied lubricant forms a thin film (protective layer) between the tool and workpiece, thereby reducing friction and shear stress and facilitating smoother cutting. The cooling effect also helps dissipate heat, preventing excessive temperature rise and reducing the forces exerted on the cutting tool. In addition, the application of MQL alleviates the variability in cutting forces throughout the part (top to bottom regions) despite variations in hardness, as seen in Figure 8a. The reduced variability of cutting forces across the build direction reduces the chances of force jumps and helps prolong cutting tool life. However, the employment of MQL does not significantly alter the machinability in the top region. In particular, the cutting forces remain comparable with those of dry cutting. This observation is vital as it suggests that for LMD of IN718, machining of the part closer to the substrate, i.e., in the initial layers of deposition, has a much stronger impact on the overall machinability of the fabricated part. Based on this outcome, the results presented hereafter focus on the machining aspects of the bottom region.

3.3.2. Process Parameter–Cutting Environment Interplay

Figure 9a–c presents the variation in cutting forces and surface roughness with changes in LMD process parameters, viz., laser power, powder feed rate, and scanning speed, respectively. The variation is evaluated for machining of the bottom region of the LMDed IN718 walls in both dry and MQL cutting environments. With an increase in laser power, a simultaneous increase in cutting forces is observed for dry cutting (Figure 9a). The correlation between increased forces with laser power can be attributed to the rise in hardness values with laser power, as depicted in Figure 7. However, with the use of MQL the cutting forces remain constant despite an increase in laser power. Despite an increase in cutting forces with laser power for dry cutting, the surface roughness values remain relatively constant with an increase in laser power. The exceptionally high surface roughness value at a lower laser power of 600 W is a result of chatter. The chatter is attributable to the highly irregular as-deposited surface produced at a laser power of 600 W (see Figure 5a). Irrespective of the cutting environment, the machining of surfaces deposited at low laser power (600 W) triggers chatter. However, further studies are needed to examine the technical reasons behind this phenomenon in detail. Under the dry cutting environment, an increase in the scanning speed resulted in a slight decrease in cutting forces (Figure 9c).

At the same time, a more significant reduction was observed with an increase in the powder feed rate (Figure 9b). The slight decrease in the cutting forces achieved by increasing the scanning speed had a remarkably positive effect on surface quality, allowing for a substantial reduction in surface roughness. Similarly, lower forces at a higher powder feed rate improved the surface quality as expected, indicating that surface quality is enhanced at a higher powder feed rate and scanning speed. It is, however, crucial to note that this observed improvement is a mid-value of laser power (800 W). Both powder feed rate and scanning speed determine the quantity of the powder delivered to the molten pool and the Intensity of the laser energy absorbed by the powder in unit time. These two parameters dominate the height and width of the deposited layers and, thereby, the surface irregularity. It is, however, very important to keep these values under a certain limit as beyond threshold values, the laser irradiation energy is not enough to entirely melt the powder and form a stable molten pool [38]. With MOL, both the power feed rate and scanning speed had little to no effect on both the cutting forces and surface roughness. There is, however, a jump in the cutting forces when the powder feed rate is increased from 8 to 10 m/min and the scanning speed is increased from 0.4 to 0.6 m/min, both at a constant laser power of 800 W. Under MQL, a significant improvement in terms of surface roughness is observed at the highest powder feed rate of 16 g/min (scanning speed 0.6 m/min). In LMD, the increased interaction of the powder feed and laser beam has the most beneficial effect on the deposited surface quality. These two parameters control the catchment, i.e., the ratio of powder feed rate and scanning speed in LMD. At a higher powder feed rate, the amount of catchment is high, and the interaction between powder and laser beam increases [39], which consequently leads to a surface with improved quality and inherently eases the machining.

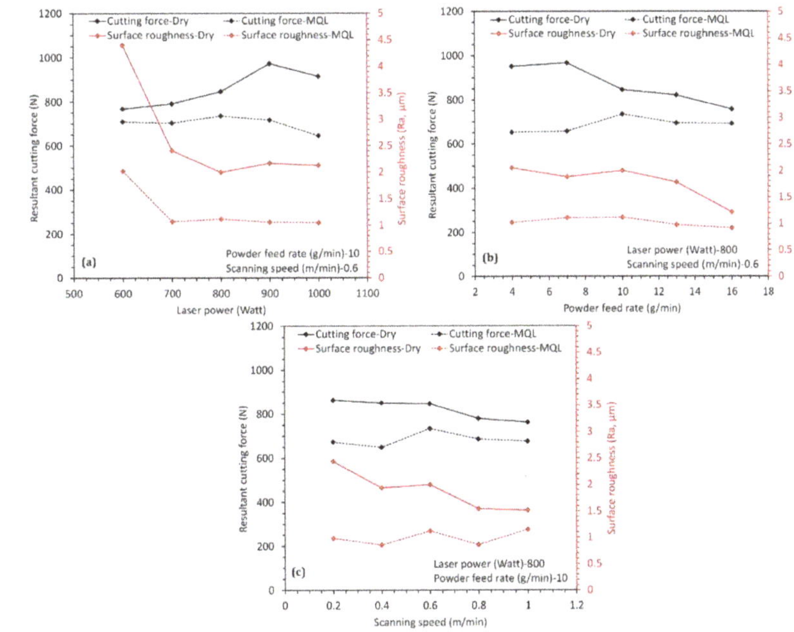

Figure 9. Effect of (**a**) laser power, (**b**) powder feed rate, and (**c**) scanning speed on the resultant cutting forces and surface roughness for machining of the bottom region of the IN718 wall under dry and MQL cutting environments.

The employment of cutting fluid during machining may lead to numerous health risks and environmental effects. Complete elimination of cutting fluids through dry machining is challenging for difficult-to-machine IN718 alloy as observed from the results in Figures 8 and 9. MQL is a suitable alternative to both dry and wet machining (i.e., flood cooling) as it not only lowers the consumption of cutting fluids, thereby reducing the occupational health hazards on the shop floor [40], without making any compromise on the surface quality of the machined surface (refer to Figures 8 and 9). MQL is not only an environmentally friendly machining process [41] but also economical as it significantly cuts down on the cutting fluid usage, as well as MQL can be easily implemented with limited changes to the existing machinery.

3.3.3. Tool Wear

Tool wear is a critical failure mechanism that not only influences the quality of the final product, but also the overall sustainability of the process in terms of energy consumption and manufacturing costs. Defining and mitigating tool wear is crucial for the success of the additive–subtractive manufacturing (ASM) process chain. In this work, the different wear modes occurring on the cutting inserts in the machining experiments were analyzed. The examination of tool wear is first conducted after the machining of the top region before proceeding to the bottom region of the deposited IN718 walls. However, as no significant tool wear is observed during the machining of the top region, the final measurement of tool wear is carried out after the machining of the bottom region. During machinability evaluation experiments, it was observed that some tools broke during the cutting process, either in the middle of the cut or during the exit from the part. The breakage of cutting inserts can be directly attributed to the vibration-induced chatter that causes a sudden and significant change in the load acting on the insert. Images of the cutting inserts that broke during the machining experiments are depicted in Figure 10. At the lowest laser power of 600 W, a significant portion of the cutting insert broke, indicative of a high degree of chatter, which is also reflected in high surface roughness in Figure 9a. The abrasion resulting from the contact between the tool and the machined surface leads to chipping and catastrophic failure.

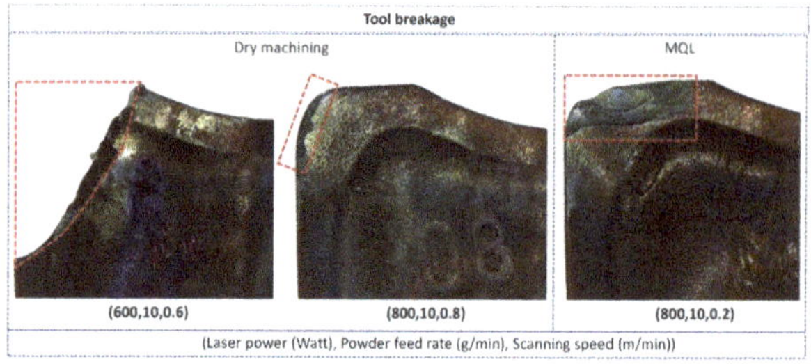

Figure 10. Images of the cutting tools that broke at the edges (shown in red boxes) during machining experiments.

Figures 11–13 compare the tool wear, i.e., the tool life for machining of LMDed IN718 wall with variation in laser power, powder feed rate, and scanning speed, respectively, under both dry and MQL cutting environments. Representative images of the cutting tools with different tool wear modes are also presented in Figures 11–13. Similar to the observed increase in cutting forces and surface roughness with an increase of laser power, the tool wear also increases when laser power is increased from 600 to 1000 W (Figure 11). Although the application of MQL reduces the overall tool wear compared to dry cutting, the tool wear still increases at higher laser powers. The steep increase in tool wear at a laser power of 900 W under MQL is caused by two simultaneous wear phenomena, i.e., coating delamination and edge chipping, as represented in Figure 11a1. At a similar laser power, the tool under dry cutting has slightly higher tool wear and the presence of microchipping at the cutting edge, as seen in Figure 11a2. The contrasting behavior of tool wear with change in cutting environment at the highest laser power of 1000 W can be accounted for by the difference in wear modes. At the highest laser power of 1000 W, there is a sharp increase in the hardness of the deposited IN718, as shown in Figure 7. In dry cutting, due to the absence of sufficient cooling combined with the steep increase in hardness for the sample at 1000 W, the cutting insert is subjected to thermal cracking, as evident from Figure 11a4 where significant edge chipping is observed. The deleterious effect of increased hardness (at a laser power of 1000 W) on tool wear is alleviated in MQL due to effective cooling, the same being evident in (Figure 11a3), where no visible defect is observed in the cutting tool. This is also in accordance with the observed reduction in cutting forces and surface roughness for machining of the LMDed IN718 wall (fabricated at a laser power of 1000 W) under MQL (see Figure 9a). The chipping could be due to the detachment of fragments of the cutting insert edge, initiated by the adhesive wear mechanism [24]. A detailed study of the underlying wear mechanisms is the subject of future work. It can be stated that despite the increase in hardness with higher laser powers, the use of MQL enables the production of surfaces with higher quality and lower cutting forces. The lower cutting forces lead to a decrease in the cutting temperature (see Figure 8) and contribute to the tool wear reduction.

Figure 12 presents the variation in the tool wear with changes in the powder feed rate. Unlike the decrease in cutting forces and surface roughness with an increase in powder feed rate, the tool wear does not follow a particular trend. For powder feed rate, until 7 g/min, the tool wear remains constant for both dry and MQL, with dry cutting registering higher wear. A comparison of the wear modes in dry and MQL environments at a powder feed rate of 7 g/min is presented in Figure 12a1,a2, respectively. While MQL machining resulted in a tool with no visible defects (Figure 12a1), the dry cutting resulted in microchipping (Figure 12a2) and correspondingly higher wear. With an increase of powder feed rate to

10 g/min, the tool wear increased in dry cutting, which does not correspond to the decrease in cutting force and roughness observed previously. Although further experiments are required to elucidate the exact reason, the post-machining analysis of the tool reveals edge chipping (Figure 12a4), which can contribute to the observed jump in the wear. The inspection of the tool used for machining the IN718 wall fabricated at the same powder feed rate (10 g/min) under MQL reveals no defects (Figure 12a3), explaining the lower wear as compared to the dry cutting. Above the powder feed rate of 10 g/min, the tool wear reduces, which possibly is a result of the increase in the powder catchment and its positive effect on surface quality, as explained in the preceding section.

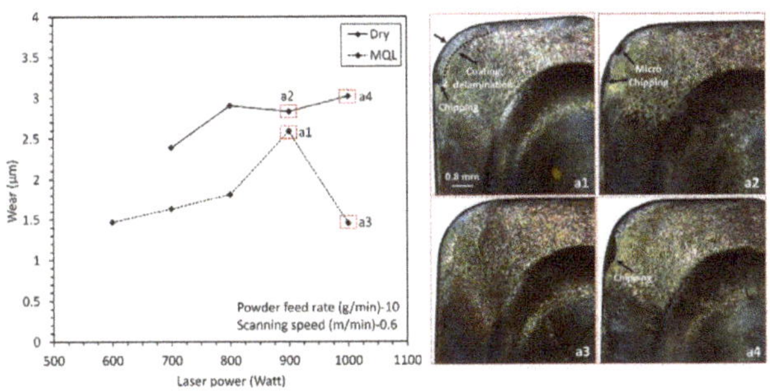

Figure 11. Variation in tool wear with change in laser power. Representative images (**a1–a4**) depict different tool wear modes corresponding to the process conditions highlighted in the graph.

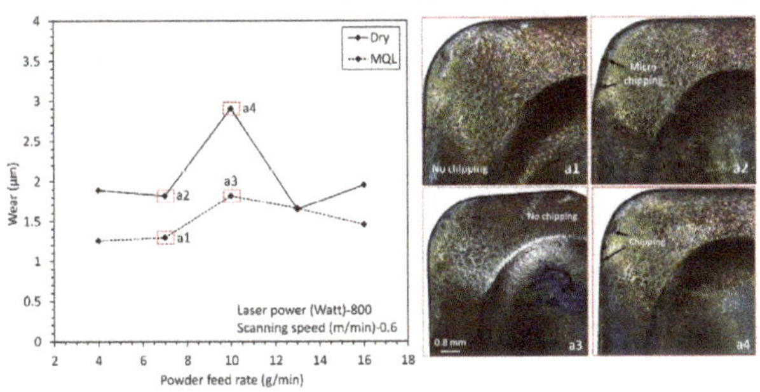

Figure 12. Variation in tool wear with a change in powder feed rate. Representative images depicting different tool wear modes are also presented. Representative images (**a1–a4**) depict different tool wear modes corresponding to the process conditions highlighted in the graph.

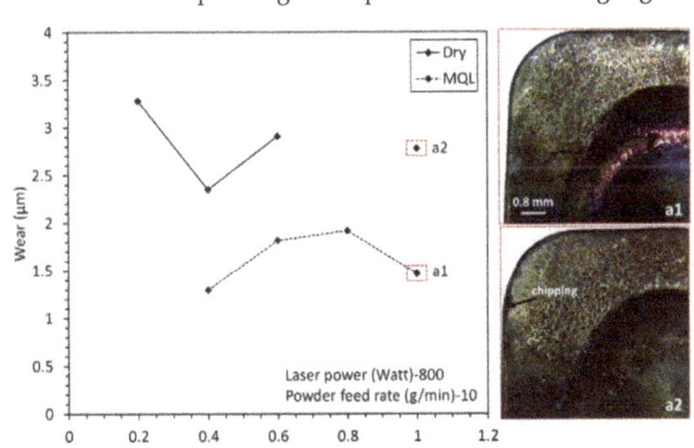

Figure 13. Variation in tool wear with a change in scanning speed. Representative images depicting different tool wear modes are also presented. Representative images (**a1**,**a2**) depict different tool wear modes corresponding to the process conditions highlighted in the graph.

The scanning speed works in collusion with the powder feed rate in determining the surface quality of the as-deposited part, which in turn controls the tool wear. For a constant powder feed rate, the powder catchment starts decreasing beyond a certain scanning speed. In other words, the laser irradiation that is responsible for creating the melt pool may not be enough, leading to the uncompleted melting of powder particles and, consequently, an irregular surface. Machining of an irregular surface is expected to produce higher tool wear, as seen in Figure 13. The trend is prominent with MQL machining up to a scanning speed of 0.8 m/min. Representative images of tools post-machining of IN718 walls at the highest scanning speed of 1 m/min under MQL and dry environment are presented in Figure 13a1,a2. The almost double-reported wear in dry cutting is discernable from the tool wear modes, with the tool under MQL displaying no defects, whereas the tool used in dry cutting produced chipping at the cutting edge. Considering the observations, it is recommended to use a combination of a higher powder feed rate and a lower scanning speed with MQL to reduce the tool wear and ease the machining of IN718 walls.

One of the primary reasons for the reduced machining performance of IN718 alloy is the tool wear [14], and hence, an adequate understanding of the tool wear is critical from the industrial perspective. Due to the low thermal conductivity of IN718, the majority of the machining-generated heat is transferred to the cutting tool, leading to high tool tip temperatures and excessive tool wear [42]. In this work, the application of MQL for machining of LMDed IN718 walls reduced the surface temperatures (Figure 8c), leading to lower tool wear (Figures 11–13) compared to dry machining. However, as tool wear is a complex interplay of several mechanisms (abrasive, adhesion, diffusion, oxidation, debonding), the influence of MQL on tool wear is not as significant compared to cutting forces and surface quality. Although under MQL machining, the tool wear remained lower compared to dry machining, the variability in tool wear with changes in LMD parameters remains, and in many cases, the tool wear in dry and MQL environments are comparable (Figures 11–13). Nevertheless, further optimization of the MQL machining for LMDed IN718 will greatly benefit the industrial applications, as reduced tool wear consequently brings down the cost and waste during the production chain.

3.3.4. Energy Consumption

The energy consumption during machining is influenced by several factors, including the workpiece and tool materials, cutting conditions, and the cutting fluids or lubricants used [43]. Evaluating the effects of the cutting environment on energy consumption is essential for achieving a sustainable and high-quality ASM process chain. The analysis presented in Figure 14 aims to help understand the energy requirements for machining different regions of LMDed IN718 walls under dry and MQL cutting environments. It is quite clear from Figure 14 that the energy consumption while machining the top region of the IN718 wall remains significantly lower than the bottom region, irrespective of the cutting environment. This observation is valid across the entire range of experimental process parameters. While MQL positively impacts machinability in terms of reduced cutting forces, surface roughness, surface temperature, and tool wear, particularly in the bottom region, the use of MQL setup does not adversely affect the hourly energy consumption. From Figure 14, it is apparent that the hourly energy consumption remains comparable to dry cutting even while machining the bottom region. Moreover, MQL machining consumes lower energy while machining the bottom region of the IN718 walls fabricated at the highest laser powers of 900 and 1000 W. This is critical as higher laser powers not only increase the hardness of the part, but also lead to higher cutting forces (Figure 9) and tool wear (Figure 11). The steep energy consumption observed for MQL machining for walls fabricated at laser power 600 W correlates with the chatter phenomena, which also resulted in higher surface roughness and catastrophic failure of the tool, as discussed earlier in Sections 3.3.2 and 3.3.3, respectively. These outcomes hold paramount significance for the ASM process chain. The potential economic benefits of MQL machining in terms of tool longevity (refer to Section 3.3.3) without a substantial increase in energy costs are significant, with positive implications for incorporating MQL in LMD and several other advanced AM and conventional manufacturing processes.

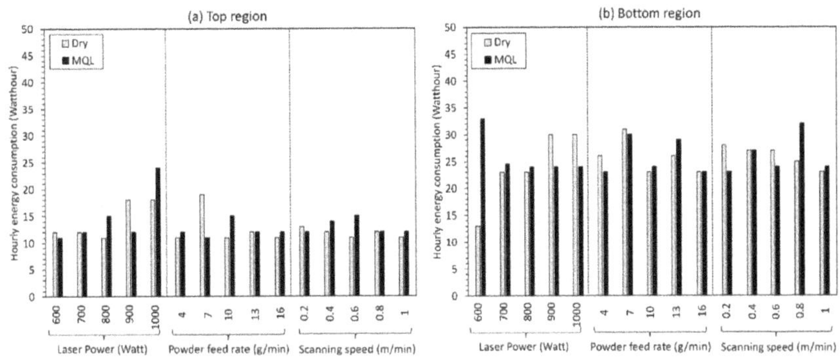

Figure 14. Comparative analysis of hourly energy consumption under different cutting environments for the machining of (**a**) the top region and (**b**) the bottom region.

4. Conclusions and Future Directions

In this work, a comprehensive assessment of the machinability of laser metal deposited (LMDed) IN718 walls in dry and minimum quantity lubrication (MQL) cutting environments was performed. An understanding of the interplay of LMD process variables (laser power, powder feed rate, and scanning speed) and the cutting environment and its effect on the machining aspects, including cutting forces, tool wear, surface quality, and energy consumption, was developed. Based on the obtained results, the study reached the following major conclusions:

1. The machinability of the LMDed IN718 wall is location-dependent owing to the build direction hardness heterogeneity. The bottom region of the IN718 wall, i.e., the initially deposited layers, with considerably higher hardness than the top region, makes machining of the deposited wall closer to the substrate notably more difficult than away from the substrate.
2. Machining of the bottom region leads to substantially higher cutting forces, surface roughness, and temperatures compared to the top region. In addition, while in the bottom region, the variation in these aspects across the entire LMD processing range is large, the variation is minimal in the top region.
3. The machinability of LMDed IN718 walls under a dry-cutting environment is inferior compared to MQL machining along the entire build direction. MQL greatly improves machining across all processing parameters regardless of the machining location; however, the effect is more pronounced in the bottom region.
4. While MQL positively impacts machinability and reduces tool wear to a great extent, the hourly energy consumption remains comparable to dry cutting. This finding holds significance for the ASM process chain, as the negligible increase in hourly energy consumption can largely compensate for the cost of the MQL setup and other accessories.
5. Laser power is identified as the parameter that most influences the processing performance. The variation in powder feed rate and scanning speed has little-to-no effect on the cutting force, whereas the increase in laser power significantly increases the cutting forces. Higher laser powers are detrimental, as they contribute to higher hardness and lead to higher surface roughness. In comparison, a combination of a higher power feed rate and lower scanning speed is essential for ease of machining.

AM using LMD is a potential route for successfully fabricating several multi-material combinations. Although this investigation provides applicable solutions for post-processing, the fabricated IN718 parts, challenges such as chatter and adhesive behavior affecting tool wear and surface quality must also be addressed. Research should be directed toward optimizing the post-processing with a larger view of improved efficiency and sustainability. Even though the LMD AM of IN718 looks promising, challenges like low manufacturing speed and build volumes hinder its growth. However, these challenges can be outweighed by the inherent ability of the LMD process to form intricate shapes with high design flexibility, reshape existing objects, and the potential for repair applications that can lead to tremendous savings in terms of material, cost, and lead time.

Author Contributions: Conceptualization—Y.S. and A.S.; Formal Analysis—O.C.O., A.K.; Funding acquisition—T.S. and A.S., Investigation—O.C.O., A.K., Y.S., Y.H., K.I., S.K. and M.B.; Methodology—O.C.O., A.K., Y.S., Y.H., K.I., S.K., M.B. and A.S.; Project Administration—A.S., T.S. and M.T.; Resources—A.S. and M.T.; Supervision—A.S.; Visualization—A.K. and O.C.O.; Writing—Original Draft—O.C.O. and A.K.; Writing—Review and Editing—S.K., M.B. and A.S. All authors have read and agreed to the published version of the manuscript.

Funding: This research received no external funding.

Data Availability Statement: The raw/processed data required to reproduce these findings cannot be shared as the data also forms part of an ongoing study.

Acknowledgments: The authors would like to acknowledge the support of research group members for help in conducting experiments.

Conflicts of Interest: The authors declare that they have no known competing financial interest or personal relationships that could have influenced the work reported in this paper.

Nomenclature

Abbreviations:	
LMD	Laser metal deposition
DED	Directed energy deposition
AM	Additive manufacturing
MQL	Minimum quantity lubrication
SS	Stainless steel
LDED	Laser-directed energy deposition
ASM	Additive–subtractive manufacturing
LMDed	Laser metal deposited
FGM	Functionally graded material
Symbols:	
V_c	Cutting speed (m/min)
f_z	Feed per tooth (mm/tooth)
a_p	Depth of cut (mm)
a_e	Cutting width (mm)
G	Temperature gradient
R	Solidification growth rate
HI	Heat input (J/m)
P	Laser power (Watt)
V	Scanning speed (m/min)

Appendix A

For the removal of the unmelted powder particles and generating a flat surface for the machinability evaluation, all the as-deposited parts are scanned using a structural light scanner (ATOS Compact Scan, GOM Metrology). As shown in Figure A1a, with the 3D data of the scanned samples, the depth of the cut values is determined by finding the deviations. The pre-treatment, i.e., initial machining, ensured that the unmelted powder was removed without changing the effective width of the test piece. The initial machining to remove the unmelted powder particles is conducted with Mitsubishi AOMT123608PEER-M VP15TF grade PVD-coated carbide cutting inserts. As depicted in Figure A1b, the as-deposited irregular surfaces (with unmelted powder particles) are processed and made ready for the subsequent machinability evaluation study.

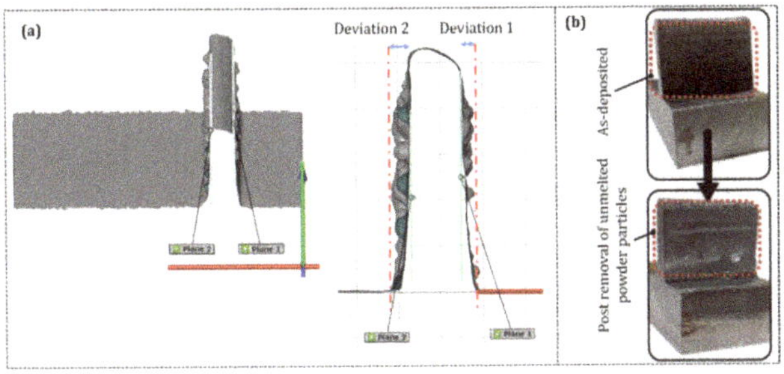

Figure A1. (**a**) Three-dimensional-scanned surface for the determination of the machining depth of cut; and (**b**) representative sample showing surfaces before and after the removal of unmelted powder particles.

References

1. Wang, X.; Jiang, J.; Tian, Y. A review on macroscopic and microstructural features of metallic coating created by pulsed laser material deposition. *Micromachines* **2022**, *13*, 659. [CrossRef] [PubMed]
2. Gu, D.D.; Meiners, W.; Wissenbach, K.; Poprawe, R. Laser additive manufacturing of metallic components: Materials, processes and mechanisms. *Int. Mater. Rev.* **2012**, *57*, 133–164. [CrossRef]
3. Arrizubieta, J.I.; Cortina, M.; Ruiz, J.E.; Lamikiz, A. Combination of laser material deposition and laser surface processes for the holistic manufacture of inconel 718 components. *Materials* **2018**, *11*, 1247. [CrossRef] [PubMed]
4. Cheng, J.; Xing, Y.; Dong, E.; Zhao, L.; Liu, H.; Chang, T.; Chen, M.; Wang, J.; Lu, J.; Wan, J. An overview of laser metal deposition for cladding: Defect formation mechanisms, defect suppression methods and performance improvements of laser-cladded layers. *Materials* **2022**, *15*, 5522. [CrossRef]
5. Mazzucato, F.; Menerini, M.; Valente, A. Laser-based Hybrid System for Inconel 718 part repairing. *Procedia CIRP* **2020**, *95*, 29–34. [CrossRef]
6. Dadbakhsh, S.; Hao, L.; Kong, C.Y. Surface finish improvement of LMD samples using laser polishing. *Virtual Phys. Prototyp.* **2010**, *5*, 215–221. [CrossRef]
7. Rombouts, M.; Maes, G.; Hendrix, W.; Delarbre, E.; Motmans, F. Surface finish after laser metal deposition. *Phys. Procedia* **2013**, *41*, 810–814. [CrossRef]
8. Ostra, T.; Alonso, U.; Veiga, F.; Ortiz, M.; Ramiro, P.; Alberdi, A. Analysis of the machining process of inconel 718 parts manufactured by laser metal deposition. *Materials* **2019**, *12*, 2159. [CrossRef]
9. Rosa, B.; Mognol, P.; Hascoët, J.Y. Laser polishing of additive laser manufacturing surfaces. *J. Laser Appl.* **2015**, *27*, S29102. [CrossRef]
10. Maffia, S.; Chiappini, F.; Maggiani, G.; Furlan, V.; Guerrini, M.; Previtali, B. Enhancing productivity and efficiency in conventional laser metal deposition process for Inconel 718–Part II: Advancing the process performance. *Int. J. Adv. Manuf. Technol.* **2023**, *129*, 279–298. [CrossRef]
11. Paulonis, D.F.; Schirra, J.J. Alloy 718 at Pratt & Whitney–Historical perspective and future challenges. *Superalloys* **2001**, *718*, 13–23.
12. Mazzucato, F.; Forni, D.; Valente, A.; Cadoni, E. Laser metal deposition of Inconel 718 alloy and as-built mechanical properties compared to casting. *Materials* **2021**, *14*, 437. [CrossRef] [PubMed]
13. Shokrani, A.; Dhokia, V.; Newman, S.T. Environmentally conscious machining of difficult-to-machine materials with regard to cutting fluids. *Int. J. Mach. Tools Manuf.* **2012**, *57*, 83–101. [CrossRef]
14. Yin, Q.; Liu, Z.; Wang, B.; Song, Q.; Cai, Y. Recent progress of machinability and surface integrity for mechanical machining Inconel 718: A review. *Int. J. Adv. Manuf. Technol.* **2020**, *109*, 215–245. [CrossRef]
15. Bagherzadeh, A.; Budak, E.; Ozlu, E.; Koc, B. Machining behavior of Inconel 718 in hybrid additive and subtractive manufacturing. *CIRP J. Manuf. Sci. Technol.* **2023**, *46*, 178–190. [CrossRef]
16. Shu, L.; Cang, X.; Zhou, J.; Heng, Z.; Wu, H.; He, W. Study on machinability and grain deformation of laser cladding manufactured and wrought IN718 alloys in dry milling process. *Mater. Today Commun.* **2023**, *34*, 105066. [CrossRef]
17. Zhao, Y.; Han, X.; Xu, Z.; Sun, Y.; Meng, W. Influence of Thermogenetic Effect on Machinability of IN718 Alloy Made by Additive–Subtractive Integrated Manufacturing. *J. Mater. Eng. Perform.* **2023**, *32*, 1–19. [CrossRef]
18. Li, B.; Zhang, R.; Malik, A.; Li, W. Machinability of partition milling stainless steel/Inconel functionally gradient material printed with directed energy deposition. *Int. J. Adv. Manuf. Technol.* **2022**, *122*, 3009–3022. [CrossRef]
19. Zhang, R.; Nagaraja, K.M.; Bian, N.; Fisher, E.; Ahmadyar, S.; Bayazitoglu, K.; Lu, H.; Li, W. Experimental studies on fabricating functionally gradient material of stainless steel 316L-Inconel 718 through hybrid manufacturing: Directed energy deposition and machining. *Int. J. Adv. Manuf. Technol.* **2022**, *120*, 7815–7826. [CrossRef]

20. Souflas, T.; Bikas, H.; Ghassempouri, M.; Salmi, A.; Atzeni, E.; Saboori, A.; Brugnetti, I.; Valente, A.; Mazzucato, A.; Stavropoulos, P. A comparative study of dry and cryogenic milling for Directed Energy Deposited IN718 components: Effect on process and part quality. *Int. J. Adv. Manuf. Technol.* **2022**, *119*, 745–758. [CrossRef]
21. Heigel, J.C.; Phan, T.Q.; Fox, J.C.; Gnaupel-Herold, T.H. Experimental investigation of residual stress and its impact on machining in hybrid additive/subtractive manufacturing. *Procedia Manuf.* **2018**, *26*, 929–940. [CrossRef]
22. Oyelola, O.; Crawforth, P.; M'Saoubi, R.; Clare, A.T. Machining of additively manufactured parts: Implications for surface integrity. *Procedia CIRP* **2016**, *45*, 119–122. [CrossRef]
23. Pereira, J.C.; Zubiri, F.; Garmendia, M.J.; Tena, M.; Gonzalez, H.; López de Lacalle, L.N. Study of laser metal deposition additive manufacturing, CNC milling, and NDT ultrasonic inspection of IN718 alloy preforms. *Int. J. Adv. Manuf. Technol.* **2022**, *120*, 2385–2406. [CrossRef]
24. Careri, F.; Umbrello, D.; Essa, K.; Attallah, M.M.; Imbrogno, S. The effect of the heat treatments on the tool wear of hybrid Additive Manufacturing of IN718. *Wear* **2021**, *470*, 203617. [CrossRef]
25. Calleja, A.; Urbikain, G.; González, H.; Cerrillo, I.; Polvorosa, R.; Lamikiz, A. Inconel® 718 superalloy machinability evaluation after laser cladding additive manufacturing process. *Int. J. Adv. Manuf. Technol.* **2018**, *97*, 2873–2885. [CrossRef]
26. Kelliger, T.; Schraknepper, D.; Bergs, T. Fundamental investigations on the machinability of additively manufactured multi-materials. *MM Sci. J.* **2021**, 5098–5105. [CrossRef]
27. Zhou, H.; Yang, Y.; Han, C.; Wei, Y.; Liu, Z.; Tai, Z.; Zhang, S.; Wang, D. Laser directed energy deposition/milling hybrid additive manufacturing of thin-walled GH4169 alloy: Effect of processing strategy on its microstructure and mechanical properties. *Mater. Sci. Eng. A.* **2023**, *882*, 145480. [CrossRef]
28. Gupta, M.K.; Boy, M.; Korkmaz, M.E.; Yaşar, N.; Günay, M.; Krolczyk, G.M. Measurement and analysis of machining induced tribological characteristics in dual jet minimum quantity lubrication assisted turning of duplex stainless steel. *Measurement* **2022**, *187*, 110353. [CrossRef]
29. Nadammal, N.; Kromm, A.; Saliwan-Neumann, R.; Farahbod, L.; Haberland, C.; Portella, P.D. Influence of support configurations on the characteristics of selective laser-melted inconel 718. *JOM* **2018**, *70*, 343–348. [CrossRef]
30. Banerjee, N.; Sharma, A. Development of a friction model and its application in finite element analysis of minimum quantity lubrication machining of Ti-6Al-4 V. *J. Mater. Process. Technol.* **2016**, *238*, 181–194. [CrossRef]
31. Jang, D.Y.; Jung, J.; Seok, J. Modeling and parameter optimization for cutting energy reduction in MQL milling process. *Int. J. Precis. Eng. Manuf.—Green Technol.* **2016**, *3*, 5–12. [CrossRef]
32. Ozaner, O.C.; Klobčar, D.; Sharma, A. Machining Strategy Determination for Single-and Multi-Material Wire and Arc Additive Manufactured Thin-Walled Parts. *Materials* **2023**, *16*, 2055. [CrossRef] [PubMed]
33. Tian, Y.; McAllister, D.; Colijn, H.; Mills, M.; Farson, D.; Nordin, M.; Babu, S. Rationalization of microstructure heterogeneity in INCONEL 718 builds made by the direct laser additive manufacturing process. *Metall. Mater. Trans. A* **2014**, *45*, 4470–4483. [CrossRef]
34. Stevens, E.L.; Toman, J.; To, A.C.; Chmielus, M. Variation of hardness, microstructure, and Laves phase distribution in direct laser deposited alloy 718 cuboids. *Mater. Des.* **2017**, *119*, 188–198. [CrossRef]
35. Li, Z.; Chen, J.; Sui, S.; Zhong, C.; Lu, X.; Lin, X. The microstructure evolution and tensile properties of Inconel 718 fabricated by high-deposition-rate laser directed energy deposition. *Addit. Manuf.* **2020**, *31*, 100941. [CrossRef]
36. Reddy, S.; Kumar, M.; Panchagnula, J.S.; Parchuri, P.K.; Kumar, S.S.; Ito, K.; Sharma, A. A new approach for attaining uniform properties in build direction in additive manufactured components through coupled thermal-hardness model. *J. Manuf. Process.* **2019**, *40*, 46–58. [CrossRef]
37. Alhuzaim, A.; Imbrogno, S.; Attallah, M.M. Controlling microstructural and mechanical properties of direct laser deposited Inconel 718 via laser power. *J. Alloys Compd.* **2021**, *872*, 159588. [CrossRef]
38. Zhang, K.; Liu, W.; Shang, X. Research on the processing experiments of laser metal deposition shaping. *Opt. Laser Technol.* **2007**, *39*, 549–557. [CrossRef]
39. Moradi, M.; Pourmand, Z.; Hasani, A.; Moghadam, M.K.; Sakhaei, A.H.; Shafiee, M.; Lawrence, J. Direct laser metal deposition (DLMD) additive manufacturing (AM) of Inconel 718 superalloy: Elemental, microstructural and physical properties evaluation. *Optik* **2022**, *259*, 169018. [CrossRef]
40. Banerjee, N.; Sharma, A. A comprehensive assessment of minimum quantity lubrication machining from quality, production, and sustainability perspectives. *Sustain. Mater. Technol.* **2018**, *17*, e00070. [CrossRef]
41. Sharma, V.S.; Singh, G.; Sørby, K. A review on minimum quantity lubrication for machining processes. *Mater. Manuf. Process.* **2015**, *30*, 935–953. [CrossRef]
42. Sugihara, T.; Enomoto, T. High speed machining of Inconel 718 focusing on tool surface topography of CBN tool. *Procedia Manuf.* **2015**, *1*, 675–682. [CrossRef]
43. Muñoz-Escalona, P.; Shokrani, A.; Newman, S.T. Influence of cutting environments on surface integrity and power consumption of austenitic stainless steel. *Robot. Comput.-Integr. Manuf.* **2015**, *36*, 60–69. [CrossRef]

lubricants

Article

Tribological Behavior of Ti-Coated Diamond/Copper Composite Coating Fabricated via Supersonic Laser Deposition

Qunli Zhang [1,2,3], Yiyun Chen [1,2,3], Bo Li [1,2,3], Changyi Wang [1,2,3], Lijuan Wu [1,2,3] and Jianhua Yao [1,2,3,*]

1 Institute of Laser Advanced Manufacturing, Zhejiang University of Technology, No. 288 Liuhe Road, Hangzhou 310023, China
2 College of Mechanical Engineering, Zhejiang University of Technology, No. 288 Liuhe Road, Hangzhou 310023, China
3 Key Laboratory of Special Purpose Equipment and Advanced Processing Technology, Ministry of Education and Zhejiang Province, Zhejiang University of Technology, No. 288 Liuhe Road, Hangzhou 310023, China
* Correspondence: laser@zjut.edu.cn

Abstract: Diamond/copper composite coating is promising for wear-resistant applications, owing to the extreme hardness of the diamond reinforcement. Ti-coated diamond/copper composite coatings with various laser powers were successfully fabricated employing the novel manufacturing technology of supersonic laser deposition (SLD). Ti-coated diamond, which was able to enhance the wettability between diamond and copper, was prepared at the optimal parameters via salt bath. Nano-spherical titanium carbides were uniformly distributed on the diamond's surface to generate a favorable interface bonding with a copper matrix though mechanical interlocking and metallurgical bonding during impact. Furthermore, the results showed that the transition layer acted as a buffer, preventing the breakage of the diamond in the coating. SLD can prevent the graphitization of the diamonds in the coating due to its low processing temperature. The coordination of laser and diamond metallization significantly improved the tribological properties of the diamond/copper composite coatings with the SLD technique. The microhardness of the diamond/copper composite coating at a laser power of 1000 W reached about 172.58 $HV_{0.1}$, which was clearly harder than that of the cold sprayed copper. The wear test illustrated that the diamond/copper composite coating at a laser power of 1000 W exhibited a low friction coefficient of 0.44 and a minimal wear rate of 11.85 $\mu m^3 \cdot N^{-1} \cdot mm^{-1}$. SLD technology shows great potential in the field of preparing wear-resistant hard reinforced phase composite coatings.

Keywords: supersonic laser deposition; Ti-coated diamond; diamond/copper composite coating; wear-resistant

Citation: Zhang, Q.; Chen, Y.; Li, B.; Wang, C.; Wu, L.; Yao, J. Tribological Behavior of Ti-Coated Diamond/Copper Composite Coating Fabricated via Supersonic Laser Deposition. *Lubricants* **2023**, *11*, 216. https://doi.org/10.3390/lubricants11050216

Received: 14 April 2023
Revised: 5 May 2023
Accepted: 10 May 2023
Published: 11 May 2023

1. Introduction

Copper and its alloys have been widely employed in the sectors of power electronics, energy, machinery, and transportation owing to its easy processing, corrosion-resistant, and electrical and thermal conductivity properties [1,2]. Due to their notoriously inferior wear resistance, copper components always undergo severe wear and erosion during daily service, which significantly shortens their service life [3]. On the contrary, the inherent ultra-hardness, large bulk modulus, and outstanding tribological properties of diamond enables it to be a superior wear-resistant material [4]. However, industrially produced diamond has limited crystal size and is challenging to machine due to its intrinsic properties. Therefore, understanding how to make full use of the advantages of diamond in industrial production has become a hot research topic.

The increased mechanical performances of diamond/copper composite indicate the remarkable improvements of the composite, which can acquire the benefits of both diamond and copper [5,6]. The ductility of the copper matrix provides the machinability of the composite, while the diamond-reinforced phase improves the wear resistance of the composite [7]. However, the poor wettability between diamond and copper causes the composite interface to have obvious defects, which seriously affect the performance of the composite [8]. Therefore, the diamond will spall in friction wear due to poor bonding. To overcome the drawback of poor interfacial bonding, construction of transition layers by chemical reactions use elements such as B [9], Cr [10], Ti [11], and W [12] to enhance the bonding between the reinforcing phase and the matrix through the wettability between carbide and copper. Carbide transition layers can effectively avoid diamond flaking in frictional wear and thus take full advantage of diamond wear resistance.

Currently, diamond/copper composites are mainly produced through powder metallurgy and metal infiltration. Powder metallurgy usually fabricates composites via pressurized sintering of copper and diamond mixed powder, such as spark plasma sintering [13], vacuum hot pressing [14], and high-temperature high-pressure process [15]. The sintering temperature of powder metallurgy is below the melting point of copper, which means that the diamond and copper powder is still linked in the form of particles with poor interface bonding. Moreover, the diamond floats during the sintering process due to its relatively low density, which leads to the uneven distribution of the two phases in the composite. On the contrary, the infiltration method [16–18] is employed to fabricate composites by melting copper at a high temperature and then infiltrating it into the preforms of diamond under pressure or no pressure conditions. The principle is mainly to cause the molten copper to enter the interstices between the diamonds using the capillary effect. The drawbacks are that the diamond is prone to ablation at temperatures of over 1000 °C, the need to provide an oxygen-free atmosphere to safeguard the composite from oxidation, and the high temperature and pressure leading to huge costs. In order to prepare diamond/copper composites with superior properties, there is an urgent need to develop a high-quality and efficient manufacturing technology that avoids the drawbacks of existing technologies.

Supersonic laser deposition (SLD) is a promising addition to manufacturing and surface modification technology [19]. During the spraying process, micron-sized powder is accelerated to 300–1400 m/s using high pressure gas through a de-Laval nozzle under laser irradiation and achieves deposition via severe plastic deformation through kinetic energy and heat energy [20]. Metals [21,22], composites [23], and even ceramics [24] have been successfully fabricated via SLD with favorable performance. SLD is dominated by kinetic energy and supplemented by thermal energy, which avoids oxidation defects, thermal residual stress, and thermal damage caused by the high-temperature preparation of composites. Furthermore, SLD can be applied to rapidly prepare various structures, coatings on any surface, and efficiently repair damaged parts [25]. As a brittle and hard material, diamond has a high critical deposition rate in commercial cold spraying, which is prone to internal fracture and cannot be effectively plastically deformed to achieve deposition. By constructing a buffer layer on the surface of diamond and introducing laser thermal softening via SLD, diamond composites can be deposited. Previous studies have rarely focused on the manufacturing and the wear resistance of core-shell diamond/copper composite coatings.

In this paper, diamond is coated with Ti via salt bath to bridge the large gap of acoustic impedance and enhance the bonding between the diamond and copper. The novel SLD additive manufacturing process is applied to prepare machinable Ti-coated diamond/copper composite with excellent wear-resistant performance. The composition, microstructure, and tribological characteristics of the diamond/copper composite coatings were thoroughly investigated after preparation at various laser power. The purpose of this paper is to improve the wettability between diamond and copper by diamond surface modification, while introducing laser to enhance interfacial bonding. It ensures that the diamond can play an excellent wear resistance role in the diamond/copper composite coating without flaking.

2. Materials and Methods

2.1. Composite Fabrication

The feedstock for SLD included irregular copper powder (Zhongke Yanuo Technology Co., Ltd., Beijing, China) ranging from 15 to 53 μm in size and irregular diamond powder (Zhongyuan Superhard Co., Ltd., Tuocheng, Henan, China) ranging from 35 to 42 μm in size after commercially available spherical Ti ranging from 15 to 48 μm in size plating via salt bath. Figure 1 shows the morphology of the original powder. Ti-coated diamonds are prepared using the salt bath method at 900 °C for 1 h to build a uniform density of carbide transition layer.

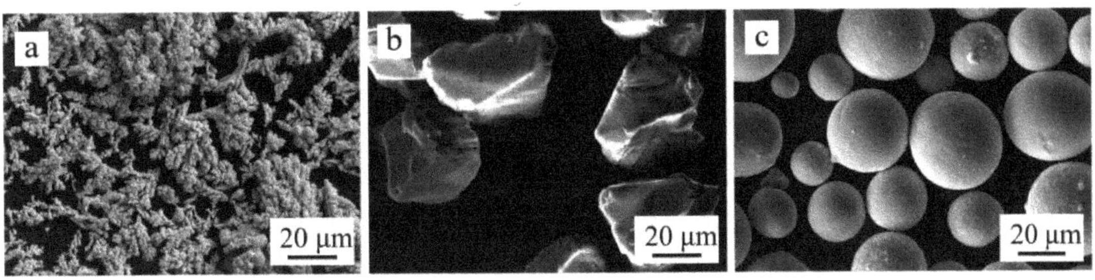

Figure 1. Morphology of the (**a**) copper, (**b**) diamond, and (**c**) Ti used in this study.

The Ti-coated diamond/copper composite coatings were fabricated onto the copper substrate using an in-house SLD system. The system was mainly composed of a six-axis mechanical arm, powder feeder, high pressure nitrogen station, water cooling system, gas heating device, and laser system (LDF6000-100 VGP, Laserline, Koblenz, Germany) [7]. The schematic of the fabrication via SLD is shown in Figure 2. The spraying nozzle had a total length of 278.0 mm with a throat diameter of 2.8 mm and an exit diameter of 6.0 mm. The produced coatings and SLD parameters are presented in Table 1. The copper and 50 vol% Ti-coated diamond powder were mechanically mixed before spraying. Through previous preparatory experiments, we observed that the composite coating suffered from poor interfacial bonding at a laser power of about 500 W, while the interfacial bonding was favorable at 1000 W. Too high a laser power leads to problems of diamond graphitization and local thermal stress concentration. Therefore, 0 W, 500 W, 1000 W, 1500 W were chosen as experimental parameters to study the effect of laser power and its different effects on the composite coatings. For facilitating the following discussion, each composite coating was marked using an abbreviation. The abbreviation 'C' means the as-deposited composite coating, while '0', '500', '1000', and '1500' indicate the laser power used to produce the composite coating. The laser and the spray gun were fixed on the mechanical arm and moved with the same traversal speed. All composite coatings were prepared at a stand-off distance of 30.0 mm and 3.0 MPa nitrogen pressure with a traversal speed of 10 mm·s^{-1}. The speed of the powder feeder was set to 2 RPM. The diamond particles were estimated 435–563 m·s^{-1} at a powder-heating temperature of 500 °C [20]. The Ti-coated diamond/copper composite coatings were fabricated onto the sand-blasted copper substrate using 24 μm corundum.

Table 1. Abbreviations of the prepared coatings and the SLD conditions.

Annotation	Laser Power [W]	Temperature [°C]
C_0	0	500
C_{500}	500	500
C_{1000}	1000	500
C_{1500}	1500	500

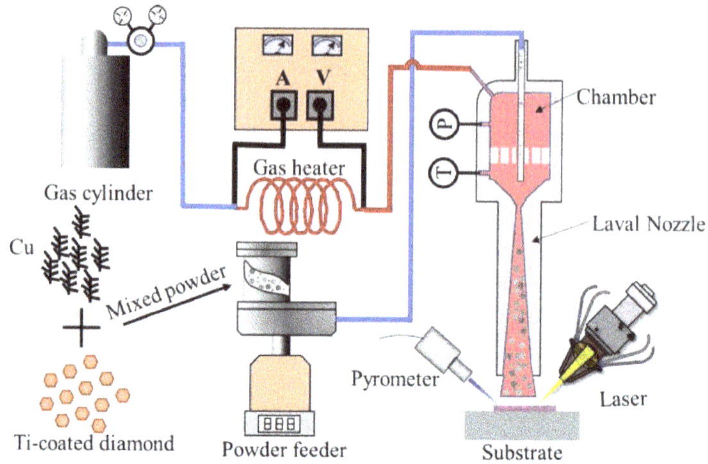

Figure 2. Schematic of the fabrication of the Ti-coated diamond/copper composite coatings via SLD.

2.2. Material Characterizations

The X-ray diffractometer (D8 Advance, Bruker, Billerica, MA, USA) was employed to investigate the phase composition of the samples using SLD with the copper (λ = 1.542 Å) source at a current of 40 mA, a voltage of 40 kV, and scan step of 0.02°. In order to prevent the metallographic procedure from damaging the morphology of the diamond in the coating, the fracture morphology of the coating was investigated. Scanning electron microscopy (SEM, EVO 18, Carl Zeiss, Stuttgart, Germany) was employed to analyze the morphology of the samples, and the elemental composition was determined using affiliated energy-dispersive spectroscopy (EDS, Nano X-flash Detector 5010, Bruker).

2.3. Wear Test

The Vickers hardness tester (HMV-2, SHIMADZU, Kyoto, Japan) was used to test the microhardness of the samples with a load of 0.98 N for 15 s. A small amount of ultra-high hardness was caused by the indentation on the diamond. The hardness of the samples was determined by averaging 10 hardness values in order to be accurate. A ball-on-disc tribometer (HT-600, Lanzhou Institute of Chemical Physics, Lanzhou, China) was employed to measure the tribological properties of the samples at room temperature. The samples were polished with a 2.5 μm metallographic polishing agent before the test to ensure the accuracy of the wear test. A silicon nitride ball with a diameter of 5 mm was used as a counter-grinding part to contact the fixed sample on the disc. The disk rotated at a speed of 350 $r \cdot min^{-1}$ under a load of 5N. The wear mechanism was investigated using SEM, EDS, and LSCM. The wear rate was obtained by calculating the volume of wear per unit length under unit load. The mean COF (μ), wear radius (r), and wear cross-sectional area (S) of each specimen was measured. The wear volume (V) was calculated as follows:

$$V = 2 \cdot \pi \cdot r \cdot S \quad (1)$$

3. Results and Discussion

3.1. Composition and Morphology of Metallized Diamond

Figure 3 exhibits the surface morphology of the diamond after Ti plating via salt bath. Compared with the original diamond in Figure 1b, it is obvious that almost all the diamonds were coated with a shell as seen in Figure 3a. The surface of the shells with significantly improved roughness was undulating and had a granular bulge, which contributed to the subsequent SLD coating preparation. In order to better observe the effect of metallized diamond via salt bath, the edge of an uncoated diamond is shown in Figure 3b. It was observed that the thickness of the transition layer closely combined with diamond was only 2–3 μm, which was thinner than the 6 μm copper-nickel layer of the diamond [26]. This was mainly due to the chemical reaction between Ti and carbon on the surface of the diamond to generate chemical bonding, which would protect the diamond and improve the wettability with copper in coating preparation using SLD. The titanium element reacted with the diamond in the molten salt through the following equations [7]:

$$Ti^{4+} + Ti \rightarrow 2Ti^{2+} \quad (2)$$

$$2TiCl_2 + C\ (diamond) \rightarrow TiC + TiCl_4 \quad (3)$$

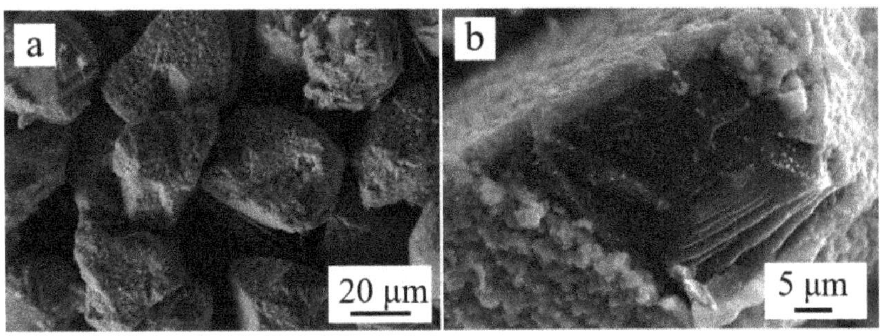

Figure 3. SEM images of (**a**) the Ti-coated diamond powder and (**b**) the cross-section of typical Ti-coated diamond.

The molten salt provided a flow atmosphere at a high temperature, which caused the diamond surface to react completely.

The XRD diffractions of diamond before and after salt bath plating are shown in Figure 4. Although the diamonds were almost completely coated as shown in Figure 3, the diamond's characteristic peaks were still detected from the diffraction of the Ti-coated diamond, so it can be concluded that the layer was thin. The characteristic peaks of TiC indicate that TiC successfully coated the surface of the diamond [27]. The characteristic peaks of Ti clearly showed that metal titanium also existed on the diamond's surface. The many miscellaneous peaks were mostly different types of titanium carbide, indicating that the transition layer presented gradient carbide changes.

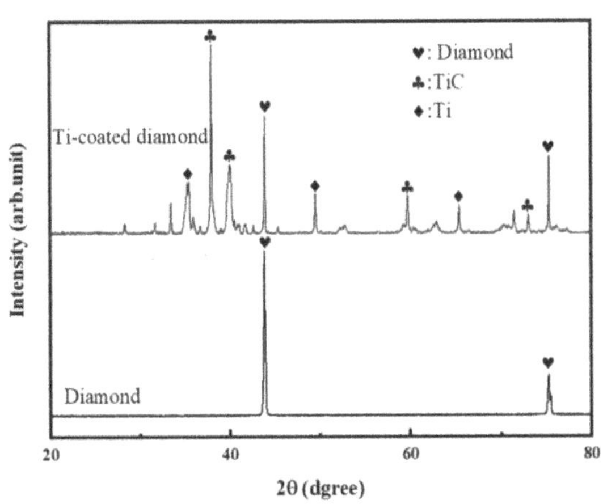

Figure 4. XRD profiles of the diamond powder before and after Ti coating.

3.2. Composites Prepared via SLD with Different Laser Power

The XRD spectra of Ti-coated diamond/copper composite coatings at various laser powers are depicted in Figure 5. Due to the relatively small content of diamond, the characteristic peaks of copper were mainly detected. When the laser power was high, the diamond deposition efficiency was improved, as can be seen in the dashed frame, which confirmed the positive effect of laser power. A small number of the characteristic peaks of diamond and titanium carbide were observed. No graphite peaks appeared at 26.5°, indicating that the diamond phase in the composite maintained its original properties with SLD [28].

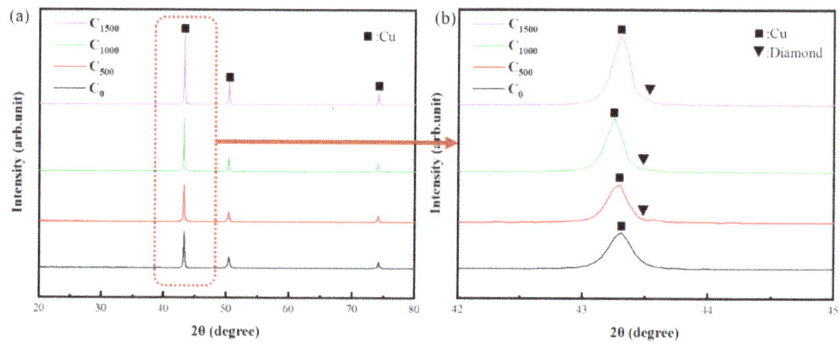

Figure 5. XRD patterns (**a**) and their partially enlarged detail (**b**) of the Ti-coated diamond/copper composites at various laser power.

The SEM images of the fracture cross-section of the Ti-coated diamond/copper composite coatings fabricated via SLD at various laser powers are shown in Figure 6. The deposition efficiency of diamond was significantly improved as the laser power increased. The pits left by the diamond shedding after the fracture were formed due to poor interface bonding under cold spraying without laser irradiation, as can be seen in Figure 6a. However, it was observed that the transition layer on the surface of the diamond peeled off in Figure 6b, which was due to the breakage of the transition layer during impact at a low laser power. As the laser power reached 1000 W, the diamond with high deposition efficiency retained most of the transition layer to protect the brittle diamonds. This was due to the better particle softening and copper plastic deformation capability of the laser and the resultant lower particle critical impact velocity. Due to the softening effect of laser irradiation on particles, the phenomenon of diamond breakage or rebound was reduced, and the deposition efficiency was improved [20]. As the laser power reached 1500 W, excessive thermal stress leads to a large number of tiny pores at the particle interface during cooling, which affected the performance of the composite coating.

The magnified SEM images of Ti-coated diamond/copper composite coatings at various laser powers are exhibited in Figure 7. As seen in Figure 7a, without the laser irradiation, the cracks in the diamond and the pores at the edge were obvious in the coating, which lead to the spalling of the diamond. At the laser power of 500 W, as shown in Figure 7b, a small number of cracks could still be seen in the coating, but the interfacial bonding was greatly improved compared to the coating without laser addition. When the laser power was 1000 W, the cracks at the interface no longer appeared (as in Figure 7c) due to the softening of the interface using laser energy. The shell of the Ti-coated diamond was deformed during the impact process, which reduced the brittle fracture of the diamond and may have produced the interpenetration of interface elements due to its wettability with the copper matrix. However, when the plastic flow of the copper matrix was large with a high laser power (in Figure 7d), the mutual impact between diamonds resulted in cracks during the deposition process.

Figure 6. SEM images of the Ti-coated diamond/copper composite coatings at various laser power: (**a**)—C_0, (**b**)—C_{500}, (**c**)—C_{1000}, (**d**)—C_{1500}.

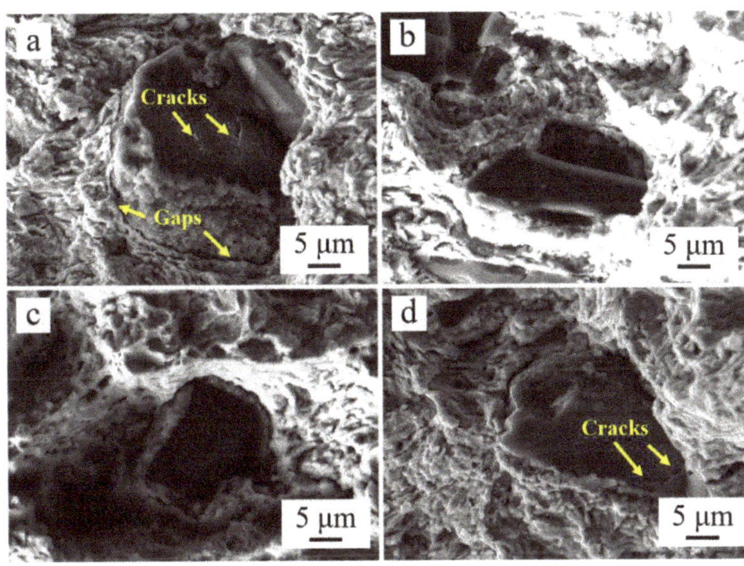

Figure 7. The magnified SEM images of the Ti-coated diamond/copper composite coatings at various laser powers: (**a**)—C_0, (**b**)—C_{500}, (**c**)—C_{1000}, (**d**)—C_{1500}.

To study the element distribution at the interface between diamond and copper, the typical characteristics of Ti-coated diamond in C_{1000} is analyzed in Figure 8. The titanium element around the diamond can be clearly observed, which was connected to the copper element. A line scan at the interface showed a 0.5 μm overlap between copper and titanium, and a 0.8 μm overlap between titanium and carbon. The transition layer of the diamond was infiltrated with copper elements under the action of the laser, which proved the excellent combination of diamond in the composite. In the SLD process, the deposition mechanism of core-shell diamond particles was mainly a cushioning protection mechanism for the shell structure under the high speed impact. Without laser irradiation, the cracks on the surface of the diamond and the broken transition layer can be seen in Figure 7a after deposition. The sample prepared at a lower laser power due to work hardening was not able to avoid the peeling of the titanium carbide or even the fracture of the diamond. However, too high a laser power resulted in the same appearance due to the impact between diamonds. The high-speed impact diamond particles were deposited by the buffer of the transition layer to protect the diamond. Additionally, the transition layer and the copper element penetrated each other to achieve an effective combination at an optimal laser power.

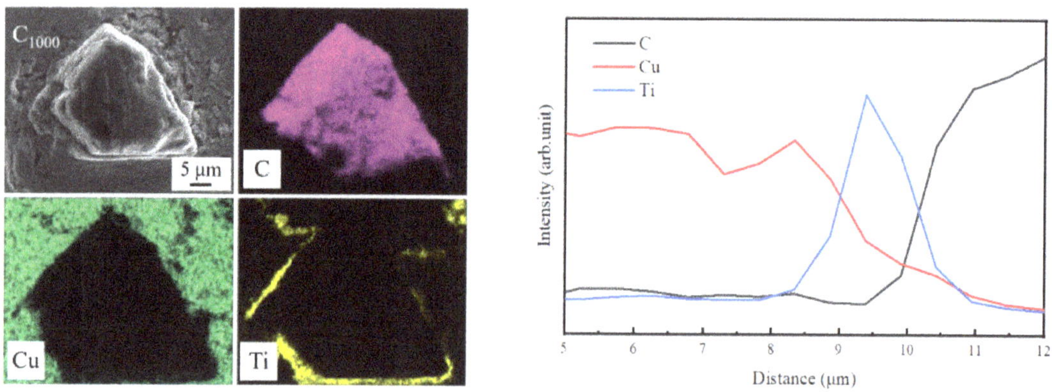

Figure 8. EDS mapping and line scanning at the interface of the Ti-coated diamond for C_{1000}.

3.3. Wear-Resistant Property

The microhardness of the Ti-coated diamond/copper composite coatings at various laser powers is shown in Figure 9. The substrate hardness of the composite was 93 $HV_{0.1}$. Clearly, the microhardness of all the laser-irradiated samples was harder than that of the cold sprayed copper due to the hammer effect of the diamond and the softening effect of laser irradiation [29]. In addition, although the diamond content in the composites enhanced with the increase of laser power, the hardness increased first and then decreased. Laser irradiation played different roles at various laser powers—with a low laser power, the microhardness was mainly caused by the hammer effect of the diamond and the work hardening caused by the plastic deformation of the copper. The maximal hardness reached 172.58 ± 4.05 $HV_{0.1}$ at 500 W. Laser irradiation may have caused tempering, resulting in a reduction of hardness [30]. As the laser power increased, the composite material was tempered and the grains at the interface were re-crystallized, which refined the grains and lead to the release of residual stresses, thus reducing the hardness. The diamond content no longer played a major role, and the microhardness decreased to 128.70 ± 8.72 $HV_{0.1}$ at 1500 W.

Figure 10 displays the friction curves of C_0, C_{500}, C_{1000}, and C_{1500} at the load of 5 N for 60 min. The average friction coefficients of C_0, C_{500}, C_{1000}, and C_{1500} were 0.65, 0.49, 0.44, and 0.39, respectively. The friction coefficients of all samples rose swiftly in the early stage of sliding friction and became stable with small fluctuations after 5 min, except for C_0. C_0 abruptly fluctuated up and down for 10 min, owing to the friction instability caused by the peeling of the diamond in the composite without the laser. C_0 had the highest friction coefficient due to the poorly bonded particles with small plastic deformation, which lead to abrasive wear. The friction coefficients of C_{500}, C_{1000}, and C_{1500} decreased with the increase of laser power. Due to the thermal stress concentration in the local area at 1500 W laser power, the stress released in friction lead to microcracks, which enhanced the lubrication effect of the copper in the unit contact area.

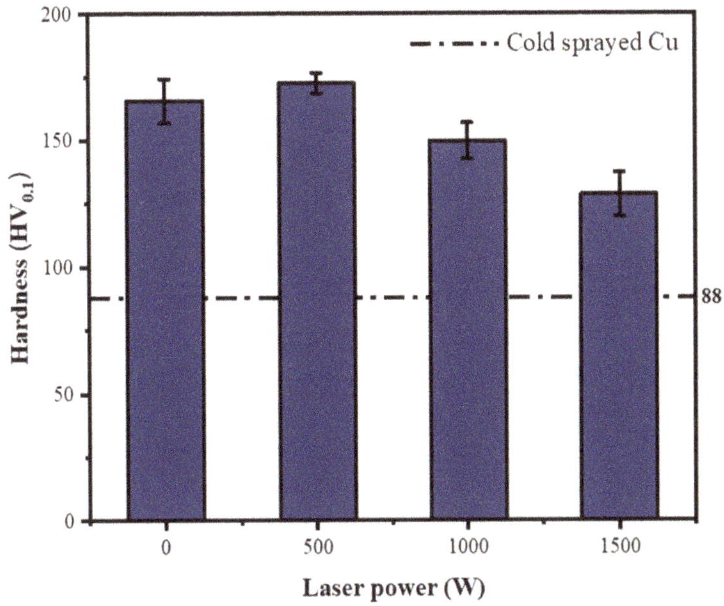

Figure 9. Microhardness of Ti-coated diamond/copper composite coatings at various laser powers.

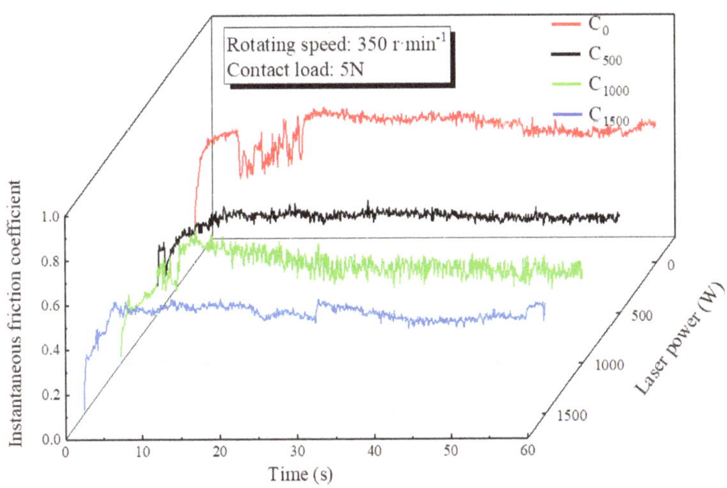

Figure 10. Evolution of the instantaneous friction coefficient at various laser powers.

An LSCM examination was conducted on the surface of the samples to evaluate surface wear. As shown in Figure 11, the roughness change of the worn track in the radial direction are illustrated. The surface roughness of C_{500} (Ra: 9.935 μm), C_{1000} (Ra: 6.438 μm), and C_{1500} (Ra: 8.403 μm) was smaller than that of C_0 (Ra: 16.770 μm). By comparing the roughness, the C_{1000} exhibited a smoother worn surface than that of the others, with a maximum between peak and valley of 12.875 μm. The wear rates of the diamond/copper composite coatings after the wear experiment are shown Table 2. The C_{1000} composite coating with a wear rate of 11.85 $\mu m^3 \cdot N^{-1} \cdot mm^{-1}$ exhibited better wear resistance performance than that of the C_0, C_{500}, and C_{1000} due to their much higher diamond content.

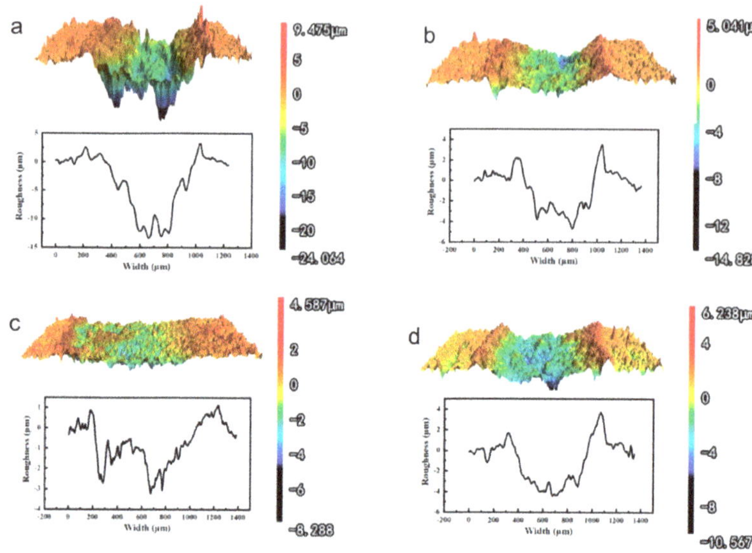

Figure 11. LSCM images of the composite coatings after the ball-on-disk test: (**a**)—C_0, (**b**)—C_{500}, (**c**)—C_{1000}, (**d**)—C_{1500}.

Table 2. Wear rates ($\mu m^3 \cdot N^{-1} \cdot mm^{-1}$) of the composite coatings after the ball-on-disk test.

Sample	C_0	C_{500}	C_{1000}	C_{1500}
Wear rate	44.59	14.81	11.85	15.89

In the wear experiment, the diamond in the coating was continuously compacted with time under the action of load. The wear mechanism of the diamond/copper composites was studied by observing the surface of the samples after the wear experiments, shown in Figure 12. The uniformly distributed diamond particles were observed in the four samples after the wear test, indicating that the diamond/copper composite prepared using SLD had promising wear resistance potential. With the increase in wear time, the diamond was embedded in the copper matrix under the protection of the transition layer, which had excellent interface bonding, and thus protected the wear of the composite coating. Due to the low hardness of the copper matrix, the copper gradually adhered to the periphery of the diamond under the action of the hard grinding part with the increase in wear time. Due to the lack of a laser, the poor interface bonding of particles of C_0 resulted in large spalling during friction, as seen in Figure 12a. After the introduction of the laser, the surface peeling of C_{500} was reduced, but it still did not fully play the role of a diamond. As shown in Figure 12c, due to the high content of diamond in C_{1000}, the interaction of diamond particles in wear lead to partial spalling and aggregation under the action of the grinding part. Some pits and grooves appeared in the worn surface. The wear track of C_{1000} showed island adhesion and spalling with a high diamond content, which possessed the best friction and wear properties. Further increasing the laser power leaded to a downtrend in the hardness of the coating, which resulted in large-scale adhesive wear, as seen in Figure 12d. The lamellar structure was prone to flaking, resulting in a high wear rate.

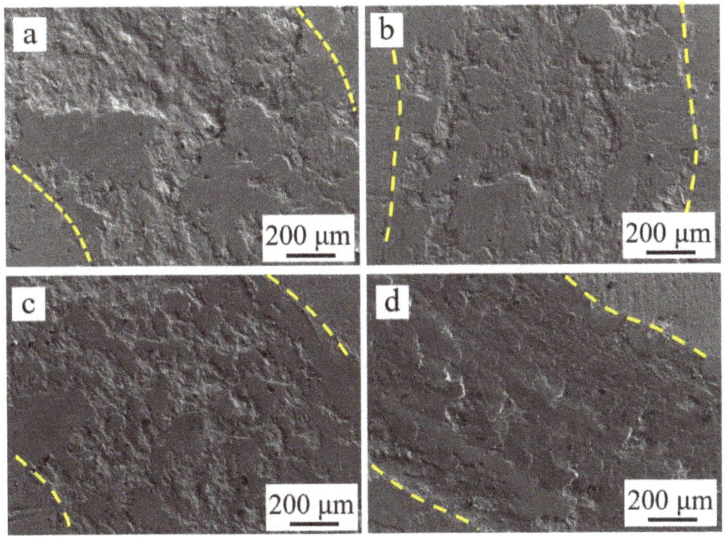

Figure 12. SEM images of the composite coatings after the ball-on-disk test: (**a**)—C_0, (**b**)—C_{500}, (**c**)—C_{1000}, (**d**)—C_{1500}.

The element distribution on the surface after wear was analyzed by observing C_{1000} with EDS energy spectrum, as shown in Figure 13. The main elements of the worn surface were Cu, O, and C. As the sliding proceeded, the copper chips converged in the local area, while the other part of the copper was oxidized at a high temperature after forming adhesive wear. As the wear time increased, the oxide layer appeared to begin fatigue spalling. The diamond was still evenly distributed throughout the matrix. The presence of the diamond provided a wear-resistant frame for the composite coating, which provided an effective obstacle to the matrix during the wear process and prevented the rapid wear of the matrix.

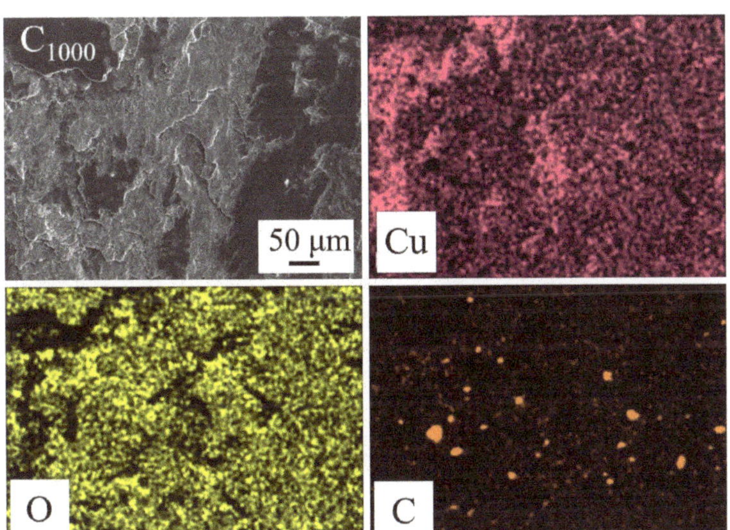

Figure 13. EDS mapping of the composite coating for C_{1000} after the ball-on-disk test.

The wear mechanisms of the diamond/copper composites prepared via SLD mainly included adhesion, oxidation, and ploughing [31]. The relatively low hardness and high ductility of the copper matrix caused the coating to be mainly adhesive wear. During the sliding process, the rough surface in the initial stage caused the copper matrix to wear continuously. As the diamond began to play a role, it protected the copper matrix from continuous wear and formed plough wear. With the increase in time, the area outside the furrow formed an oxidation lubrication grinding at a high temperature. From Figure 12c, it is evident that a large amount of diamond hindered the plastic deformation of the copper. The element of the transition layer of the Ti-coated diamond and the copper element continuously infiltrated each other during cyclic wear, as shown in Figure 14. Under the hindrance of the diamond, the copper chips were observed to be ground into nano-scale, which helped to improve the wear resistance of the composite coating [32]. The transition layer on the surface of the diamond was worn, but the diamond was still tightly embedded in the matrix and tightly bonded. The element distribution diagram also revealed that the diamond had a strong combination with the copper matrix through the metallized transition layer after friction and the wear test.

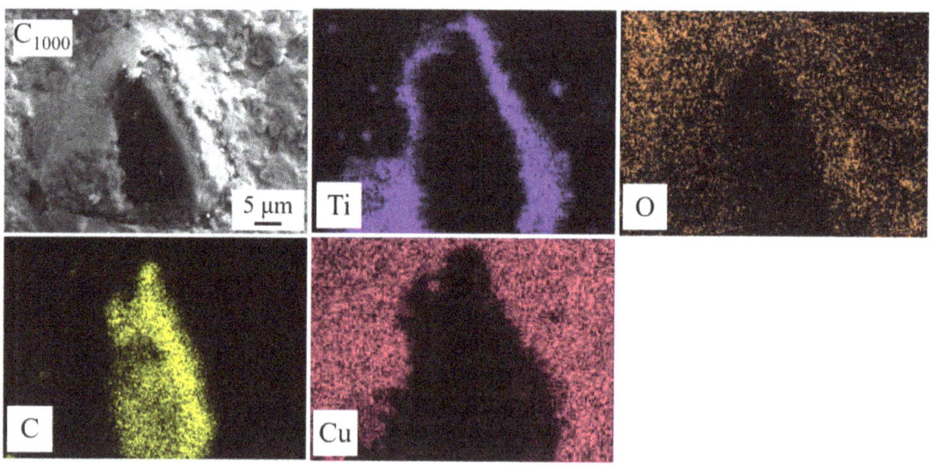

Figure 14. EDS mapping of Ti-coated diamond for C_{1000} after wear.

4. Conclusions

In summary, Ti-coated diamond/copper composite coatings were fabricated employing solid-state SLD technology using metalized diamond and copper composite powder as feedstock. The SLD technology used a high energy laser beam to heat the deposited powder and substrate simultaneously during the cold spray process. It softened the powder and the matrix, improving the plastic deformation ability and promoting the deposition efficiency, compactness, and interfacial bonding strength of the coating. Through experimental characterization and analysis, we drew the following conclusions:

(1) The Ti-coated diamond improved the bonding with the copper matrix and played an important role in friction and wear, showing great potential as feedstock for SLD.
(2) The synergistic effect of laser irradiation and diamond metallization was able to improve the interfacial bonding to ensure that the diamond did not peel off. The interfacial bonding of the Ti-coated diamond/copper composite coating prepared at a laser power of 1000 W was the best without cracks at the interface.
(3) The Ti-coated diamond/copper composite coatings fabricated using SLD had superior wear-resistant properties, and the coating with higher diamond content had better wear resistance. The microhardness of the Ti-coated diamond/copper composite

coating at a laser power of 1000 W reached about 172.58 $HV_{0.1}$, which exhibited the lowest friction coefficient of 0.44 and the minimal wear rate of 11.85 $\mu m^3 \cdot N^{-1} \cdot mm^{-1}$.

(4) The adhesive wear and oxidation wear were the main wear mechanisms with a small amount of ploughing wear. The oxidation of the copper provided lubrication and improved the wear resistance property.

Author Contributions: Methodology, B.L.; Formal analysis, C.W. and L.W.; Writing—original draft, Y.C.; Writing—review & editing, Q.Z. and J.Y. All authors have read and agreed to the published version of the manuscript.

Funding: This project is support by the National Natural Science Foundation of China (52075495), the Zhejiang Provincial Natural Science Foundation (LY22E050017), and the Science and Technology Project of State Grid Ningxia Electric Power Co., Ltd. (5229CG200069).

Data Availability Statement: No new data were created or analyzed in this study. Data sharing is not applicable to this article.

Conflicts of Interest: The authors declare no conflict of interest. The authors declare that this study received funding from the National Natural Science Foundation of China (52075495), the Zhejiang Provincial Natural Science Foundation (LY22E050017), and the Science and Technology Project of State Grid Ningxia Electric Power Co., Ltd. (5229CG200069). The funders were not involved in the study design, collection, analysis, interpretation of data, the writing of this article, or the decision to submit it for publication.

References

1. Zuo, J.; Lin, Y. An investigation of thermal-mechanical interaction effect on PVD coated tool wear for milling Be/Cu alloy. *Vacuum* **2019**, *167*, 271–279. [CrossRef]
2. Shaik, M.A.; Golla, B.R. Development of highly wear resistant Cu–Al alloys processed via powder metallurgy. *Tribol. Int.* **2019**, *136*, 127–139. [CrossRef]
3. Kumar, L.; Singh, H. Effect of nanostructured Cu on microstructure, microhardness and wear behavior of Cu-xGnP composites developed using mechanical alloying. *J. Compos. Mater.* **2021**, *55*, 2237–2248. [CrossRef]
4. Erdemir, A.; Martin, J.M. Superior wear resistance of diamond and DLC coatings. *Curr. Opin. Solid State Mater. Sci.* **2018**, *22*, 243–254. [CrossRef]
5. Liu, D.; Tian, H. Microstructure, mechanical and elevated temperature tribological behaviors of the diamond/Cu composites prepared by spark plasma sintering method. *Diam. Relat. Mater.* **2019**, *91*, 138–143. [CrossRef]
6. Hayat, M.D.; Singh, H. Enhanced interfacial bonding in copper/diamond composites via deposition of nano-copper on diamond particles. *JOM* **2022**, *74*, 949–953. [CrossRef]
7. Wu, L.J.; Zhang, G. Study on microstructure and tribological performance of diamond/Cu composite coating via supersonic laser deposition. *Coatings* **2020**, *10*, 276. [CrossRef]
8. Hagio, T.; Park, J.H. Electrodeposition of nano-diamond/copper composite platings: Improved interfacial adhesion between diamond and copper via formation of silicon carbide on diamond surface. *Surf. Coat. Technol.* **2020**, *403*, 126322. [CrossRef]
9. Bai, G.; Li, N. High thermal conductivity of Cu-B/diamond composites prepared by gas pressure infiltration. *J. Alloys Compd.* **2018**, *735*, 1648–1653. [CrossRef]
10. Wang, L.H.; Li, J.W. Combining Cr pre-coating and Cr alloying to improve the thermal conductivity of diamond particles reinforced Cu matrix composites. *J. Alloys Compd.* **2018**, *749*, 1098–1105. [CrossRef]
11. Lei, L.; Bolzoni, L. High thermal conductivity and strong interface bonding of a hot-forged Cu/Ti-coated-diamond composite. *Carbon* **2020**, *168*, 553–563. [CrossRef]
12. Pan, Y.P.; He, X.B. Optimized thermal conductivity of diamond/Cu composite prepared with tungsten-copper-coated diamond particles by vacuum sintering technique. *Vacuum* **2018**, *153*, 74–81. [CrossRef]
13. Yang, L.; Sun, L. Thermal conductivity of Cu-Ti/diamond composites via spark plasma sintering. *Diam. Relat. Mater.* **2019**, *94*, 37–42. [CrossRef]
14. Luo, F.; Jiang, X.S. Microstructures, mechanical and thermal properties of diamonds and graphene hybrid reinforced laminated Cu matrix composites by vacuum hot pressing. *Vacuum* **2023**, *207*, 111610. [CrossRef]
15. Chen, H.; Jia, C. Interfacial characterization and thermal conductivity of diamond/Cu composites prepared by two HPHT techniques. *J. Mater. Sci.* **2011**, *47*, 3367–3375. [CrossRef]
16. Jia, J.H.; Bai, S.X. Effect of tungsten based coating characteristics on microstructure and thermal conductivity of diamond/Cu composites prepared by pressueless infiltration. *Ceram. Int.* **2019**, *45*, 10810–10818. [CrossRef]
17. Chen, C.; Guo, H. Thermal conductivity of diamond/copper composites with a bimodal distribution of diamond particle sizes prepared by pressure infiltration method. *Rare Met.* **2011**, *30*, 408–413. [CrossRef]

18. Li, J.W.; Zhang, H.L. Optimized thermal properties in diamond particles reinforced copper-titanium matrix composites produced by gas pressure infiltration. *Compos. Part A Appl. Sci. Manuf.* **2016**, *91*, 189–194. [CrossRef]
19. Gorunov, A.I. Features of coatings obtained by supersonic laser deposition. *J. Therm. Spray Technol.* **2018**, *27*, 1194–1203. [CrossRef]
20. Yin, S.; Hassani, M. Unravelling the deposition mechanism of brittle particles in metal matrix composites fabricated via cold spray additive manufacturing. *Scr. Mater.* **2021**, *194*, 113614. [CrossRef]
21. Lupoi, R.; Sparkes, M. High speed titanium coatings by supersonic laser deposition. *Mater. Lett.* **2011**, *65*, 3205–3207. [CrossRef]
22. Riveiro, A.; Lusquinos, F. Supersonic laser spray of aluminium alloy on a ceramic substrate. *Appl. Surf. Sci.* **2007**, *254*, 926–929. [CrossRef]
23. Li, B.; Jin, Y. Solid-state fabrication of WCp-reinforced Stellite-6 composite coatings with supersonic laser deposition. *Surf. Coat. Technol.* **2017**, *321*, 386–396. [CrossRef]
24. Tsukamoto, K.; Uchiyama, F. Ceramic coating technique using laser spray process. *Surf. Eng.* **2013**, *6*, 45–48. [CrossRef]
25. Li, N.; Wang, Q. Research progress of coating preparation on light alloys in aviation field: A review. *Materials* **2022**, *15*, 8535. [CrossRef]
26. Yin, S.; Cizek, J. Metallurgical bonding between metal matrix and core-shelled reinforcements in cold sprayed composite coating. *Scr. Mater.* **2020**, *177*, 49–53. [CrossRef]
27. Wei, C.; Xu, X. Titanium coating on the surface of diamond particles by a novel rapid low-temperature salt bath plating method. *Chem. Phys. Lett.* **2020**, *761*, 138091. [CrossRef]
28. Wu, Y.P.; Luo, J.B. Critical effect and enhanced thermal conductivity of Cu-diamond composites reinforced with various diamond prepared by composite electroplating. *Ceram. Int.* **2019**, *45*, 13225–13234. [CrossRef]
29. Wang, Y.; Deng, N. The effect of Fe/Al ratio and substrate hardness on microstructure and deposition behavior of cold-sprayed Fe/Al coatings. *Materials* **2023**, *16*, 878. [CrossRef]
30. Daodon, W.; Saetang, V. Improvement of frictional property of AISI D2 tool steel surface against JIS SPFC 980Y advanced high-strength steel by using laser texturing process. *Lubricants* **2023**, *11*, 68. [CrossRef]
31. Zhou, H.; Yao, P. Friction and wear maps of copper metal matrix composites with different iron volume content. *Tribol. Int.* **2019**, *132*, 199–210. [CrossRef]
32. Liu, S.; Yu, X. Effect of Cu-doped carbon quantum dot dispersion liquid on the lubrication performance of polyethylene glycol. *Lubricants* **2023**, *11*, 86. [CrossRef]

lubricants

MDPI

Article

Effect of a Substrate's Preheating Temperature on the Microstructure and Properties of Ni-Based Alloy Coatings

Yu Liu [1,2], Haiquan Jin [1], Tianhao Xu [1], Zhiqiang Xu [1,3], Fengming Du [4], Miao Yu [5], Yali Gao [1,*] and Dongdong Zhang [1]

1 School of Mechanical Engineering, Northeast Electric Power University, Jilin 132012, China
2 International Shipping Research Institute, Gongqing Institute of Science and Technology, Gongqing City 332020, China
3 Beijing Electric Power Transmission and Transformation Company Limited, Beijing 102400, China
4 Marine Engineering College, Dalian Maritime University, Dalian 116026, China
5 Graduate School, Daejin University, Pocheon-si 11159, Republic of Korea
* Correspondence: dehuigyl@126.com

Abstract: Laser cladding is a new technology to fabricate a coating on the surface of a metal substrate. The properties on copper substrates are usually not very good due to the high thermal conductivity and reflectivity. The appropriate preheating temperature is helpful to fabricate coatings with good quality and properties, especially for copper substrates. In order to investigate the effect of different preheating temperatures, four coatings with different preheating temperatures (100, 200, 300 and 400 °C) were fabricated via a laser on a copper substrate. The microstructures and properties of four coatings were investigated using SEM, XRD, EDS, a Vickers microhardness meter, a wear tester and an electrochemical workstation. The results show that the elements from Ni-based alloy powder were uniformly distributed among the binding region, which obtained a good metallurgical bonding. The microstructure was mainly composited of cellular, dendrite and plane crystals, and the main reinforced phases were γ (Fe, Ni), $Cr_{0.09}Fe_{0.7}Ni_{0.21}$, WC and Ni_3B. The values of average microhardness of the four coatings were 614.3, 941.6, 668.1 and 663.1 $HV_{0.5}$, respectively. The wear rates of the four coatings were 9.7, 4.9, 12.5 and 13.3 $\times 10^{-5}$ $mm^3 \cdot N^{-1} \cdot m^{-1}$, respectively, which were less than that of the copper substrate (4.3 $\times$ 10−3 $mm^3 \cdot N^{-1} \cdot m^{-1}$). The decrease in wear rate was due to the existence of the reinforced phases, such as WC, Ni_3B, M_7C_3 (M=Fe, Cr) and $Cr_{0.09}Fe_{0.7}Ni_{0.21}$. The fine crystals in the coating preheated at 200 °C also improved the wear resistance. Additionally, the minimum values of corrosion current density were 3.26×10^{-5}, 2.34×10^{-7}, 4.02×10^{-6} and 4.21×10^{-6} $mA \cdot mm^{-2}$, respectively. It can be seen that the coating preheated at 200 °C had higher microhardness, lower wear rates and better corrosion resistance due to the existence of reinforced phases and fine and uniform crystals.

Keywords: laser cladding; Ni-based alloy; preheating temperature; friction and wear; electrochemistry

Citation: Liu, Y.; Jin, H.; Xu, T.; Xu, Z.; Du, F.; Yu, M.; Gao, Y.; Zhang, D. Effect of a Substrate's Preheating Temperature on the Microstructure and Properties of Ni-Based Alloy Coatings. *Lubricants* **2024**, *12*, 21. https://doi.org/10.3390/lubricants12010021

Received: 3 November 2023
Revised: 18 December 2023
Accepted: 19 December 2023
Published: 10 January 2024

1. Introduction

Copper alloys are widely used in the manufacture of condenser tubes for ocean platforms due to their good corrosion resistance [1–3]. However, corrosion failure often occurs during long-term service, which reduces the service life of the condenser tube [4]. In order to improve the corrosion resistance of condenser tubes, some surface modification technologies are used to fabricate coatings, such as electroplating and thermal spraying [5–7].

In recent decades, laser cladding has been widely used to fabricate coatings due to its advantages of high efficiency and low cost [8–12]. As a coating material, a single metal powder or alloy powder plays an indispensable role in laser cladding technology [13,14]. During the process of laser cladding, some self-fluxing powders, such as Fe-based, Ni-based and Co-based ones and high-entropy alloys, are widely used [15–20]. Of these, Ni-based powder is the most widely used during the process of laser cladding because of its good wear resistance, good corrosion resistance and moderate price [21]. Many researchers have prepared coatings on the surface of copper alloys via laser cladding technology. Bysakh et al. [22] used a CO_2 laser to obtain a cladding layer of a Cu-Fe-Al-Si alloy on the surface of a copper alloy; however, this coating had some defects. Li et al. [23] prepared a Ni60 alloy coating on the surface of aluminum bronze via supersonic flame spraying and then performed laser remelting at a laser power of 3.2 kW with a CO_2 laser. Although a cladding layer with good metallurgical bonding to the substrate was obtained, bubbles still appeared due to the molten pool disturbance. There are many factors affecting the quality of laser cladding, including the laser power, scanning speed, powder feeding rate and preheating temperature [24–27]. In addition, the preheating temperature is one of the key factors of laser cladding that has been studied by many scholars [28–31]. Although the preheating temperatures of different substrates have been studied, there are few studies about the preheating temperature of copper substrate. Because of the higher thermal conductivity and reflectivity of copper substrate, the selection of the preheating temperature is particularly important to fabricate a good coating using a laser.

In this paper, four different preheating temperatures (100, 200, 300 and 400 °C) were selected to fabricate a coating on a copper substrate. The macroscopic morphology and dilution rate were studied. The phases, microstructure, elemental distribution and properties of the four coatings were investigated using SEM, XRD, EDS, a Vickers microhardness meter, a wear tester and an electrochemical workstation. Meanwhile, the microstructure evolution of the four coatings is discussed.

2. Experimental Conditions

2.1. Materials

Copper alloy is used as the substrate; its size is 50 × 30 × 10 mm and its composition is shown in Table 1. Ni-based alloy powder is used as the cladding material; its composition is shown in Table 2. The mass fractions of the Fe, Cr and WC elements are 12.0%, 12.0% and 5.0%, respectively. There are small amounts of the C, B and Si elements.

Table 1. Chemical composition of the copper alloy (wt.%).

Zn	Al	Mn	Fe	Cu
29.0	6.3	3.2	1.4	60.1

Table 2. Chemical composition of the Ni-based alloy powder (wt.%).

Fe	C	B	Si	Cr	WC	Ni
12	1	2.5	1.5	12	5	Bal.

Figure 1 is an SEM micrograph of the Ni-based alloy powder. It can be seen that the shape of most particles is nearly spherical and the surface is smooth. There are a few irregular particles with the shapes resembling a block or strip. Additionally, some particles with small sizes have tended to gather together.

Figure 1. SEM micrograph of the Ni-based alloy powder.

2.2. Methods

The device was a CO_2 laser with a continuous mode and the parameters are as follows: laser power 1.4 kW, laser beam diameter 3.0 mm, scanning speed 2 mm/s and overlapping ratio 30%. The substrate preheating temperatures were 100, 200, 300 and 400 °C, respectively. After laser cladding, the coatings were cut into 10 × 10 × 10 mm samples. After the samples were buffed and polished, a ferric chloride solution was used to etch the cladding coating for 10 s. Then, the samples were observed using optical microscopy and scanning election microscopy (TESCAN MIRA). At the same time, the element distribution was analyzed via EDS. The phases were detected by X-ray diffraction (TD-3500). The microhardness was tested using a Vickers microhardness meter with a load of 50 gf at 15 s. The microhardness was measured from the top of the Ni-based coating to the copper substrate and each point was repeated five times at the same height. The friction and wear properties were tested using a reciprocating fatigue friction and wear tester (MGW-02) with a tip force of 10 N and a friction frequency 2 Hz. An experiment testing the electrochemical corrosion was also carried out to investigate the corrosion resistance of cladding coatings. A 3.5% NaCl solution was used as the electrolyte in a CHI600E electrochemical workstation.

3. Results

3.1. Macroscopic Morphology and Dilution Rate

Figure 2 shows the macroscopic morphology of coatings preheated at four temperatures (100, 200, 300 and 400 °C). The four coatings can be successively fabricated on copper substrate but there are some pores and cracks on the surface.

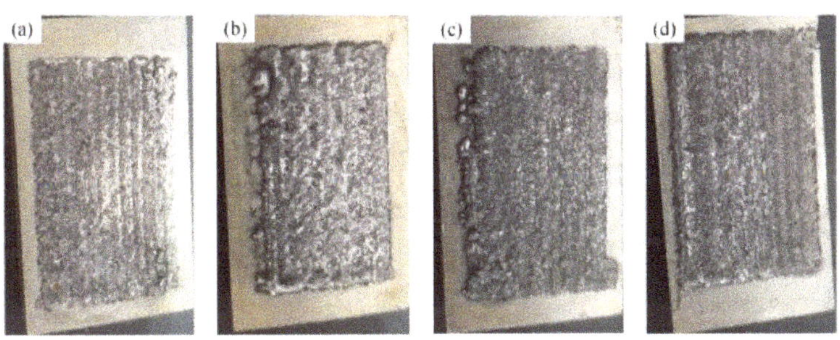

Figure 2. Macroscopic morphology of coatings preheated at four temperatures. (**a**) 100 °C; (**b**) 200 °C; (**c**) 300 °C; (**d**) 400 °C.

The dilution rate is an important index to judge the performance of a coating. A low dilution rate is helpful to improve the coating's properties. The molten pool depth and height of the coating are two parameters to measure dilution rate, as shown in Figure 3. The dilution rate can be calculated using Equation (1).

$$\eta = \frac{Depth}{Depth + Height} \times 100\% \tag{1}$$

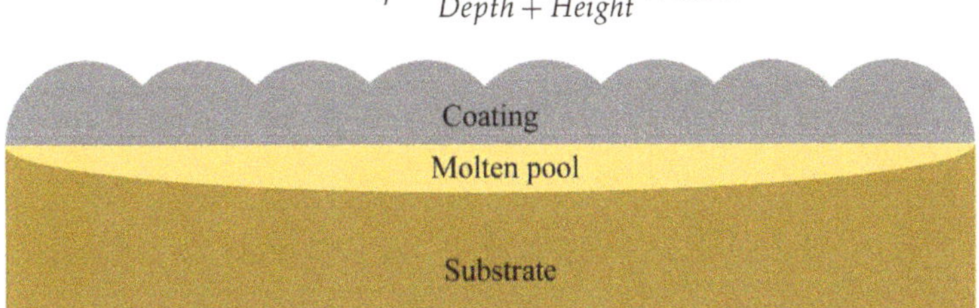

Figure 3. Schematic diagram of the molten pool and coating.

In Equation (1), η is the dilution rate, *Depth* is the average depth of the coating's molten pool, and *Height* is the average height of the coating.

Table 3 shows the height, molten pool depth and dilution rate of the four coatings with different preheating temperatures (100, 200, 300 and 400 °C). At the four different preheating temperatures, the coatings' heights are 0.27, 0.48, 0.39 and 0.27 mm, respectively. The depths of the molten pool are 0.1, 0.12, 0.16 and 0.18 mm, respectively. With the increasing preheating temperature, the coatings' heights first increase and then decrease. However, the depths of the molten pool increase gradually along with the increase in the preheating temperature. The main reason for this is that higher preheating temperatures make the substrate easier to melt and so it forms a molten pool with higher temperatures, which increases the depths of the molten pool. At the four different preheating temperatures, the dilution rates were 27%, 20%, 29% and 40%, respectively. The coating with a preheating temperature of 200 °C has the lowest dilution rate (20%), which suggests that it can maintain the properties of the coating and reduce the mixing of elements between the substrate and coating.

Table 3. Heights, molten pool depths and dilution rates of the four coatings.

Preheating Tem. (°C)	100	200	300	400
Height (mm)	0.27	0.48	0.39	0.27
Depth (mm)	0.1	0.12	0.16	0.18
Dilution rate (%)	27	20	29	40

3.2. Phases

The X-ray diffraction patterns of the coatings preheated at four different temperatures (100, 200, 300 and 400 °C) are shown in Figure 4. When the preheating temperature was 100 °C, the main phases are γ (Fe, Ni), WC, W_2C, Ni_3B, and M_7C_3 (M=Fe, Cr). When the preheating temperature was 200 °C, the higher preheating temperature increases the heat of the molten pool and prolongs the solidification time of the molten pool. Elements such as Ni, Cr and Fe have enough time to form a reinforced phase. Therefore, a new phase, $Cr_{0.09}Fe_{0.7}Ni_{0.21}$, appears. When the preheating temperatures are 300 and 400 °C, the amount of some elements decreases gradually due to the higher temperature of molten pool, which results in the decrease in WC, W_2C and boride. At the same time, a large number of substrate elements enter into the coating and form $Cu_{0.64}Zn_{0.36}$ and Ni_2W_4C phases.

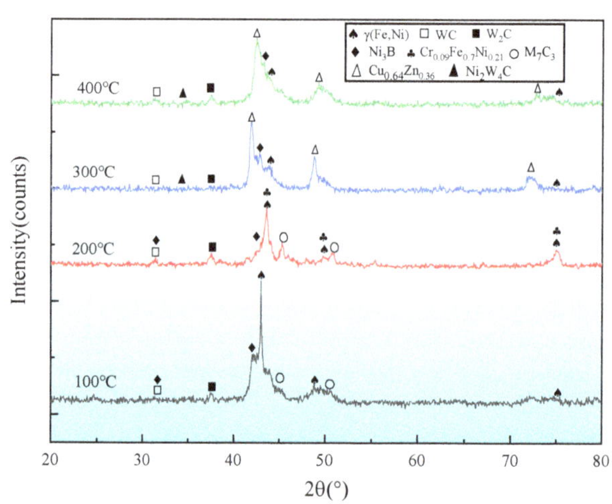

Figure 4. XRD patterns of coatings preheated at four preheating temperatures (100, 200, 300 and 400 °C).

3.3. Microstructure

Figure 5 depicts the microstructures of coatings preheated at four temperatures (100, 200, 300 and 400 °C). Figure 5(a1,b1,c1,d1) shows the microstructure of the coatings' top regions. When the preheating temperature was 100 °C, there are a large number of equiaxed crystals. When the preheating temperatures were 200, 300 or 400 °C, some cellular crystals appear in the top regions. In addition, the coating preheated at 200 °C has fine cellular crystals. As shown in Figure 5(a2,b2,c2,d2), the microstructure of the middle position is mainly a dendrite structure. In addition, when the preheating temperature was 200 °C, the dendrite crystals are fine and uniform. With the increase in preheating temperature, the secondary dendrites grow gradually on the primary dendrites, as shown in Figure 5(c2). When the preheating temperature was 400 °C, the dendrite crystals become larger. At the bottom of the four coatings, when the preheating temperature was 100 °C, some small holes appear in the bonding region, as shown in Figure 5(a3), which decreases the binding strength between the coating and the substrate. With the increase in preheating temperature, the holes gradually disappear, as shown in Figure 5(b3,c3,d3).

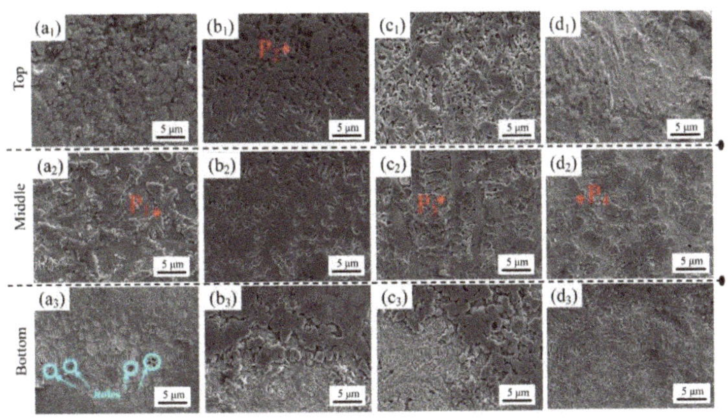

Figure 5. SEM micrographs of four coatings that were preheated at 100 (**a1–a3**), 200 (**b1–b3**), 300 (**c1–c3**) and 400 °C (**d1–d3**).

Table 4 shows the percentage of element atoms at the P1–P4 points in the four coatings with different preheating temperatures (100, 200, 300 and 400 °C). According to the percentage of element atoms at the P1 point, there are large amounts of Fe and Ni in the coating with the preheating temperature of 100 °C that form the supersaturated solid solution of the γ (Fe, Ni) phase. Some C atoms form WC and W_2C and others form the Fe_7C_3 phase. Compared with the P1 point, the percentage of element atoms at the P2 point shows that the content of Cr increases, which becomes Cr_7C_3 phase. At the same time, there are a few borides. The percentage of element atoms at the P3 and P4 points indicates that a large number of Cu and Zn elements appear at P3 and P4, which is mainly the $Cu_{0.64}Zn_{0.36}$ phase. With the increase in preheating temperature, more and more elements from the substrate enter into the molten pool. This is consistent with the results of Figure 4.

Table 4. Percentage of element atoms at the P1–P4 points in the coatings.

Point	Atomic Concentration %							
	Ni	Fe	C	W	B	Cu	Zn	Cr
P1	25.4	29.6	23.8	9.4	4.5	7.2	-	0.1
P2	39.7	11.6	35.4	5.4	2.8	1.7	-	3.4
P3	31.1	15.2	20.9	3.3	-	16.8	8.8	3.9
P4	25.5	15.4	9.8	5.4	-	23.7	18.6	1.6

Figure 6 depicts the SEM micrograph and EDS element line scanning results for the four coatings with different preheating temperatures (100, 200, 300 and 400 °C). From the substrate to the coating, the copper element decreases gradually, whereas the Ni element increases gradually. In the diffusion zone, Cu and Ni dissolve each other and form good metallurgical bonding. At the same time, the width of the diffusion zone increases with the increase in preheating temperature. When the preheating temperature was 200 °C, the coating has the least amount of Cu. More of the Cu element appears in the diffusion zone, which also decreases the relative amount of the Ni element. The other elements in the substrate and coating, such as Fe, Al and Cr, also enter into the diffusion zone. When the preheating temperature is higher, more of the Cu element enters into the coating, as shown in Figure 6c,d.

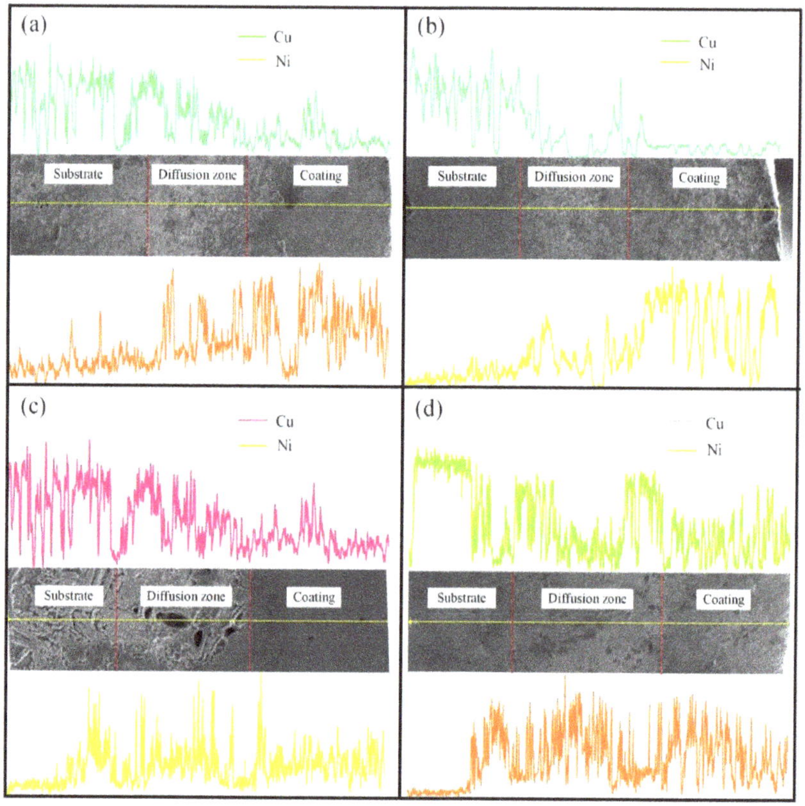

Figure 6. SEM micrograph and EDS line scan results for the four coatings with different preheating temperatures (100, 200, 300 and 400 °C). (**a**) 100 °C; (**b**) 200 °C; (**c**) 300 °C; (**d**) 400 °C.

3.4. Microhardness

Figure 7 shows the microhardnesses of the four coatings' cross sections. The maximum microhardness values of four the coatings are 680.7, 1069.3, 815.2 and 760.9 $HV_{0.5}$, respectively. The values for the average microhardness are 614.3, 941.6, 668.1 and 663.1 $HV_{0.5}$, respectively. As a whole, the microhardness in the strengthening layer is stable and the fluctuation is not very large. At the binding region, the microhardness decreases quickly. When the preheating temperature was 200 °C, the coating has the highest average microhardness (941.6 $HV_{0.5}$), which is caused by the reinforced phases M_7C_3, WC, Ni_3B and $Cr_{0.09}Fe_{0.7}Ni_{0.21}$ and the fine crystals in the top region. However, when the preheating temperature was 300 °C, some new phases, $Cu_{0.64}Zn_{0.36}$ and Ni_2W_4C, appear due to entrance of Cu and Zn elements from copper substrate, which decreases the coating's microhardness.

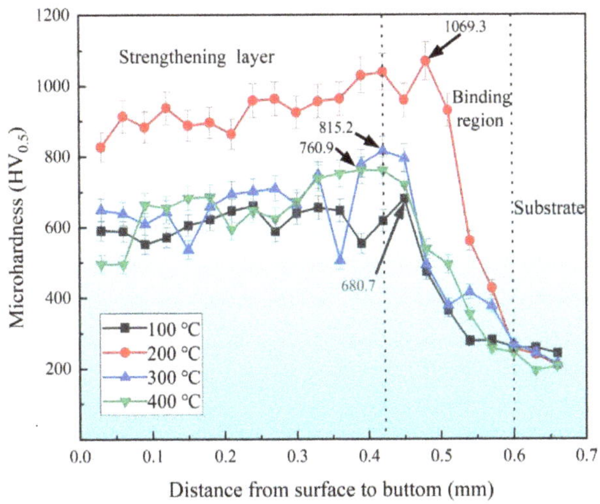

Figure 7. Microhardness of the four coatings' cross sections.

3.5. Wear Resistance

The friction coefficients of four coatings with different preheating temperatures (100, 200, 300 and 400 °C) are shown in Figure 8. The coating that had a preheating temperature of 200 °C has the smallest friction coefficient (around 0.1), which only has a small fluctuation, whereas the coating that had a preheating temperature 400 °C has the largest friction coefficient (around 0.3) and fluctuation. The friction coefficients of the coatings preheated at 100 and 300 °C fluctuate around 0.2 and also have large fluctuations. From the above analysis, the smallest friction coefficient (around 0.1) in the 200 °C preheated coating may be due to its lower surface roughness and fine crystals.

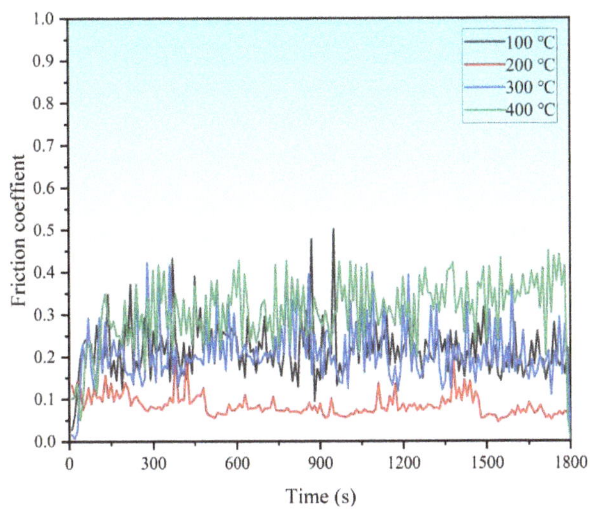

Figure 8. Friction coefficients of coatings preheated at four preheating temperatures (100, 200, 300 and 400 °C).

Figure 9 shows the wear rates of coatings preheated at different temperatures (100, 200, 300 and 400 °C). The wear rates of coatings preheated at 100, 200, 300 and 400 °C are 9.7, 4.9, 12.5 and 13.3 × 10^{-5} $mm^3 \cdot N-^1 \cdot m^{-1}$, respectively. The wear rate of the coating preheated at 200 °C is the lowest, only about one-third of the wear rate of the coating preheated at 400 °C. The wear rate of copper substrate is 4.3 × 10^{-3} $mm^3 \cdot N^{-1} \cdot m^{-1}$. Compared with the substrate, the wear resistance of the four coatings has an obvious improvement due to the existence of the reinforced phases, such as WC, Ni_3B, M_7C_3 (M=Fe, Cr) and $Cr_{0.09}Fe_{0.7}Ni_{0.21}$. Additionally, when the preheating temperature was 200 °C, the coating has more fine crystals in the top region.

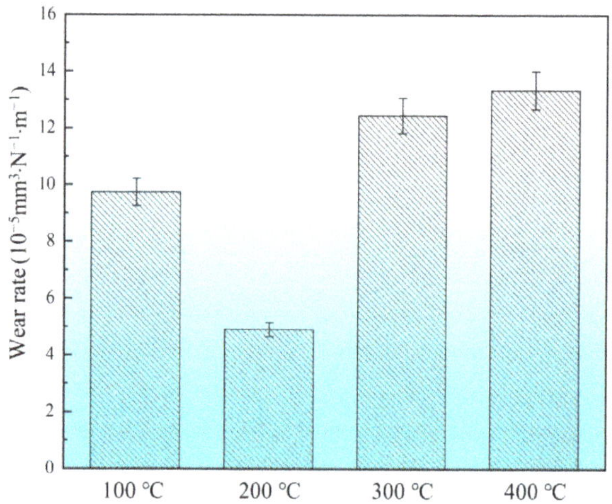

Figure 9. Wear rates of coatings preheated at four preheating temperatures (100, 200, 300 and 400 °C).

Figure 10 depicts the morphology of the friction and wear of coatings preheated at different temperatures (100, 200, 300 and 400 °C). It can be seen that the coating that had a preheating temperature 200 °C has the smallest depth and width of grinding crack. More M_7C_3, WC, W_2C and borides improve the coating hardness and reinforced phase content. Meanwhile, the fine and uniform crystals are also helpful to enhance the wear resistance, which forms abrasive wear and makes the wear volume decrease.

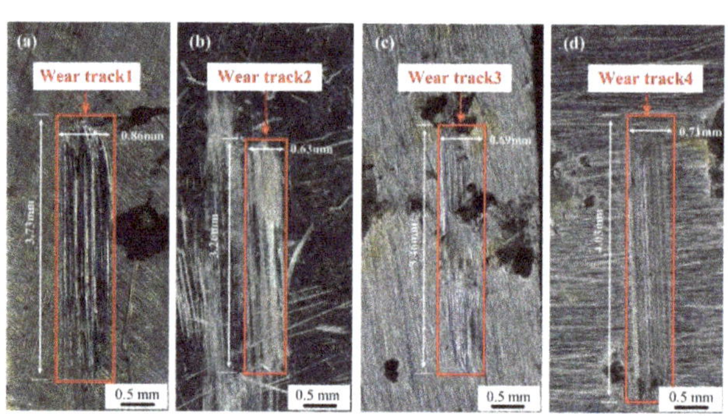

Figure 10. Morphology of friction and wear of coatings preheated at four preheating temperatures (100, 200, 300 and 400 °C). (**a**) 100 °C; (**b**) 200 °C; (**c**) 300 °C; (**d**) 400 °C.

3.6. Corrosion Resistance

Figure 11 shows the polarization curves of coatings preheated at four temperatures (100, 200, 300 and 400 °C). The polarization curves of the four coatings present a similar trend and show the certain passivation behavior. When the preheating temperatures were

100, 200, 300 and 400 °C, the free corrosion potentials are −0.708, −0.690, −0.893 and −0.817 V, respectively. The minimum values of corrosion current density are 3.26×10^{-5}, 2.34×10^{-7}, 4.02×10^{-6} and 4.21×10^{-6} mA·mm^{-2}, respectively. Moreover, the corrosion current density of the copper substrate is 1.14×10^{-6} mA·mm^{-2}, which is less than that of the coatings preheated at 100, 300 or 400 °C and more than that of the coating preheated at 200 °C. Therefore, the coating preheated at 200 °C has the lowest corrosion current density (2.34×10^{-7} mA·mm^{-2}), which indicates that the coating has the best corrosion resistance due to a large amount of Cr_7C_3. This phase improves the corrosion resistance. Although the coatings with preheating temperatures of 300 and 400 °C have a Cr_7C_3 phase, a lot of Cu and Zn elements diffuse into the coating and reduce the corrosion resistance.

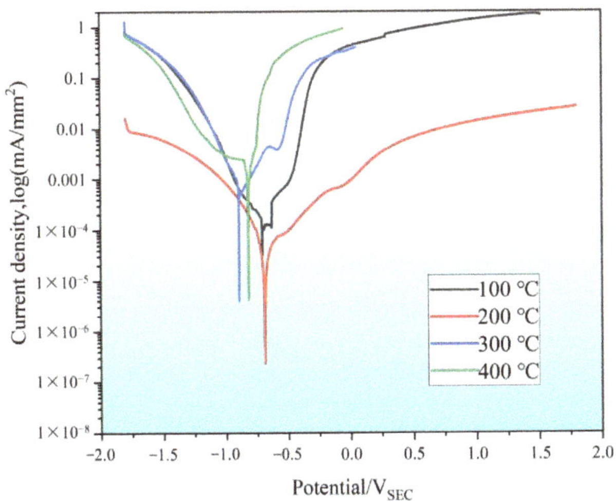

Figure 11. Polarization curves of coatings preheated at four preheating temperatures (100, 200, 300 and 400 °C).

4. Discussion

Ni-based alloy coatings with different preheating temperatures (100, 200, 300 and 400 °C) were fabricated on copper substrate. Due to the difference in preheating temperature, the microstructure, quality and properties of the coatings are different. Therefore, the phases, microstructure, microhardness, wear resistance and corrosion resistance were investigated. The microhardnesses of the coatings were 614.3, 941.6, 668.1 and 663.1 $HV_{0.5}$ under preheating at temperatures of 100, 200, 300 and 400 °C, respectively. The average hardness is highest at the preheating temperature of 200 °C, which is caused by the reinforcing phases such as M_7C_3, WC, Ni_3B and $Cr_{0.09}Fe_{0.7}Ni_{0.21}$ and the diffusion of fine grains and fewer Cu and Zn elements in the top region. The wear rates of the four coatings were 9.7, 225, 4.9, 12.5 and 13.3 × 10^{-5} mm^3·N^{-1}·m^{-1}, respectively. The 200 °C preheated coating has the lowest wear rate, which is due to the correct preheating temperature to produce finer and more uniform crystals in the top area of the coating. In the electrochemical test, the coating preheated at 200 °C has the lowest corrosion current density (2.34×10^{-7} mA·mm^{-2}), which is due to the fact that the coating is less affected by the diffusion of Cu and Zn elements and the presence of a large amount of Cr_7C_3 in the coating, resulting in the best corrosion resistance.

4.1. *Effect of Preheating Temperature on Pore and Crack*

During the laser cladding process, if the gas in molten pool does not escape in time, some pores will appear in the coating. The existence of pores will reduce the mechanical and corrosion resistance of a coating. According to the Stocks equation [32], the gas overflow in the molten pool is mainly related to the viscosity of the liquid metal. With an increase in temperature, the viscosity of the liquid metal decreases and the gas can more easily escape from the molten pool. Figure 12 shows the schematic diagram of gas overflow in the molten pool during the laser cladding process. At lower preheating temperature, the viscosity is larger, which makes it difficult for the gas to escape and results in some pores being left in the coating, as shown in Figure 12a. This is in accordance with the result shown in Figure 5(a3). In contrast, if the preheating temperature is higher, the gas can more easily escape, as shown in Figure 12b,c. Therefore, no pores appeared in the cladding coating when the preheating temperatures were 200, 300 and 400 °C.

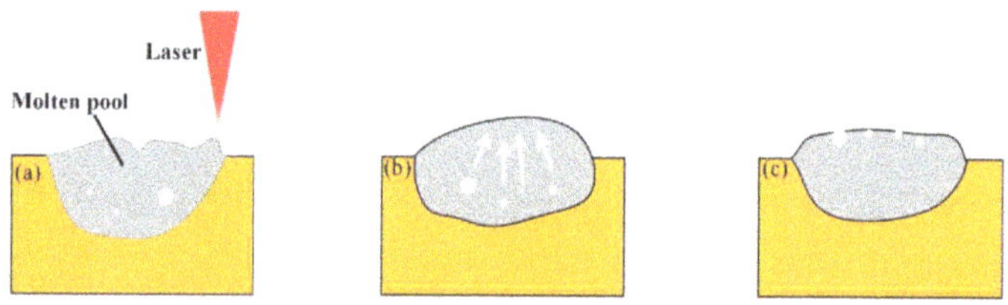

Figure 12. Diagram of the gas overflow in the molten pool during laser cladding. (**a**) lower preheating temperature; (**b**) higher preheating temperature; (**c**) the highest preheating temperature.

When the preheating temperature was 400 °C, there were some cracks at the bottom of the cladding coating, as shown in Figure 13. Higher preheating temperatures increase the heat accumulation in the molten pool, which brings larger thermal stress and shrinkage between the different phases. When the thermal stress is higher than the tensile strength of the cladding coating during the cooling of the molten pool, cracks may appear in some places, such as the crack in Figure 13.

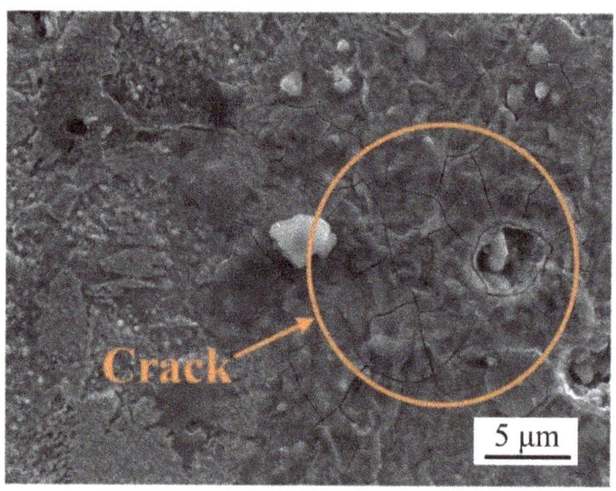

Figure 13. Crack at the bottom of the coating preheated at 400 °C.

4.2. *Effect of Preheating Temperature on Microstructure*

The microstructure evolution under four preheating temperatures is shown in Figure 14. When the preheating temperature is 100 °C, a large number of equiaxed crystals form in the top part of cladding coating, which do not grow due to the lower preheating temperature. With the decrease in preheating temperature, the crystals begin to grow as dendrite crystals in the middle of the cladding coating. When the preheating temperature is 200 °C, because of the increase in preheating temperature, the higher temperature promotes crystals to grow as fine cellular crystals in the top of the coating. When the preheating temperatures are 300 and 400 °C, although the microstructure of top regions is also cellular, the number of cellular crystals is more than that of the coating preheating at 200 °C. Because of the larger cooling rate and mutual inhibition, the cellular crystals are smaller. In the middle regions, a large amount of heat accumulates in the molten pool and the secondary dendrites grow on the basis of the primary dendrites. However, when the preheating temperature is 400 °C, the higher preheating temperature causes more substrate elements into the liquid molten pool and appear in the $Cu_{0.64}Zn_{0.36}$ phase, which destroys the growth of the primary dendrites and secondary dendrites.

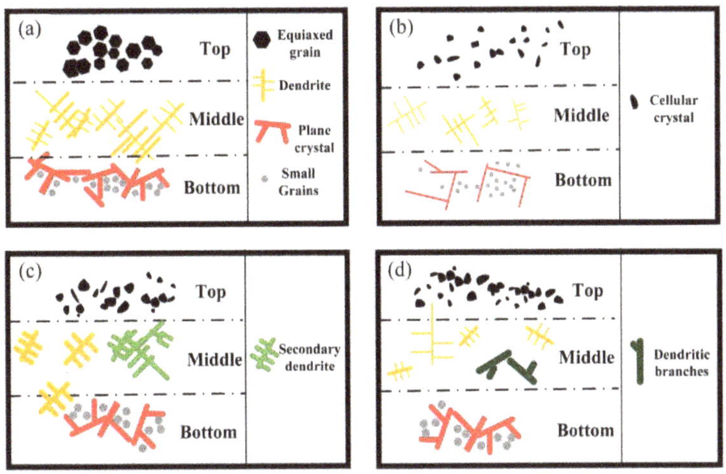

Figure 14. Microstructure evolution of four preheating temperatures. (**a**) 100 °C; (**b**) 200 °C; (**c**) 300 °C; (**d**) 400 °C.

5. Conclusions

(1) Four Ni-based alloy coatings with different preheating temperatures were successfully fabricated on the copper substrate. When the preheating temperature is 100 °C, there are some pores at the bottom of coating. When the preheating temperature is 400 °C, the effect of thermal stress is too great and a few cracks appear. At the four different preheating temperatures, the dilution rates were 27%, 20%, 29% and 40%, respectively. The coating with a preheating temperature of 200 °C has the lowest dilution rate (20%), which can maintain the property of the coating and reduces the mix of elements between the substrate and coating.

(2) When the preheating temperature is 100 °C, the main phases are γ (Fe, Ni), WC, W_2C, Ni_3B and M_7C_3 (M=Fe, Cr). When the preheating temperature is 200 °C, a new phase $Cr_{0.09}Fe_{0.7}Ni_{0.21}$ appears. When the preheating temperatures are 300 and 400 °C, a large number of substrate elements enter into the coating and form $Cu_{0.64}Zn_{0.36}$ and Ni_2W_4C phases. The microstructure of the cladding coating is mainly composed of cellular, dendrite and planar crystals. The coating preheated at 200 °C has fine and uniform crystals.

(3) At the four different preheating temperatures, the values of average microhardness for the four coatings are 614.3, 941.6, 668.1 and 663.1 $HV_{0.5}$, respectively. When the preheating temperature is 200 °C, it has the maximum average microhardness, which is caused by the reinforced phases M_7C_3, WC, Ni_3B and $Cr_{0.09}Fe_{0.7}Ni_{0.21}$ and the fine crystals in the top region. However, when the preheating temperature is 300 °C, some new phases ($Cu_{0.64}Zn_{0.36}$ and Ni_2W_4C) appear due to the entrance of Cu and Zn elements from the copper substrate, which decreases the coating's microhardness.

(4) The friction coefficients of the four coatings are about 0.2, 0.1, 0.2 and 0.3, respectively. The coating with a preheating temperature of 200 °C has the smallest friction coefficient (around 0.1), which also has a small fluctuation, whereas the coating with a preheating temperature of 400 °C has the largest friction coefficient (around 0.3) and fluctu-

ation. The wear rates of the coatings preheated at 100, 200, 300 and 400 °C are 9.7, 4.9, 12.5 and 13.3 × 10−5 $mm^3 \cdot N^{-1} \cdot m^{-1}$, respectively. The wear rate of the coating preheated at 200 °C is the lowest due to the existence of the reinforced phases and fine crystals.

(5) When the preheating temperature is 100, 200, 300 and 400 °C, the corrosion potentials are −0.708, −0.690, −0.893 and −0.817 V, respectively. Moreover, the minimum values for the corrosion current density are 3.26×10^{-5}, 2.34×10^{-7}, 4.02×10^{-6} and 4.21×10^{-6} $mA \cdot mm^{-2}$, respectively. The coating with a preheating temperature of 200 °C has the maximum corrosion potential (−0.69 V) and the lowest corrosion current density (2.34×10^{-7} $mA \cdot mm^{-2}$), which represents the best corrosion resistance. This is due to the high amount of Cr7C3 in the coating with a preheating temperature of 200 °C and the diffusion of a smaller amount of Cu and Zn elements into the coating.

Author Contributions: Software, Y.L.; Writing—original draft, T.X. and H.J.; Resources, Y.L.; Data processing, Y.G. and F.D.; Investigation, Z.X., D.Z. and Y.G.; Formal analysis, M.Y. and D.Z.; Supervision, Y.L., Y.G. and M.Y. All authors have read and agreed to the published version of the manuscript.

Funding: This research was funded by the Science and Technology Development of Jilin Province, grant number 20230101335JC. This thesis work is also supported by Northeast Electric Power University.

Data Availability Statement: Data are contained within the article.

Conflicts of Interest: Author Z.X. was employed by the company Beijing Electric Power Transmission and Transformation Company Limited. The remaining authors declare that the research was conducted in the absence of any commercial or financial relationships that could be construed as a potential conflict of interest.

References

1. Wang, Y.L.; Zhuo, L.C.; Yin, E.H. Challenges and potentials/trends of tungsten-copper (WCu) composites/pseudo-alloys: Fabrication, regulation and application. *Int. J. Refract. Met. H* **2021**, *100*, 105648. [CrossRef]
2. Ebrahimi, M.; Par, M. Twenty-year uninterrupted endeavor of friction stir processing by focusing on copper and its alloys. *J. Alloys Compd.* **2019**, *781*, 1074–1090. [CrossRef]
3. Zhang, H.; Fu, H.; Zhu, S.; Yong, W.; Xie, J. Machine learning assisted composition effective design for precipitation strengthened copper alloys. *Acta Mater.* **2021**, *215*, 117118. [CrossRef]
4. Rao, T.S.; Bera, S. Protective layer dissolution by chlorine and corrosion of aluminum brass condenser tubes of a nuclear power plant. *Eng. Fail. Anal.* **2021**, *123*, 105307.
5. Zhou, H.-H.; Liao, Z.-W.; Fang, C.-X.; Li, H.-X.; Feng, B.; Xu, S.; Cao, G.-F.; Kuang, Y.-F. Pulse electroplating of Ni-W-P coating and its anti-corrosion performance. *Trans. Nonferrous Met. Soc. China* **2018**, *28*, 88–95. [CrossRef]
6. Li, T.; Dang, N.; Liang, M.; Guo, C.; Lu, H.; Ma, J.; Liang, W. TEM observation of general growth behavior for silver electroplating on copper rod. *Appl. Surf. Sci.* **2018**, *451*, 148–154. [CrossRef]
7. Zang, J.; Li, H.; Sun, J.; Shen, Y.; Su, N.; Feng, X. Microstructure and thermal conductivity of Cu-Cu2AlNiZnAg/diamond coatings on pure copper substrate via high-energy mechanical alloying method. *Surf. Interfaces* **2020**, *21*, 100742. [CrossRef]
8. Liu, Y.; Xu, T.; Liu, Y.; Gao, Y.; Di, C. Wear and heat shock resistance of Ni-WC coating on mould copper plate fabricated by laser. *J. Mater. Res. Technol.* **2020**, *9*, 8283–8288. [CrossRef]
9. Sciacca, G.; Sinico, M.; Cogo, G.; Bigolaro, D.; Pepato, A.; Esposito, J. Experimental and numerical characterization of pure copper heat sinks produced by laser powder bed fusion. *Mater. Des.* **2022**, *214*, 110415. [CrossRef]
10. Guan, W.; Gao, M.; Lv, H.; Yuan, J.; Chen, D.; Zhu, T.; Fang, Y.; Liu, J.; Wang, H.; Tang, Z.; et al. Laser cladding of layered Zr/Cu composite cathode with excellent arc discharge homogeneity. *Surf. Coatings Technol.* **2021**, *421*, 127454. [CrossRef]
11. Liu, Y.; Li, Z.Y.; Li, G.H.; Tang, L. Friction and wear behavior of Ni-based alloy coatings with different amount of WC-TiC ceramic particles. *J. Mater. Sci.* **2023**, *58*, 1116–1126. [CrossRef]
12. Feng, X.; Cui, X.; Zheng, W.; Lu, B.; Dong, M.; Wen, X.; Zhao, Y.; Jin, G. Effect of the protective materials and water on the repairing quality of nickel aluminum bronze during underwater wet laser repairing. *Opt. Laser Technol.* **2019**, *114*, 140–145. [CrossRef]
13. Harooni, A.; Iravani, M.; Khajepour, A.; King, J.M.; Khalifa, A.; Gerlich, A.P. Mechanical properties and microstructures in zirconium deposited by injected powder laser additive manufacturing. *Addit. Manuf.* **2018**, *22*, 537–547. [CrossRef]
14. Ramakrishnan, A.; Dinda, G. Microstructural control of an Al–W aluminum matrix composite during direct laser metal deposition. *J. Alloys Compd.* **2020**, *813*, 152208. [CrossRef]

15. Zhang, H.; Wang, L.; Zhang, S.; Wu, C.; Zhang, C.; Sun, X.; Chen, J. An investigation on wear and cavitation erosion-corrosion characteristics of the TiC modified Fe-based composite coating via laser cladding. *J. Mater. Res. Technol.* **2023**, *26*, 8440–8455. [CrossRef]
16. Wang, H.; Sun, Y.; Qiao, Y.; Du, X. Effect of Ni-coated WC reinforced particles on microstructure and mechanical properties of laser cladding Fe-Co duplex coating. *Opt. Laser Technol.* **2021**, *142*, 107209. [CrossRef]
17. He, B.-M.; Liu, X.-B.; Zhang, F.-Z.; Liu, Z.-Y.; Zhang, S.-H. Tribological and oxidation behaviors of TiN reinforced Co matrix composite coatings on Inconel718 alloy by laser cladding. *Tribol. Int.* **2023**, *188*, 108781. [CrossRef]
18. Liu, Y.; Wu, Y.; Ma, Y.; Gao, W.; Yang, G.; Fu, H.; Xi, N.; Chen, H. High temperature wear performance of laser cladding Co06 coating on high-speed train brake disc. *Appl. Surf. Sci.* **2019**, *481*, 761–766. [CrossRef]
19. Naizabekov, A.; Samodurova, M.; Bodrov, E.; Lezhnev, S.; Samoilova, O.; Trofimov, E.; Mikhailov, D.; Litvinyuk, K.; Trofimova, S.; Latfulina, Y.; et al. Use of laser cladding for the synthesis of coatings from high-entropy alloys reinforced with ceramic particles. *Case Stud. Constr. Mater.* **2023**, *19*, e02541. [CrossRef]
20. Wu, T.; Yu, L.T.; Chen, G.; Wang, R.Q.; Xue, Y.P.; Lu, Y.H.; Luan, B.L. Effects of Mo and Nb on the microstructure and high temperature oxidation behaviors of CoCrFeNi-based high entropy alloys. *J. Mater. Res. Technol.* **2023**, *27*, 1537–1549. [CrossRef]
21. Zhu, L.; Xue, P.; Lan, Q.; Meng, G.; Ren, Y.; Yang, Z.; Xu, P.; Liu, Z. Recent research and development status of laser cladding: A review. *Opt. Laser Technol.* **2021**, *138*, 106915. [CrossRef]
22. Bysakh, S.; Chat, K.; Maiwald, T. Microstructure evolution in laser alloyed layer of Cu-Fe-Al-Si on Cu substrate. *Mater. Sci. Eng. A* **2004**, *375*, 661–665. [CrossRef]
23. Li, Y.M.; Song, J.G.; Wang, L. Structure and Wear Performance of Laser Cladding Ni-Based Coatings on Aluminum Bronze Surface. *Mater. Prot.* **2012**, *45*, 30–32+2.
24. Su, Z.; Li, J.; Shi, Y.; Ren, S.; Zhang, Z.; Wang, X. Effect of process parameters on microstructure and tribological properties of Ni60A/Cr_3C_2 laser cladding on $60Si_2Mn$ steel. *Surf. Coatings Technol.* **2023**, *473*, 130005. [CrossRef]
25. Kumar, V.; Kumar, R.; Das, A.K. Comprehensive study of process parameters and their effect on tribological properties of laser cladded Ti6Al4V alloy: A review. *Mater. Today Proc.* **2023**, *23*, 2214–7853. [CrossRef]
26. Chen, T.; Wu, W.; Li, W.; Liu, D. Laser cladding of nanoparticle TiC ceramic powder: Effects of process parameters on the quality characteristics of the coatings and its prediction model. *Opt. Laser Technol.* **2019**, *116*, 345–355. [CrossRef]
27. Liu, Y.; Xu, T.; Li, G.; Li, Z.; Du, F.; Chen, G. A multi-objective optimization of laser cladding process of Ni-based coating on the preheated copper alloy. *Mater. Today Commun.* **2023**, *35*, 105614. [CrossRef]
28. Lai, Q.; Abrahams, R.; Yan, W.Y.; Qiu, C.; Mutton, P.; Paradowska, A.; Fang, X.Y.; Soodi, M.; Wu, X.H. Effects of preheating and carbon dilution on material characteristics of laser-cladded hypereutectoid rail steels. *Mater. Sci. Eng. A.* **2018**, *712*, 548–563. [CrossRef]
29. Bidron, G.; Doghri, A.; Malot, T.; Fournier-Dit-Chabert, F.; Thomas, M.; Peyre, P. Reduction of the hot cracking sensitivity of CM-247LC superalloy processed by laser cladding using induction preheating. *J. Am. Acad. Dermatol.* **2020**, *277*, 116461. [CrossRef]
30. Weisheit, A.; Rittinghaus, S.-K.; Dutta, A.; Majumdar, J.D. Studies on the effect of composition and pre-heating on microstructure and mechanical properties of direct laser clad titanium aluminide. *Opt. Lasers Eng.* **2020**, *131*, 106041. [CrossRef]
31. Zhou, L.; Chen, S.; Wei, M.; Liang, J.; Liu, C.; Wang, M. Microstructure and properties of 24CrNiMoY alloy steel prepared by direct laser deposited under different preheating temperatures. *Mater. Charact.* **2019**, *158*, 109931. [CrossRef]
32. Hastie, J.C.; Koelblin, J.; Kartal, M.E.; Attallah, M.M.; Martinez, R. Evolution of internal pores within AlSi10Mg manufactured by laser powder bed fusion under tension: As-built and heat treated conditions. *Mater. Des.* **2021**, *204*, 109645. [CrossRef]

lubricants

MDPI

Article

Laser-Fabricated Micro-Dimples for Improving Frictional Property of SKH51 Tool Steel Surfaces

Chansovannkumpheak Phun [1], Witthaya Daodon [2,*], Kamthon Septham [3], Peerapong Kumkhuntod [3], Hao Zhu [4] and Viboon Saetang [1,*]

1 Department of Production Engineering, Faculty of Engineering, King Mongkut's University of Technology Thonburi, Bangkok 10140, Thailand; chansovankumpheak.ph@rmuti.ac.th
2 Department of Industrial Engineering, Faculty of Engineering and Technology, Rajamangala University of Technology Isan, Nakhon Ratchasima 30000, Thailand
3 Department of Mechanical Engineering, Faculty of Engineering, King Mongkut's University of Technology Thonburi, Bangkok 10140, Thailand; kamthon.sep@mail.kmutt.ac.th (K.S.)
4 School of Mechanical Engineering, Jiangsu University, Zhenjiang 212013, China
* Correspondence: witthaya.da@rmuti.ac.th (W.D.); viboon.tan@kmutt.ac.th (V.S.); Tel.: +66-4423-3000 (ext. 3330) (W.D.); +66-2470-9194 (V.S.); Fax: +66-2872-9081 (V.S.)

Abstract: Friction involved in metal-forming processes typically leads to the wear of tool and die surfaces, and in turn shortens the tool's service life. A thriving need for reducing surface friction requires the tool surface to be modified. This paper presents the surface modification of SKH51 tool steel, on which the hexagonal array of micro-dimples is fabricated by a nanosecond pulse laser. Using the average laser power of 25 W can create decent dimples for trapping lubricant and enabling hydraulic pressure at the surfaces in contact. The effect of dimple density and sliding speed on the coefficient of friction was examined in this study through the pin-on-disc test, in which a stainless steel pin was applied against the tool steel disc with a constant load. The laser-textured tool steel surface with a dimple density of 35% had a friction coefficient of 0.087, which was lower than that of the untextured surface by 12.6% when using a sliding speed of 15 cm/s. In addition to friction reduction, there was no substantial wear found on the laser-textured surface compared to the untextured sample. The findings of this study can be a processing guideline and benefit the treatment of tool and die surfaces for friction and wear reduction in metal-forming and related processes.

Keywords: laser; dimple; tribology; friction; surface; SKH51

Citation: Phun, C.; Daodon, W.; Septham, K.; Kumkhuntod, P.; Zhu, H.; Saetang, V. Laser-Fabricated Micro-Dimples for Improving Frictional Property of SKH51 Tool Steel Surfaces. *Lubricants* **2023**, *11*, 456. https://doi.org/10.3390/lubricants11110456

Received: 22 September 2023
Revised: 12 October 2023
Accepted: 21 October 2023
Published: 24 October 2023

1. Introduction

The friction and wear of metal-forming tools are significant factors affecting the surface quality of the formed parts and tool life in metal-forming processes [1]. Coating the tool surface with low-friction coatings and/or elevating lubrication effectiveness are vital in reducing friction and wear in the processes [2–4]. The roughness of tool surfaces is also a key parameter that influences the kinetic friction force between the surfaces in contact. Sigvant et al. [5] noted that the decrease in tool surface roughness substantially decreases the coefficient of friction. In addition to these approaches, micro-scale surface texturing has recently become a method for enhancing lubrication and reducing friction in tribological pairs [6–8]. The surface texturing can be achieved by using electrochemical machining [9,10], electrical-discharge micromachining [11], and laser-texturing processes [12–15]. Among these techniques, laser is the most preferred as it offers a high material removal rate, high texturing resolution, and fully automated processing with good reliability. Two texturing structures, i.e., micro-groove and micro-dimple, are normally fabricated on the tool surface to help reduce friction. The different sizes and geometries of the texturing structures have been found to cause different tribological properties [16]. Therefore, this area of interest is still open and requires in-depth investigations to further reveal the underlying mechanisms and influences of the texturing parameters on friction and wear improvement.

Micro- or submicro-features are created on the tool surface through the vaporization mechanism induced by the laser surface texturing process. The influence of laser-textured surfaces and speed-load parameters on the transition from a boundary to a hydrodynamic lubrication regime was studied by Kovalchenko et al. [17]. Laser surface texturing extends the range of speed-load parameters for hydrodynamic lubrication and lowers the friction coefficient of surfaces in contact. At greater speeds, higher loads, and with higher viscosity oil, the advantages of laser-textured surfaces are more noticeable. Thrust bearings, journal bearings, and mechanical seals are examples of friction pairs where surface texturing can substantially improve their tribological characteristics [18–20]. The textured surface containing micro-dimples has the ability to generate micro-hydrodynamic pressure, retain and distribute lubricant [21], increase lubricant thickness [22], reduce contact area, and capture wear debris [23]. This directly extends the tool's service life in addition to the friction force reduction.

Abe et al. [24] employed TiCN-based cermet dies with micro-dimples to enhance wear resistance during stainless steel cup ironing. The dimples enable lubricant flow and reduce friction through liquid transfer at the die–cup interface. The influence of surface texturing on friction reduction in a silicon nitride ceramic tool was clarified by Wakuda et al. [25]. Their results show that micro-dimple size and dimple density have a significant impact on tribological behaviors. However, dimple shape has less influence on the change in friction coefficient. The study suggests a dimple diameter of around 100 μm with a dimple density ranging from 5 to 20% to provide a significant reduction in friction. Daodon and Saetang [12] applied a laser texturing process to reduce the surface friction of a cold-work steel tool sliding against an advanced high-strength steel. The laser-textured surface with a dimple density of 5.6% has a lower coefficient of friction compared to the untextured surface. In addition, the wettability of the textured surface still remains unaffected, so the lubricant can wet the surface and effectively facilitate friction reduction. Schneider et al. [26] studied the influence of dimple aspect ratio, dimple density, and arrangements on friction in mixed lubrication. The lowest friction is obtained when using 10% dimple density and 0.1 aspect ratio associated with a hexagonal arrangement of dimples on the work surface. Abe et al. [27] demonstrated that laser-textured dies can enhance seizure resistance in aluminum alloy sheet and stainless steel cup ironing. The micro-dimples arranged in a grid array increase the load-carrying capacity of the lubricant and improve the ironing limits. According to many past studies, the dimple diameter and aspect ratio of about 100 μm [12,25,28–32] and 0.1 [26,30,32], respectively, are usually suggested to be fabricated on the tool surface for improving its tribological properties. However, there have been few discussions on the effect of dimple density on friction reduction. In addition, most of the past studies examined the tribological performance of textured surfaces with a dimple density less than 20%. An increase in dimple density greater than this value can be of a high potential for further decreasing the coefficient of friction. This issue needs to be elaborated to provide a better understanding of the relationship between the dimple density and friction improvement.

This paper aims to investigate the effect of dimple density on the friction coefficient. Due to the high pulse energy, low photon cost, and localized ablation with a small heat-affected zone offered by nanosecond pulse lasers, this type of laser was employed in this study to texture micro-dimples with different dimple densities on SKH51 tool steel surface. This tool steel is a common grade for tools and dies used in metal-forming processes. AISI304 stainless steel representing a workpiece material in the forming processes pressed against the textured tool steel surface in a pin-on-disc test for determining the coefficient of friction of the two surfaces in contact. The findings of this study will further unveil the tribological performance of micro-dimpled textured surfaces and provide a guideline for friction reduction in metal forming as well as cutting processes.

2. Materials and Methods

2.1. Micro-Dimple Fabrication

A nanosecond pulse laser (IPG YLP-1-100, IPG Laser GmbH & Co. KG, Burbach, Germany) emitting a wavelength of 1064 nm and a constant pulse duration of 100 ns was applied in this study to create micro-dimples on the surface of SKH51 tool steel. The hardness of the metal was 64 HRC. To achieve this, the steel was heat treated in a vacuum heat treatment furnace, first preheated to temperatures of 850 and 1050 °C for holding times of 60 and 30 min, respectively, heated to a hardening temperature of 1220 °C for a short period of 15 min and then quenched in nitrogen gas at a pressure of 280 kPa. The workpieces were subsequently tempered at 560 °C for 90 min and cooled to room temperature under a nitrogen atmosphere at a pressure of 80 kPa. The thermophysical properties and chemical composition of the tool steel are listed in Table 1. The distribution of laser beam profile was Gaussian with the beam quality factor (M^2) of 1.65. The collimated laser beam was focused by an f-theta lens having a focal length of 100 mm. The laser beam irradiated on the top surface of workpiece at the off-focused position where the beam diameter at $1/e^2$ (d_b) of 100 μm was attained as shown in Figure 1a. This beam diameter was expected to produce a dimple diameter of about 100 μm, which is recommended by previous studies [12,25,28–32] as a proper size of dimple for trapping lubricant and enabling the hydraulic pressure at the contact surface to reduce the friction. The laser pulse repetition rate (f) and irradiation time (t) for each dimple ablation were kept constant at 100 kHz and 0.1 s, respectively. According to past literature, a suitable aspect ratio of micro-dimples is about 0.1 [26,30,32], so the expected depth of the dimple having the diameter of 100 μm is 10 μm. Thereby, the determination of average laser power (P) for achieving this dimple depth was performed in this study through a set of experiments. After a number of trial tests, the average laser power ranging from 10 to 50 W was able to provide a micro-dimple on the tool steel surface without the excessive melting of workpiece and substantial formation of recast structures around the laser-ablated dimple. The laser parameters used in the fabrication of micro-dimples are summarized in Table 2.

After the laser ablation, the workpiece surface was ground by emery papers with grit numbers of 500, 800, 1000, 1500, 2000, and 2500 to remove bulges of recast structures depositing at and around the dimple edge. The diameter and depth of dimples obtained under the different laser power were measured by a 3D laser confocal microscope (Olympus OLS5000, Olympus Corp., Tokyo, Japan). The laser power that caused the dimple aspect ratio of about 0.1 was then selected and applied in the surface texturing of tool steel for friction test. The micro-dimples were arranged in the hexagonal pattern as shown in Figure 1b. This pattern has been proven by past studies to help the reduction of surface friction [26,33]. The dimples were evenly spaced according to the hexagonal pattern. The closer the distance between the adjacent dimples is employed, the higher the dimple density is obtained. The dimple density (ρ_d) is calculated by using:

$$\rho_d = \frac{\pi}{2\sqrt{3}}\left(\frac{d}{p_d}\right)^2 \quad (1)$$

$$L = \frac{p_d}{2\tan 30^\circ} \quad (2)$$

where d, p_d, and L are the diameter of dimple, horizontal and vertical distances between the adjacent dimples, respectively. In this study, four levels of dimple density, i.e., 5%, 15%, 25%, and 35%, were made on the tool steel surface whose coefficient of friction was subsequently quantified by a tribometer. The micrograph of laser-textured surfaces with the different levels of dimple density taken by an optical microscope is presented in Figure 1c–f.

Table 1. Thermophysical properties and chemical composition of SKH51 tool steel.

Thermophysical Properties						Value		
Density, ρ (kg/m^3)						8138		
Specific heat capacity, c_p (kJ/kg °C)						0.46		
Melting temperature, T_m (°C)						1430		
Vaporization temperature, T_v (°C)						2861		
Thermal conductivity, k (W/m °C)						24		
Chemical composition (wt%)								
C	Si	Mn	P	S	Cr	Mo	W	V
0.80–0.88	0.45	0.40	0.030	0.030	3.80–4.50	4.70–5.20	5.90–6.70	1.70–2.10

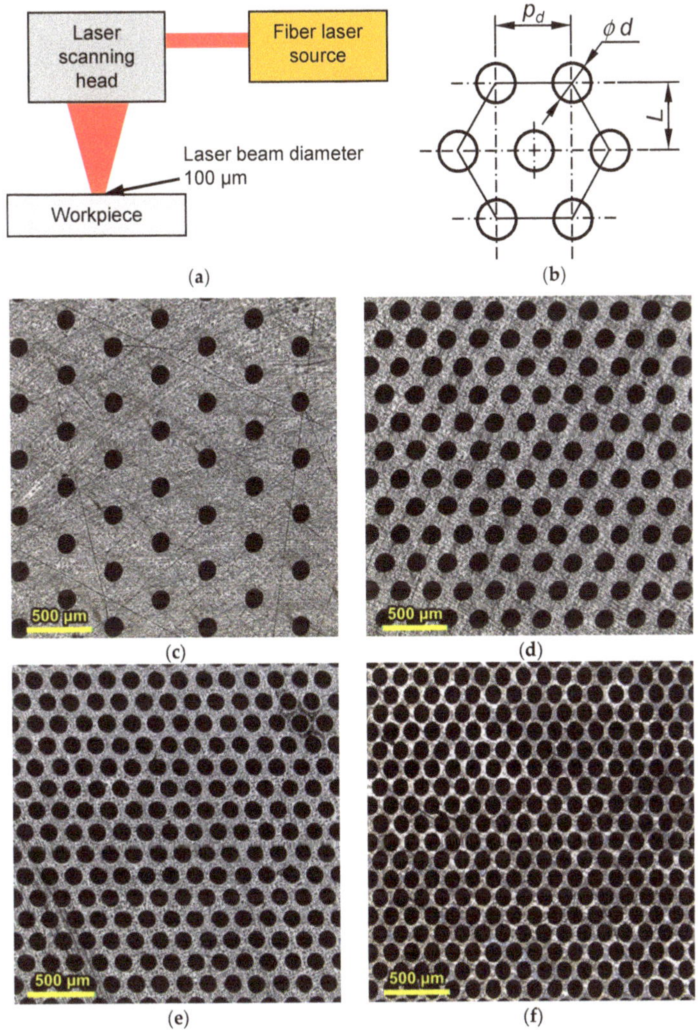

Figure 1. Schematic of (**a**) experimental setup and (**b**) the hexagonal array of micro-dimples; (**c**) 5% dimple density; (**d**) 15% dimple density; (**e**) 25% dimple density; (**f**) 35% dimple density.

Table 2. Laser parameters used for fabricating micro-dimples.

Parameter	Value
Laser wavelength (nm)	1064
Average laser power, P (W)	10, 15, 20, 25, 30, 35, 40, 45 and 50
Laser pulse repetition rate, f (kHz)	100
Irradiation time, t (s)	0.1
Laser beam diameter at $1/e^2$, d_b (μm)	100

2.2. Measurement of Friction

The pin-on-disc test was carried out in this study to determine the friction coefficient of the untextured and laser-textured SKH51 tool steel surfaces by using a tribometer. The tool steel was assigned to be the disc having the diameter and thickness of 41 mm and 4 mm, respectively. AISI304 stainless steel with the hardness of 15 HRC was applied as the flat-ended pin, on which the diameter of the flatten area of 3 mm and the average surface roughness (R_a) of 0.04 μm were prepared. The surface of tool steel disc possessing the dimple density of 0% (untextured surface), 5%, 15%, 25%, and 35% were tested under a normal load of 10 N against the 304 stainless steel pin as shown in Figure 2. Before the test, the laser-textured surfaces were ground by emery papers with grit numbers of 500, 800, 1000, 1500, 2000, and 2500 to remove the recast depositing around the dimples. The R_a of the final surface after the 2500-grit polishing was about 0.04 μm. The samples were cleaned by acetone for 10 min in an ultrasonic cleaner and then dried. The stainless steel pin was fixed in a stationary position, while the disc was rotated at a constant speed. Two sliding speeds, i.e., 5 and 15 cm/s, were conducted in the test of each sample. The sliding speed was changed by adjusting the sliding radius. By selecting sliding radii of 5 mm and 15 mm, consistent sliding velocities of 5 cm/s and 15 cm/s were achieved, respectively. This setting allowed the test to maintain a stable rotational speed, thereby ensuring that only the linear speed at the contact point varied. This approach was chosen to minimize potential variations in temperature that could arise from changing the rotational frequency, and the sliding distance was 500 m. The pin-on-disc test was performed at room temperature of 25 °C and lubricated by Castrol Iloform TDN81 (BP–Castrol (Thailand) Ltd., Bangkok, Thailand), whose kinematic viscosity at 40 °C is 157–180 cSt.

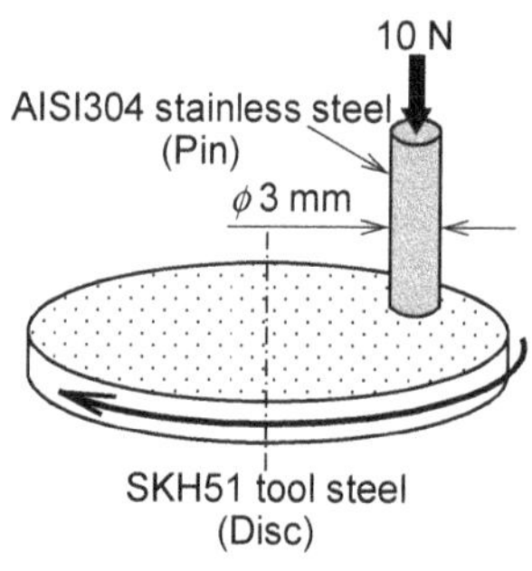

Figure 2. Rotating SKH51 tool steel disc and a 304 stainless steel pin employed in the pin-on-disc test.

3. Results and Discussion

3.1. Effects of Laser Power on the Diameter and Depth of Micro-Dimples

In this study, a micro-dimple was created by irradiating a laser beam for 0.1 s duration on the surface of SKH51 tool steel. The nine levels of average laser power ranging from 10 to 50 W were employed in the experiments, and the dimple dimensions were measured accordingly. Figure 3 presents the 3D morphology of micro-dimples fabricated by using the different laser powers, and the cross-sectional profile of the dimples is shown in Figure 4a. The increase in the laser power apparently increased the dimple size. The maximum depth of the dimples was located at its center where the laser intensity was highest. In addition to the enlarging of the dimples, the recast structures and heat-affected zone were found inside the dimple and around its edge as shown in Figure 4b. Furthermore, although this study did not specifically evaluate the composition and microstructural changes, our previous research noted that increased laser power caused a reduction in hardness near the edge of the pocket. Such a variation in hardness may be indicative of microstructural changes resulting from retempering [12]. The deposition of recast at the edge of the dimples was a result of molten flow induced by the recoil pressure and Marangoni effect. The molten metal was forced to flow and/or eject out of the laser-irradiated region and then solidified at the dimple edge. This thereby led to the protrusion of dimple rim due to the recast deposition. Although the most recast was able to be removed by the surface polishing prior to the dimple measurements, some still remained around the dimple edge. The total removal of recast protruding from the metal surface is possible by increasing the polishing time. However, some dimples are subjected to over polishing which makes them too shallow or even disappeared. A trade-off between the recast removal and maintaining the dimple depth as well as the existence of the dimple array was made in the surface polishing step, in which a small number of recasts with a protrusion height of a few microns were allowed to remain on the metal surface. The amount of recasts was found to increase with the laser power applied in the ablation. This also affected the surface polishing time and effort involved in the recast removal.

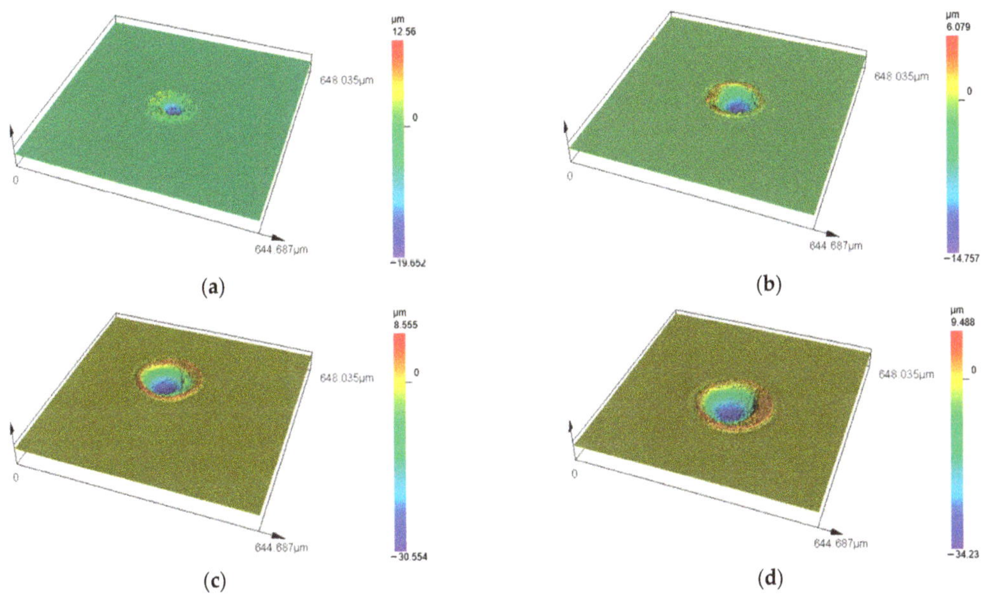

Figure 3. Morphology of micro-dimples produced by using the average laser power of: (**a**) 15 W; (**b**) 25 W; (**c**) 35 W; (**d**) 45 W.

The diameter, depth, and aspect ratio of micro-dimples caused by the different laser powers are presented in Figure 5a–c. The dimple diameter was about 80 µm when using the laser power of 10 W, and it linearly increased with the laser power. The relationship between laser power and dimple depth was also linear until the laser power reached 40 W. The depth sharply increased with a large deviation when the laser powers of 45 and 50 W were applied. Using higher laser power typically promotes a thicker molten layer where its high flow velocity and high recoil pressure substantially push a large portion of molten material out of the dimple. This thereby results in a deeper dimple with more recasts depositing at its edge. Since the behavior of the molten layer becomes more dynamic under higher laser power, a high deviation of dimple depth can thus be expected. According to the results, the transition from the static to the dynamic regimes of laser ablation occurred at the average laser power of 40 W. In the static regime, the dimple depth can be predicted by a heat conduction model associated with the conservation of energy. The temperature field induced by the irradiation of a Gaussian laser beam with the average power P and spot radius ω_b is expressed as:

$$T(r,z) = \frac{(1-R)P}{2\pi kz}\exp\left(-\frac{2r^2}{\omega_b^2}\right) + T_o \quad (3)$$

where R, k, r, z, and T_o are the reflectivity (0.7) and thermal conductivity of the metal, the radius and depth of the temperature field, and room temperature, respectively. When considering the vaporization temperature of the metal (T_v) to be the ablation point, the dimple depth (z) is predicted by using:

$$z(r) = \frac{(1-R)P}{2\pi k(T_v - T_o)}\exp\left(-\frac{2r^2}{\omega_b^2}\right) \quad (4)$$

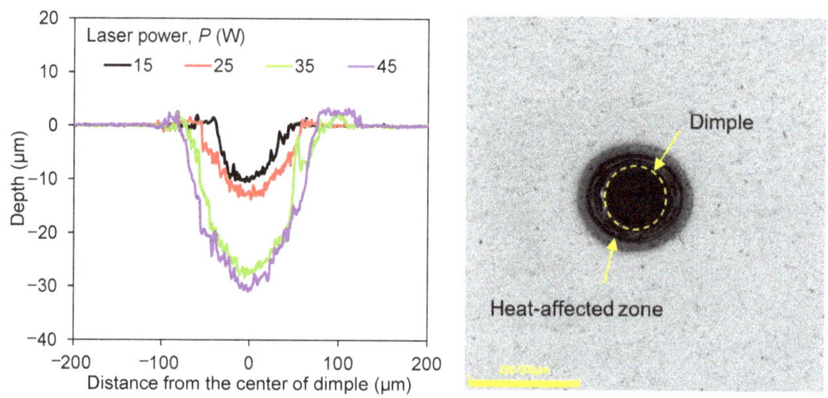

(**a**) (**b**)

Figure 4. (**a**) Profile of micro-dimples caused by different laser powers; (**b**) Micrograph of micro-dimple fabricated by 25 W laser power.

The predicted dimple depth subjected to the vaporization point is also plotted in Figure 5b. The prediction provided a good agreement with the measured depth under the static regime, in which the applied laser power was less than 40 W. A comparison between the measured and predicted profile of micro-dimples caused by the laser power of 25 W is also shown in Figure 5d. Since the dimple formation is more induced by the outward flow of the molten layer due to the recoil pressure and thermocapillary effect in the dynamic regime of ablation ($P > 40$ W), the melting point of the metal (T_m), which is able to indicate the depth of the molten layer, can be applied in Equation (4) to approximate the dimple depth. The predicted depth regarding the melting point is also shown in Figure 5b. The prediction was found to be in line with the dimple depth caused by the 45 and 50 W laser powers.

According to the suggested dimple dimensions [25,26], the diameter and aspect ratio of micro-dimples should be about 100 μm and 0.1, respectively, to introduce the effective reduction of surface friction. Among the laser ablation conditions performed in this study, the average laser power of 25 W was selected since it was able to provide the desired dimple dimensions with fewer recast structures, good repeatability, and a small deviation of dimples across the textured area. This processing condition was further applied to texture the hexagonal array of micro-dimples on the SKH51 tool steel surface for the friction test using the pin-on-disc method.

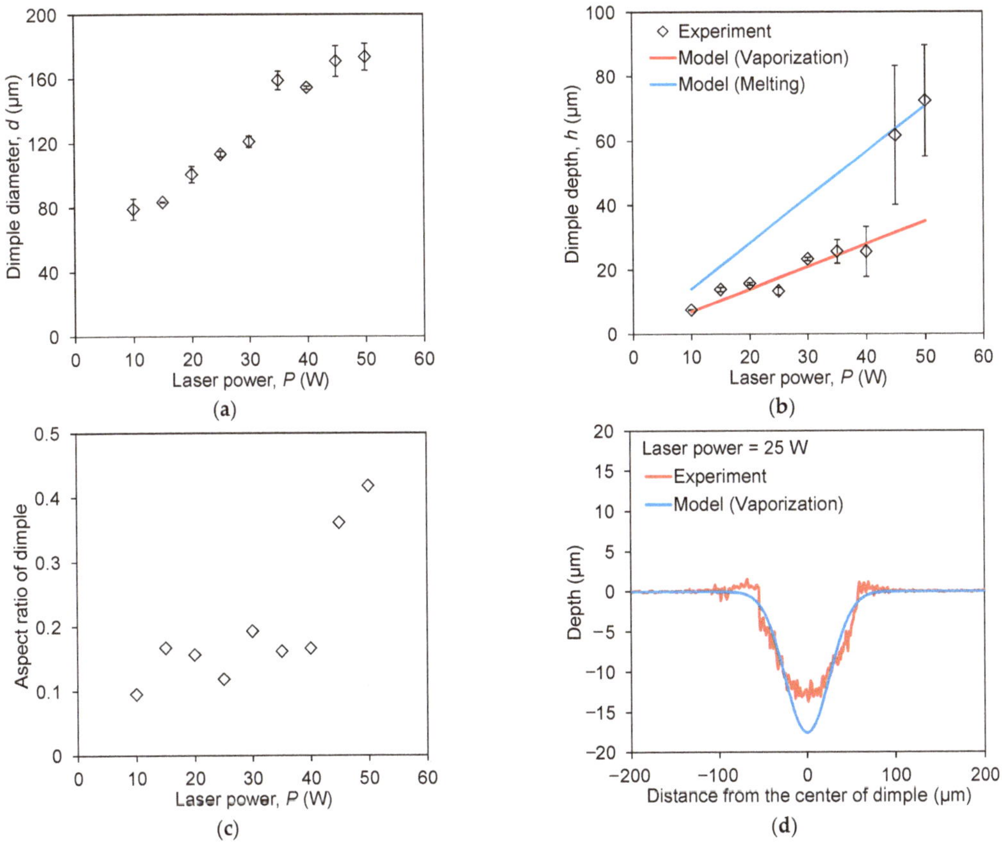

Figure 5. Effects of laser power on the (**a**) diameter, (**b**) depth, and (**c**) aspect ratio of micro-dimples; (**d**) measured and predicted profiles of micro-dimples when using 25 W laser power.

3.2. Effects of Micro-Dimple Density on the Friction and Wear of SKH51 Tool Steel Surfaces

Five levels of dimple density including the untextured surface and two levels of sliding speed were carried out in the pin-on-disc test. The plots of the friction coefficient reading from the tribometer against the sliding distance under the test speed of 5 and 15 cm/s are presented in Figure 6a,b. The friction coefficients of all surfaces exhibited a sharp increase at the initial stage, followed by a subsequent decline. This phenomenon is attributed to the initial roughness of the fresh friction pair that is high during the initial running-in period. The coefficient of friction obtained at the sliding speed of 5 cm/s showed more fluctuation than at the other speed. This is owing to the unstable friction force at the contact surface [34], where the adhesion between the two metals likely occurs and interrupts. This situation is due to the transition between boundary lubrication and mixed lubrication regimes. At lower speeds, the contact time between the surfaces increases, causing more intimate contact and adhesive interactions between the tool steel and stainless steel surfaces. This thereby results in stick–slip behavior and increased friction fluctuations [34,35]. In addition, the coefficient of friction taken at 5 cm/s sliding speed was higher than that at 15 cm/s by 8.3% on average regardless of the percentage of dimple density used. The high surface friction at slow sliding speed is caused by a diminished degree of separation facilitated by the lubricant. The lubrication is typically under a mixed condition consisting of both hydrodynamic and boundary lubrication regimes [36]. The relative motion between the two surfaces is minimal when using slow sliding speed, and the balance between the two lubrication regimes is altered. The lubrication tends to be dominated by the boundary lubrication regime, where a direct contact between asperities on the surfaces becomes more prominent and the lubricant is unable to separate the surfaces effectively. The micro-dimples, which are designed to retain lubricant and reduce direct metal-to-metal contact, are therefore ineffective for the surfaces in contact under the boundary lubrication condition. This finding also corresponds to the study of Vilhena et al. [37], noting that the friction tends to be high and to fluctuate when a low sliding speed is applied. On the other hand, as the sliding speed increases, the shear forces acting on the lubricant become stronger. This induces a better distribution of lubricant within the micro-dimples, thus promoting the mixed lubrication condition. The hydrodynamic effect involved in the mixed lubrication contributes to the separation of the surfaces in contact to a greater extent. This substantially leads to friction reduction [17,38,39]. Mizuno and Okamoto [40] explored the influence of lubricant properties and sliding velocity on lubrication. Employing a surface with micro-dimples to capture lubricant proves advantageous for friction reduction during a metal-forming process. The boundary film thickness can also be enhanced by maximizing the lubricant entrapment within the interface in addition to the use of high-viscosity lubricants.

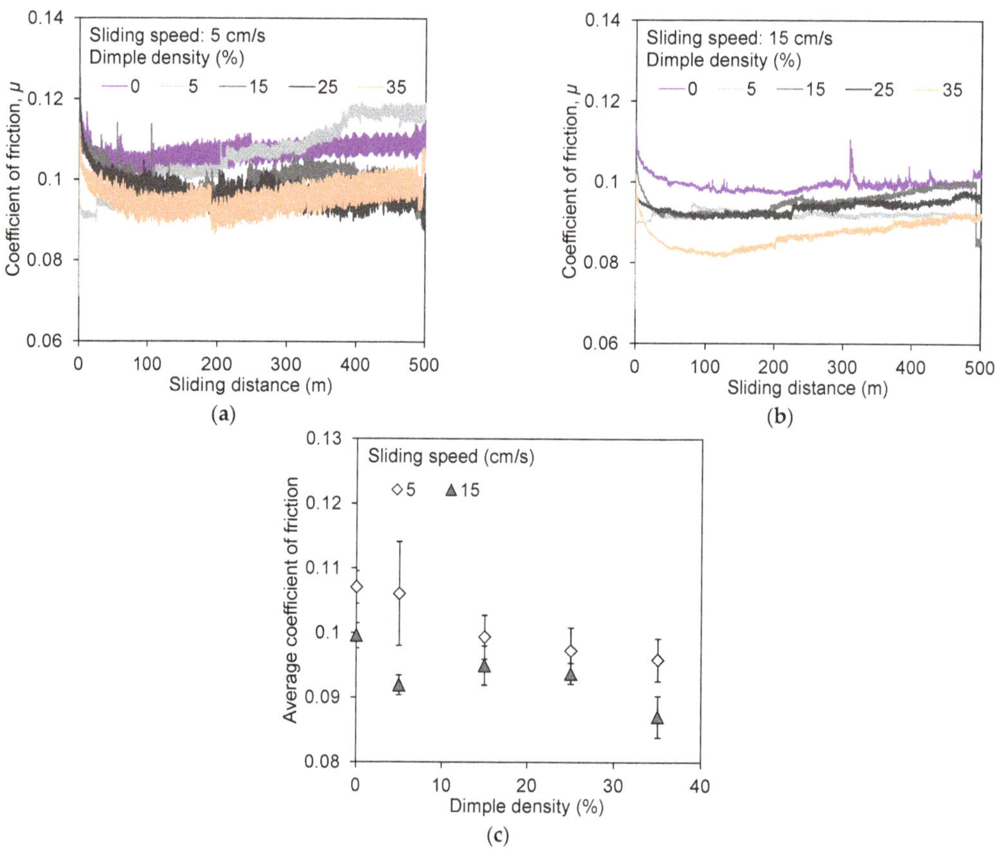

Figure 6. Friction coefficient of the sample surfaces having the different dimple densities and tested at the sliding speed of (**a**) 5 and (**b**) 15 cm/s, and (**c**) the average coefficient of friction of each test.

The average coefficient of friction of each sample is shown in Figure 6c, indicating a decreasing trend with the increased dimple density for both tested sliding speeds. The friction coefficient of the untextured surface (0% dimple density) subjected to the sliding speed of 5 and 15 cm/s was 0.107 and 0.099. The lowest coefficient of friction obtained at both sliding speeds was 0.096 and 0.087 when the 35% dimple density was used. This is equivalent to the reduction of 10.5% and 12.6%, respectively. With the aid of regression analysis, the decreasing rate of the friction coefficient against the dimple density at the sliding speed of 5 and 15 cm/s was about 0.0003 and 0.0002. This implies that the contribution of the dimple density to friction reduction is about the same rate for both sliding speeds tested in this study.

Comparing the friction coefficient of the laser-textured surface to that of the untextured surface, it was observed that the friction values of all textured surfaces, except for those with a dimple density of 5% tested at a sliding speed of 5 cm/s, were significantly lower than those of the untextured surface. The decreased friction coefficient induced by the increase in dimple density is caused by micro-hydrodynamic lubrication [41,42]. This lubricating mechanism is apparent when the lubricant is entrapped, pressurized, and extracted from the micro-dimples. The presence of an asymmetric hydrodynamic pressure distribution or wedging effect above the dimples provides load-carrying forces [43], which facilitate the interface separation [37,43]. The wedging effect occurs when the pressure distribution in the lubricant film generates positive hydrodynamic pressure in the convergent part of the micro-dimples and negative hydrodynamic pressure in the divergent part as shown in Figure 7. This leads to a situation in which the magnitude of the positive pressure is greater than that of the negative pressure and thus results in a net positive pressure within the micro-dimple unit. The positive pressure offers an extra carrying force, which helps increase the load-carrying capacity of the friction pair. The numerical simulation performed by Abe et al. [27] also reveals the change in pressure distribution of lubricant flow on the dimpled surface. A higher dimple density or shorter flat portion length between the adjacent dimples likely leads to an adequate supply of lubricants into the interface, thus enhancing the micro-hydrodynamic lubrication. Our finding also corresponds to the studies of Shimizu et al. [22], Liu et al. [36], and Brizmer et al. [44], noting that a shorter length of the flat portion or a higher dimple density promotes more lubricant transfer between the adjacent dimples. This, in turn, increases the load-carrying capacity and reduces the coefficient of friction accordingly.

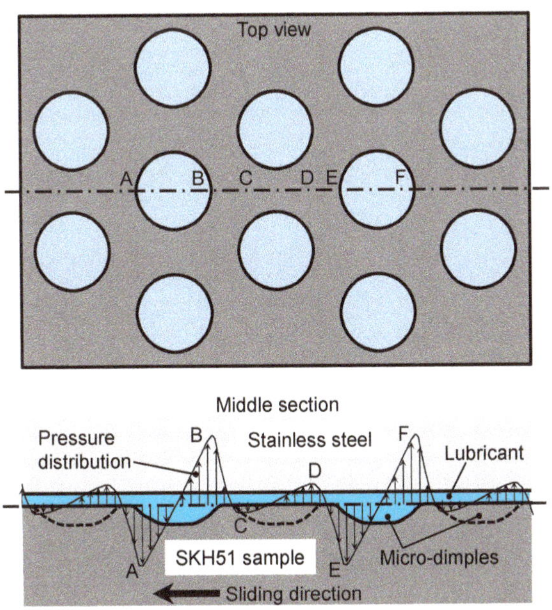

Figure 7. Schematic diagram of pressure distribution along the middle plane of micro-dimples.

Following the pin-on-disc test using the tribometer at the sliding speed of 5 and 15 cm/s, the wear scars on the sliding track were macroscopically observed for the textured and untextured surfaces as illustrated in Figure 8. The observed difference in surface color in Figure 8a compared to the other images could be attributed to factors such as oxidation. It should be noted that any surface oxidation in Figure 8a might affect the reported frictional properties when compared with the other pictures. Although the AISI304 stainless steel pin is softer than the SKH51 tool steel disc, its surface is naturally covered by a passive layer primarily consisting of chromium oxide (Cr_2O_3). This passive layer is harder than the base metal. When some elements of stainless steel are detached from its surface during the sliding test, these hard chromium oxide particles behave like abrasive agents (third bodies) between the contact pairs, leading to abrasive wear on the sliding track. As the surfaces continue to slide against each other, the hard chromium oxide particles cause material removal and wear on both stainless steel and tool steel surfaces. However, no substantial seizures were found on any of the surfaces, and only thin scratches were observed. At the sliding speed of 5 cm/s, where the boundary lubrication tended to dominate, the untextured and textured surfaces with a dimple density of 5% struggled to provide the effective lubrication for separating the surfaces in contact. In the boundary lubrication regime, the lubricant film is not well developed and a direct metal-to-metal contact is likely occurred, thus leading to the formation of shallow scratches as shown in Figure 8a,b. This is also reflected in the friction coefficient-sliding distance curves, which show high friction coefficients and increasing trends for the tested surfaces (Figure 6a). At the sliding speed of 15 cm/s, the textured surface having a dimple density of 15% exhibited thin scratches as shown in Figure 8c. Although using a higher sliding speed likely affects the lubrication regime, the dimple density of 15% may not be sufficient to provide a decent lubrication. This could lead to a situation that the lubricant film is unable to effectively prevent the couple surfaces in contact and the thin scratches are thereby obtained as a result. The laser-textured surfaces with the dimple densities of 25% and 35% were of high potential for promoting the ingress of the lubricant into the interface of the mating components. This significantly improved the lubrication performance and only tiny traces of wear are apparent on the sliding track as shown Figure 8d,e.

A comparison between the morphology of laser-textured surfaces before and after the pin-on-disc test is shown in Figure 9, indicating that the shape of the micro-dimples remained unchanged even after testing for a distance of 500 m. This is apparently a positive outcome, as the micro-dimples are able to provide the lubrication-enhancing function over an extended period of use. Higher dimple densities can hold a greater volume of lubricant within the dimples. As a result, a robust and continuous lubricating film is formed, preventing direct contact and minimizing wear.

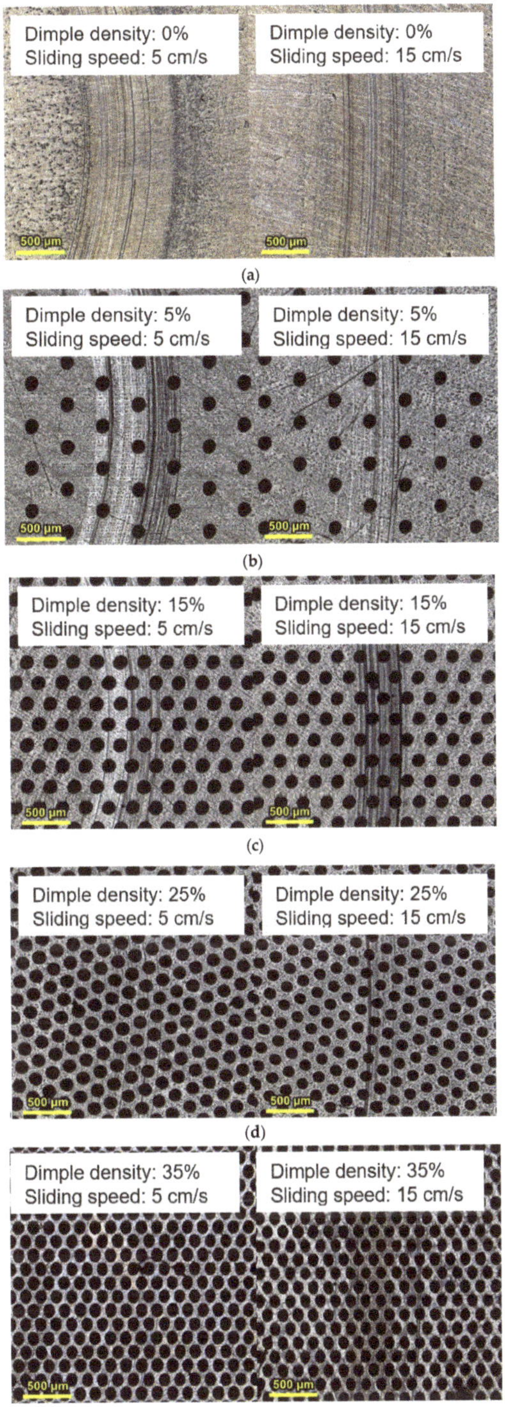

Figure 8. (**a**) Untextured (0% dimple density) and (**b–e**) laser-textured surface of samples after the pin-on-disc test at the sliding speed of 5 and 15 cm/s.

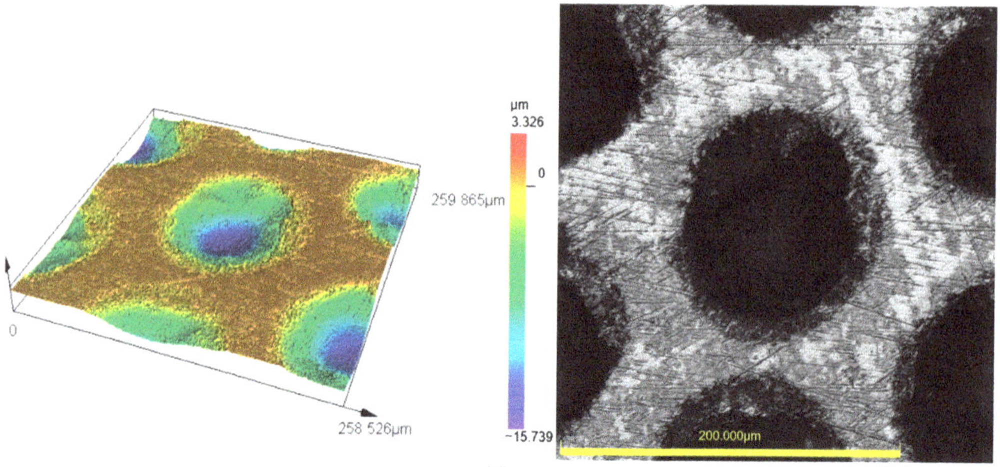

(**a**)

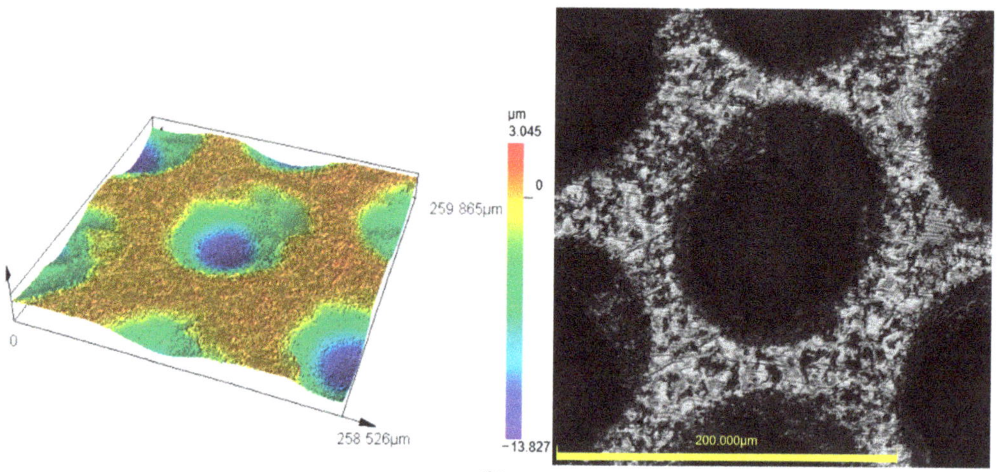

(**b**)

Figure 9. Morphology of sample surfaces with the dimple density of 35%: (**a**) before and (**b**) after the pin-on-disc test at the sliding speed of 5 cm/s.

SEM micrograph of the worn track on the untextured and textured surfaces is shown in Figure 10. The thin scratches are noticeable on the untextured surface when using the sliding speed of 15 cm/s as depicted in Figure 10a. This is because of the breakdown of the thin lubricant film that was unable to support the load, so the direct contact between the friction pairs occurred accordingly. Regarding the chemical composition listed in Table 3, a greater amount of Ni is found on the untextured surface compared to the other after the pin-on-disc test. Since Ni is not present in the composition of SKH51 tool steel as noted in Table 1, it can be implied that a number of Ni were transferred from the AISI304 stainless steel counterpart to the tool steel surface during the sliding test. This thereby indicates a combination of adhesion and abrasion during the sliding contact, and the adhesion of Ni occurred more on the untextured surface. In contrast, the textured surface exhibits only slight wear with a smooth appearance in the plateau area as shown in Figure 10b,c. The dimples remained intact after the sliding test. This is attributed to the lubricating oil entrapped in the dimples that is continuously supplied to the tribo-contacts during sliding. Therefore, the presence of an oil film on the textured surface facilitates wear reduction to a great extent. Additionally, the textured surface can capture wear debris in the micro-dimples, preventing surface scratching due to the metallic debris. The EDS mapping of the textured surface before the sliding test shown in Figure 10d reveals a high oxygen content in the micro-dimples, where surface oxidation is typically induced by laser. The oxygen content is higher after the pin-on-disc test (Figure 10e), where the debris of metal oxides is trapped. This can limit the amount of wear particles as well as oxide debris generated under the mixed lubrication conditions to circulate in the tribo-contact. By trapping wear debris in the micro-dimples, surface scratching caused by the metallic debris is thus minimized. In addition to reducing surface friction through micro-hydrodynamic lubrication, the textured surface containing the micro-dimples has the ability to capture wear debris and limit surface scratching.

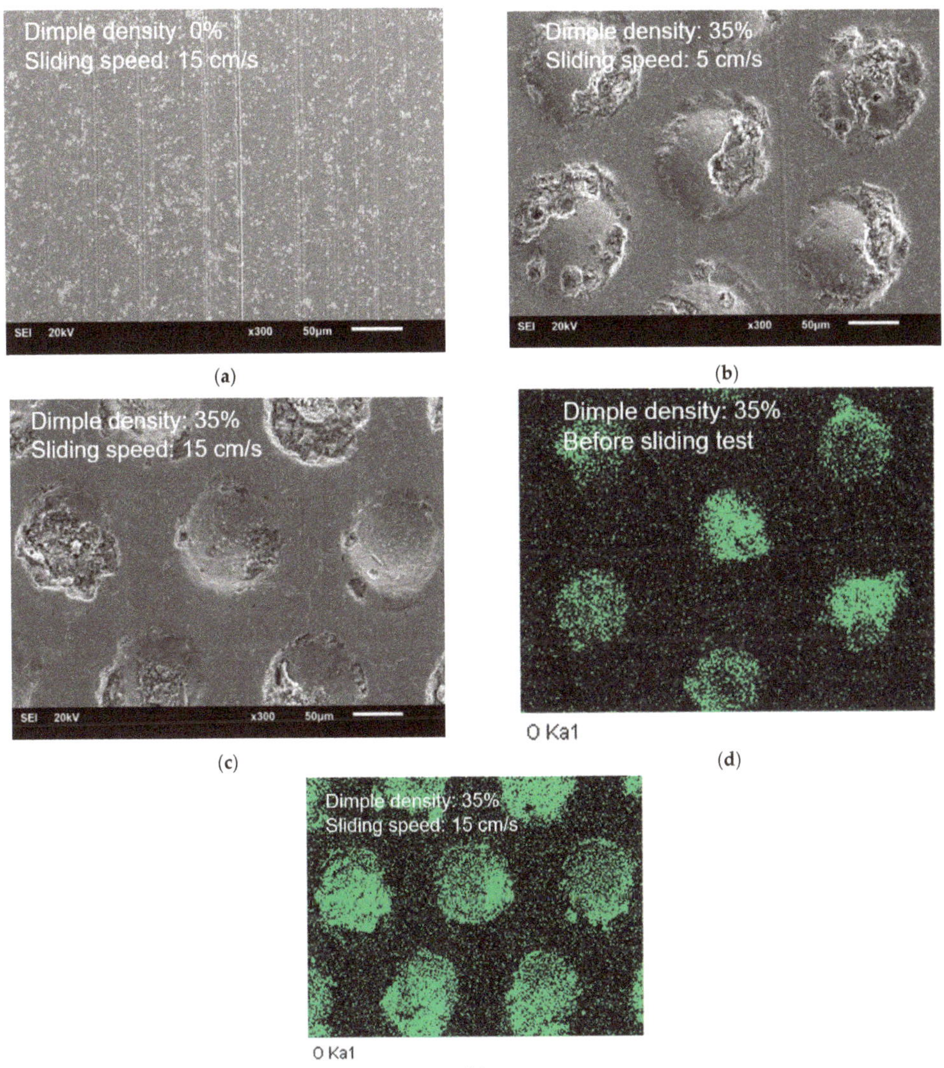

Figure 10. SEM image of (**a**) the untextured and (**b**,**c**) laser-textured surfaces after the pin-on-disc test at the sliding speed of (**b**) 5 and (**a**,**c**) 15 cm/s; (**d**) EDS mapping for oxygen on the laser-textured surface before and (**e**) after sliding test.

Table 3. Chemical composition taken from the EDS mapping of untextured and laser-textured surfaces.

Element (wt%)	O	V	Cr	Fe	Ni	Mo	W
Untextured surface	0.80	2.17	4.31	78.73	0.15	5.44	8.41
Textured surface	5.66	1.35	2.80	76.27	0.02	6.34	7.55

4. Conclusions

A nanosecond pulse laser was employed in this study to fabricate micro-dimples on the surface of SKH51 tool steel in an attempt to reduce its friction coefficient when sliding against AISI304 stainless steel. The effects of average laser power on the dimple sizes were examined and analyzed together with the laser ablation model. The untextured and micro-dimple textured surfaces having the different dimple densities underwent the pin-on-disc test to determine their coefficient of friction. The major findings and implications of this study can be summarized as follows:

(1) The diameter and depth of the micro-dimples were found to increase with the laser power applied in the texturing process. However, a clean dimple with less recast deposition was obtained when using the laser power less than 40 W. The dimple diameter of about 100 μm with the aspect ratio of 0.1 and less formation of recast structures was achievable by using the laser power of 25 W.

(2) Regarding the laser ablation model, the predicted dimple profile had a good agreement with the measured profile. The prediction was, however, accurate when the flow of the molten layer was minimal. The proposed model can be of help in defining the laser texturing conditions to create the desired dimple dimensions.

(3) The friction coefficient of the surfaces in contact was reduced from 0.099 to 0.087 and from 0.107 to 0.096 at the sliding speed of 5 and 15 cm/s, respectively, when the tool steel surface was textured with the 35% dimple density. The micro-dimples can induce the positive pressure on the surfaces in contact, where the load-carrying force assists the separation of the surfaces through the hydrodynamic effect and reduces the friction accordingly.

(4) In addition to friction reduction, there was no substantial wear found on the micro-dimple textured surfaces. The micro-dimples are able to continuously supply lubricant into the contact interface and also trap wear debris to prevent surface scratching during sliding. According to the findings, it is apparent that the texturing of tool and die surfaces with high density of micro-dimples can reduce friction and wear on the surfaces. This leads to the prolongation of the service life of tools and dies employed in metal forming and other related manufacturing and mechanical applications, e.g., cutting tools used in machining processes, devices in material handling, and transmission systems.

Author Contributions: Conceptualization, W.D. and V.S.; methodology, C.P., W.D. and V.S.; validation, C.P., W.D., K.S., P.K., H.Z. and V.S.; formal analysis, C.P., W.D., K.S., H.Z. and V.S.; investigation, C.P., W.D., P.K. and V.S.; writing—original draft preparation, C.P.; writing—review and editing, W.D., K.S., H.Z. and V.S.; visualization, C.P., W.D. and V.S.; supervision, W.D., K.S. and V.S.; project administration, V.S.; funding acquisition, H.Z. and V.S. All authors have read and agreed to the published version of the manuscript.

Funding: The authors received support from the Petchra Pra Jom Klao Master's Degree Research Scholarship from King Mongkut's University of Technology Thonburi; Thailand Science Research and Innovation (TSRI) under Fundamental Fund 2023; Research Grant for New Scholar (Grant No. NKR2564INC005) from Rajamangala University of Technology Isan; and National Natural Science Foundation of China (No. 52275431).

Data Availability Statement: Not applicable.

Acknowledgments: The authors would like to express their appreciation to King Mongkut's University of Technology Thonburi, Rajamangala University of Technology Isan, and Jiangsu University for supporting this research.

Conflicts of Interest: The authors declare that they have no known competing financial interests or personal relationships that could have appeared to influence the work reported in this paper.

References

1. Nielsen, C.V.; Bay, N. Review of friction modeling in metal forming processes. *J. Mater. Process. Technol.* **2018**, *255*, 234–241. [CrossRef]
2. Bay, N.; Azushima, A.; Groche, P.; Ishibashi, I.; Merklein, M.; Morishita, M.; Nakamura, T.; Schmid, S.; Yoshida, M. Environmentally benign tribo-systems for metal forming. *CIRP Ann. Manuf. Technol.* **2010**, *59*, 760–780. [CrossRef]
3. Sulaiman, M.H.; Farahana, R.N.; Bienk, K.; Nielsen, C.V.; Bay, N. Effects of DLC/TiAlN-coated die on friction and wear in sheet-metal forming under dry and oil-lubricated conditions: Experimental and numerical studies. *Wear* **2019**, *438–439*, 203040. [CrossRef]
4. Prieske, M.; Hasselbruch, H.; Mehner, A.; Vollertsen, F. Friction and wear performance of different carbon coatings for use in dry aluminium forming processes. *Surf. Coat. Technol.* **2019**, *357*, 1048–1059. [CrossRef]
5. Sigvant, M.; Pilthammar, J.; Hol, J.; Wiebenga, J.H.; Chezan, T.; Carleer, B.; van den Boogaard, T. Friction in sheet metal forming: Influence of surface roughness and strain rate on sheet metal forming simulation results. *Procedia Manuf.* **2019**, *29*, 512–519. [CrossRef]
6. Gachot, C.; Rosenkranz, A.; Hsu, S.M.; Costa, H.L. A critical assessment of surface texturing for friction and wear improvement. *Wear* **2017**, *372–373*, 21–41. [CrossRef]
7. Dhage, S.; Jayal, A.D.; Sarkar, P. Effects of surface texture parameters of cutting tools on friction conditions at tool-chip interface during dry machining of AISI 1045 Steel. *Procedia Manuf.* **2019**, *33*, 794–801. [CrossRef]
8. Guo, D.; Wang, D.; Wu, S.; Qi, H.; Saetang, V. Investigation on turning of Inconel 718 using differently coated microtextured tools. *Proc. Inst. Mech. Eng. Part E J. Process Mech. Eng.* **2023**. [CrossRef]
9. Byun, J.W.; Shin, H.S.; Kwon, M.H.; Kim, B.H.; Chu, C.N. Surface texturing by Micro ECM for friction reduction. *Int. J. Precis. Eng. Manuf.* **2010**, *11*, 747–753. [CrossRef]
10. Zhu, H.; Jiang, Z.; Han, J.; Saetang, V.; Xu, K.; Liu, Y.; Zhang, Z.; Huang, S.; Zhou, J. Fabrication of oxide-free dimple structure on germanium via electrochemical jet machining enhanced by opposing laser irradiation. *J. Manuf. Process.* **2023**, *85*, 623–635. [CrossRef]
11. Singh, M.; Jain, V.K.; Ramkumar, J. Micro-texturing on flat and cylindrical surfaces using electric discharge micromachining. *J. Micromanufacturing* **2020**, *4*, 127–137. [CrossRef]
12. Daodon, W.; Saetang, V. Improvement of frictional property of AISI D2 tool steel surface against JIS SPFC 980Y advanced high-strength steel by using laser texturing process. *Lubricants* **2023**, *11*, 68. [CrossRef]
13. Kasem, H.; Stav, O.; Grützmacher, P.; Gachot, C. Effect of low depth surface texturing on friction reduction in lubricated sliding contact. *Lubricants* **2018**, *6*, 62. [CrossRef]
14. Etsion, I. State of the art in laser surface texturing. In *Advanced Tribology*; Luo, J., Meng, Y., Shao, T., Zhao, Q., Eds.; Springer: Berlin/Heidelberg, Germany, 2010; pp. 761–762.
15. Mao, B.; Siddaiah, A.; Liao, Y.; Menezes, P.L. Laser surface texturing and related techniques for enhancing tribological performance of Engineering Materials: A Review. *J. Manuf. Process.* **2020**, *53*, 153–173. [CrossRef]
16. Vishnoi, M.; Kumar, P.; Murtaza, Q. Surface texturing techniques to enhance tribological performance: A Review. *Surf. Interfaces* **2021**, *27*, 101463. [CrossRef]
17. Kovalchenko, A.; Ajayi, O.; Erdemir, A.; Fenske, G.; Etsion, I. The effect of laser texturing of steel surfaces and speed-load parameters on the transition of lubrication regime from boundary to hydrodynamic. *Tribol. Trans.* **2004**, *47*, 299–307. [CrossRef]
18. Etsion, I.; Halperin, G.; Brizmer, V.; Kligerman, Y. Experimental investigation of laser surface textured parallel thrust bearings. *Tribol. Lett.* **2004**, *17*, 295–300. [CrossRef]
19. Lu, X.; Khonsari, M.M. An experimental investigation of dimple effect on the Stribeck curve of journal bearings. *Tribol. Lett.* **2007**, *27*, 169–176. [CrossRef]
20. Etsion, I.; Burstein, L. A model for mechanical seals with regular microsurface structure. *Tribol. Trans.* **1996**, *39*, 677–683. [CrossRef]
21. Kitamura, K.; Makino, T.; Nawa, M.; Miyata, S. Tribological effects of punch with micro-dimples in blanking under high hydrostatic pressure. *CIRP Ann.* **2016**, *65*, 249–252. [CrossRef]
22. Shimizu, T.; Kobayashi, H.; Vorholt, J.; Yang, M. Lubrication analysis of micro-dimple textured die surface by direct observation of contact interface in sheet metal forming. *Metals* **2019**, *9*, 917. [CrossRef]
23. Geiger, M.; Popp, U.; Engel, U. Excimer laser micro texturing of cold forging tool surfaces-influence on tool life. *CIRP Ann. Manuf. Technol.* **2002**, *51*, 231–234. [CrossRef]
24. Abe, Y.; Mori, K.; Hatashita, F.; Shiba, T.; Daodon, W.; Osakada, K. Improvement of seizure resistance in ironing of stainless steel cup with cermet die having fine lubricant pockets. *J. Mater. Process. Technol.* **2016**, *234*, 195–207. [CrossRef]

25. Wakuda, M.; Yamauchi, Y.; Kanzaki, S.; Yasuda, Y. Effect of surface texturing on friction reduction between ceramic and steel materials under lubricated sliding contact. *Wear* **2003**, *254*, 356–363. [CrossRef]
26. Schneider, J.; Braun, D.; Greiner, C. Laser Textured Surfaces for Mixed Lubrication: Influence of Aspect Ratio, Textured Area and Dimple Arrangement. *Lubricants* **2017**, *5*, 32. [CrossRef]
27. Abe, Y.; Sugiura, M.; Ando, T.; Kumkhuntod, P.; Septham, K.; Daodon, W.; Mori, K.-i. Improvement of seizure resistance in ironing of aluminum alloy sheets and stainless steel cups by utilizing laser textured die having lubricant pockets. *Metals* **2023**, *13*, 803. [CrossRef]
28. Zhu, Y.; Chen, J.; Du, J.; Fan, Y.; Zheng, J. Tribological behavior of laser textured nodular cast iron surface. *Ind. Lubr. Tribol.* **2019**, *71*, 949–955. [CrossRef]
29. Paul Joshua, S.; Dinesh Babu, P. Effect of laser textured surface with different patterns on tribological characteristics of bearing material AISI 52100. *J. Cent. South Univ.* **2020**, *27*, 2210–2219. [CrossRef]
30. Qi, X.; Wang, H.; Dong, Y.; Fan, B.; Zhang, W.; Zhang, Y.; Ma, J.; Zhou, Y. Experimental analysis of the effects of laser surface texturing on tribological properties of PTFE/Kevlar fabric composite weave structures. *Tribol. Int.* **2019**, *135*, 104–111. [CrossRef]
31. Prem Ananth, M.; Ramesh, R. Tribological improvement of titanium alloy surfaces through texturing and TiAlN coating. *Surf. Eng.* **2014**, *30*, 758–762. [CrossRef]
32. Wang, S.; Yan, F.; Chen, A. Tribological effects of laser surface texturing and residual stress. *Ind. Lubr. Tribol.* **2018**, *70*, 126–132. [CrossRef]
33. Gyawali, G.; Joshi, B.; Tripathi, K.; Kim, S.-H.; Wohn Lee, S. Effect of microtexturing on tribological performance of Ni/Ni–sic composite coatings. *Surf. Eng.* **2015**, *31*, 701–707. [CrossRef]
34. He, Y.; Yang, J.; Wang, H.; Gu, Z.; Fu, Y. Micro-dimple and micro-bulge textures:influence of surface topography types on stick–slip behavior under starved lubrication. *Appl. Surf. Sci.* **2022**, *585*, 152501. [CrossRef]
35. Bay, N.; Olsson, D.D.; Andreasen, J.L. Lubricant test methods for sheet metal forming. *Tribol. Int.* **2008**, *41*, 844–853. [CrossRef]
36. Liu, W.; Ni, H.; Chen, H.; Wang, P. Numerical simulation and experimental investigation on tribological performance of micro-dimples textured surface under hydrodynamic lubrication. *Int. J. Mech. Sci.* **2019**, *163*, 105095. [CrossRef]
37. Vilhena, L.M.; Podgornik, B.; Vižintin, J.; Možina, J. Influence of texturing parameters and contact conditions on tribological behaviour of laser textured surfaces. *Meccanica* **2011**, *46*, 567–575. [CrossRef]
38. Zenebe Segu, D.; Hwang, P. Friction control by multi-shape textured surface under pin-on-disc test. *Tribol. Int.* **2015**, *91*, 111–117. [CrossRef]
39. Kovalchenko, A.; Ajayi, O.; Erdemir, A.; Fenske, G. Friction and wear behavior of laser textured surface under lubricated initial point contact. *Wear* **2011**, *271*, 1719–1725. [CrossRef]
40. Mizuno, T.; Okamoto, M. Effects of lubricant viscosity at pressure and sliding velocity on lubricating conditions in the compression-friction test on sheet metals. *Trans. ASME J. Lubr. Technol.* **1982**, *104*, 53–59. [CrossRef]
41. Vermeulen, M.; Scheers, J. Micro-hydrodynamic effects in EBT textured steel sheet. *Int. J. Mach. Tools Manuf.* **2001**, *41*, 1941–1951. [CrossRef]
42. Li, K.; Jing, D.; Hu, J.; Ding, X.; Yao, Z. Numerical investigation of the tribological performance of micro-dimple textured surfaces under hydrodynamic lubrication. *Beilstein J. Nanotechnol.* **2017**, *8*, 2324–2338. [CrossRef] [PubMed]
43. Tang, W.; Zhou, Y.; Zhu, H.; Yang, H. The effect of surface texturing on reducing the friction and wear of steel under lubricated sliding contact. *Appl. Surf. Sci.* **2013**, *273*, 199–204. [CrossRef]
44. Brizmer, V.; Kligerman, Y.; Etsion, I. A laser surface textured parallel thrust bearing. *Tribol. Trans.* **2003**, *46*, 397–403. [CrossRef]

lubricants

MDPI

Article

Anti-Wear Property of Laser Textured 42CrMo Steel Surface

Hua Zhang *, Xinfeng Pei and Xindong Jiang

School of Mechanical Engineering, Nantong University, Nantong 226019, China; peixinfeng221@163.com (X.P.); jiangxd2@sany.com.cn (X.J.)
* Correspondence: ntuzhh@ntu.edu.cn; Tel.: +86-0513-85012674

Abstract: In this work, laser processing technology was utilized to fabricate micro-textures on the surface of 42CrMo steel to improve its wear resistance under high load conditions and provide an effective method to solve the wear of tooth plates in oil drilling wellhead machinery. Firstly, the friction process of the textured components was conducted by finite element analysis. Additionally, various forms of textures were compared and measured by this method to optimize the shape and parameters of the patterns. Secondly, three types of texture shapes, such as micro-dimples, micro-grooves, and reticular grooves, were created on the surface of 42CrMo steel. Lastly, the tribological characteristics of the micro-textures were analyzed in the dry friction experiments. Compared with the untextured surface, the wear resistance of the textured 42CrMo steel has been improved, and the anti-wear property of the micro-dimples was better than micro-grooves and reticular grooves. Along the direction of friction sliding, the wear of the front end is more worn than the rear end. Micro-dimples with a diameter of 0.8 mm, a spacing of 1.2 mm, and an area occupancy of 34.8% were fabricated at an output power of 200 W and a frequency of 5 Hz. The wear of the textured surface has been reduced by more than 80% in the process of ring-block dry friction with a load of 50 N, a rotation speed of 35 r/min, and a time of 15 min. The wear mechanism is mainly abrasive wear. The results showed that the hardness of the surface could be improved by laser hardening. In addition, micro-dimples on 42CrMo steel can store abrasive particles, mitigate the formation of furrows and reduce the abrasive wear of tooth plates.

Keywords: laser processing; textured surface; wear resistance; friction

1. Introduction

The surface texturing technique has been considered a promising method in tribology applications because it could improve the performance of the friction system, including the wear resistance, the lifespan, and the bearing capacity due to the ability of surface textures to function as micro-containers for lubricant and particles [1]. Therefore, surface texturing has been used to manufacture mechanical components, for example, thrust bearings, cutting tools, piston rings, and seals [2–7]. Electric discharge machining (EDM), micro-grinding, and laser surface texturing are applied to fabricate textures on specimen surfaces researchers [8–12]. Compared with the other techniques, laser surface texturing has the advantages of high automation, high repeatability, and fine-scale processing [13].

At present, a large number of research on surface texturing has been expected to improve the tribological properties at the interface by reducing the contact area and increasing the lubrication pocket area on the surface. Ran Duan et al. created micro-textures on an Al_2O_3/TiC ceramic cutting tool and the textured tool to machine the AISI H13 steel in the conventional cooling condition [14]. The results showed that the structures of the surface textures performed with good integrity at different cutting speeds. Youqiang Xing et al. fabricated micro-textures on the surface of an Al_2O_3/TiC ceramic specimen to investigate the wear mechanism by using a ball-on-disk tribometer. The micro-textures were used in combination with different solid lubricants [15]. Results showed that laser surface textures combined with WS_2 solid lubricants could decrease wear and friction significantly.

Citation: Zhang, H.; Pei, X.; Jiang, X. Anti-Wear Property of Laser Textured 42CrMo Steel Surface. *Lubricants* **2023**, *11*, 353. https://doi.org/10.3390/lubricants11080353

Received: 20 July 2023
Revised: 16 August 2023
Accepted: 17 August 2023
Published: 21 August 2023

Chi-Wai Chan et al. explored the effects of laser surface treatment on enhancing the properties of commercial Ti6Al4V and CoCrMo alloy implant materials [16]. The results indicated that the laser-treated Ti6A4V surfaces exhibited a noticeable reduction in adhesion. Such properties were attributable to the combined effects of reduced hydrophobicity, thicker and stable oxide films, and the presence of laser-induced textures. Tao Wang et al. produced micro-textures on end faces of two-phase mechanical face seals by using laser surface texturing [17]. They found that the textures with suitable parameters could attain a higher speed in comparison with the seal with a plain end face. Etsion I. et al. investigated the effects of partially laser surface textured piston rings on the fuel consumption and exhaust gas composition of a compression-ignition IC engine [18]. It was found that t the fuel consumption of the partial LST piston rings decreased by 4%. Julius Caesar Puoza et al. fabricated micro-dimples on the surface of the cylinder bore by Nd:YAG laser texturing [19]. The results showed that the cylinder bore with surface textures could effectively decrease the reverse drag power and torque. The fuel consumption was also reduced significantly in comparison with the standard one. Karam Kang researched the effects of surface textures on the punch in the Aluminum can manufacturing process [20]. The micro-textured punches with Ra in the range of 0.15–0.3 μm were preferred over polished and grounded punches to obtain fine surface roughness of the can body sheet. Tatsuhiko Aizawa et al. investigated the Engineering durability of surface texture by exposure testing to air at the atmospheric condition [21]. The testing results showed that modification of the AISI304 surface had sufficient stability for its long-term usage in the air. Georg Schnell et al. created micro-textures on the half-bearing shells of the journal bearings to research the wear mechanism [22].The findings showed that the journal bearings with micro-textures formed a thicker lubricant film than the untextured journal bearings. Jian Zhan et al. manufactured dimples on the cylinder wall to investigate the wear mechanism by using a cylinder liner-piston ring wear tester [23]. The results showed that the distribution angle of the dimples had a significant effect on the tribological characteristics of the cylinder liner-piston ring system. The wear resistance of the system was effectively improved when the dimple distribution angle was 60°. Xingyu Liang et al. studied the effect of partial laser surface texturing on piston rings by simulation and experiment investigation [24]. The results of the study indicated that the micro-dimple textures on both sides of the ring surface can improve the friction coefficient. Kafayat Eniola Hazzan et al. investigated the influence of the laser parameters and the current limitations of laser processing [25]. Laser parameters could affect ect heat-affected zone on the surface of the specimen. The heat-affected zone was related to the surface finish. They found the optimized multi-pass scanning speeds could deliver sufficient energy to process textures to the required depth and profile with minimal defects. Kairui Zheng et al. investigated the effects of micro-textured tools on cutting performances [26]. The results showed that the cutting force was effectively reduced by manufacturing suitable micro-textures on the surface of the cutting tool. Mourier et al. manufactured micro-cavities by laser surface texturing and carried out EHL tests with a tribometer [27]. The results showed that the micro-cavities, whose depth was 300 nm, could get an increase in the thickness of the lubricant film. Shunchu Liu et al. investigated the effects of laser surface textures and lubrication with graphene on the tribological and dynamical performance of non-conformal sliding contacts [28]. The results showed that laser surface textures have a more significant effect on the reduction of frictional vibrations and noise generated from sliding contacts, and laser textures enhanced the frictional reduction performance of lubrication. Zhaoqiang Wang et al. fabricated three kinds of micro-textures on the surface of the valve plates to investigate the effects of textures with different parameters on the tribological characteristics [29]. The experimental results showed that surface textures had significant effects on pressure, load-carrying capacity, and elastic deformation. The square textures were the preferable shape, and the tribological characteristics were improved when the depths of textures were from 20 μm to 50 μm. Akshay Gaikwad et al. investigated the effects of laser texturing on the tribological properties of Ti6Al4V in contact with a ceramic ball [30]. Results showed that wear and friction were reduced for all the textured samples as compared to an untextured surface. Chen Li et al. fabricated surface micro-textures on the electrosurgical blades by the long-pulse fiber laser [31]. The results showed that the electrosurgical blades with surface micro-textures exhibited good performances of adhesion and anti-friction, which minimized the potential danger for the patients. Caraguay et al. created micro-textures on the AISI-A30 carbon steel to study the adhesion strength by using a pull-off test and a shaft load blister test [32]. The results showed that the textured area had a significant effect on the adhesion mechanism of the carbon steel. Tooth plates are the key components of the oil drilling wellhead machinery, which are commonly made of 42CrMo steel. During the operation of the oil drilling wellhead machinery, tooth plates are prone to wear under high pressure. They need to be replaced frequently, which would reduce production efficiency. In this paper, how to reduce wear has been a vital problem. A model of the micro-textured surface friction pairs was established and simulated by the finite element analysis software. Nd: YAG laser was used to fabricate three types of textures on the surface of 42CrMo alloy steel, which were micro-dimples, micro-grooves, and reticular grooves. Then the dry friction experiments were carried out to analyze the influence of surface textures on the wear resistance of 42CrMo steel surface.

2. Finite Element Simulation

2.1. A Model of Friction Pairs

During the operation of the wellhead machinery, the motion pairs consisted of the tooth plate and the drill pipe, which contacted each other and produced relative motion under the external load. Therefore, the friction contact between the tooth plate and the drill pipe could be considered linear sliding friction pairs. Figure 1 shows the schematic diagram of the tooth plate and the drill pipe. It can be seen from Figure 1 that the drill pipe made contact with the tooth plate under load conditions. The load applied on the drill pipe is P. The radius of the drill pipe is R. Elastic moduli of the friction pairs are E_1 and E_2, respectively. Their Poisson ratios are ε_1 and ε_2. E' is the equivalent elastic modulus. b is the half-width of the contact area.

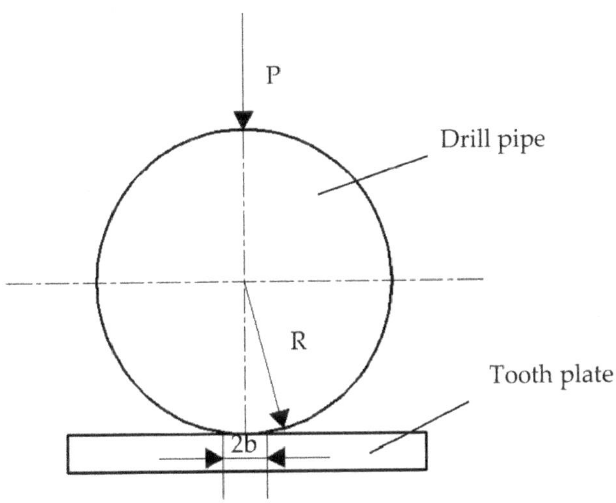

Figure 1. Schematic diagram of the contact between the tooth plate and the drill pipe.

According to the Hertz contact theory, the half-width of the contact area can be calculated as follows.

$$b = \left(\frac{4PR}{\pi E'}\right)^{1/2} \quad (1)$$

The equivalent elastic modulus can be calculated as follows.

$$\frac{1}{E'} = \frac{1-\varepsilon_1{}^2}{E_1} + \frac{1-\varepsilon_2{}^2}{E_2} \tag{2}$$

Actually, there are eight tooth plates of the same size (100 mm × 40 mm × 20 mm) in the oil drilling wellhead machinery. The drill pipe and the tooth plate are both made of 42CrMo steel. The elastic modulus and the Poisson ratio are 210 GPa and 0.3, respectively. The diameter of the drill pipe is in the range of 88 to 160 mm, and the torque value is in the range of 81,350 to 108,465 N·m. The load is in the range of 508,437.5 to 1,232,556.8 N. It can be calculated that the contact half width of a single tooth plate is in the range of 0.66 to 1.01 mm.

The friction pairs were composed of upper and lower sliding blocks. There are three kinds of micro-textures on the friction surface of the upper sliding blocks. The surface textures included micro-dimples, micro-grooves, and reticular grooves. The models of friction pairs are shown in Figure 2. The sizes of the upper sliding block and the lower sliding block were 10 mm × 10 mm × 5 mm and 30 mm × 10 mm × 10 mm, respectively.

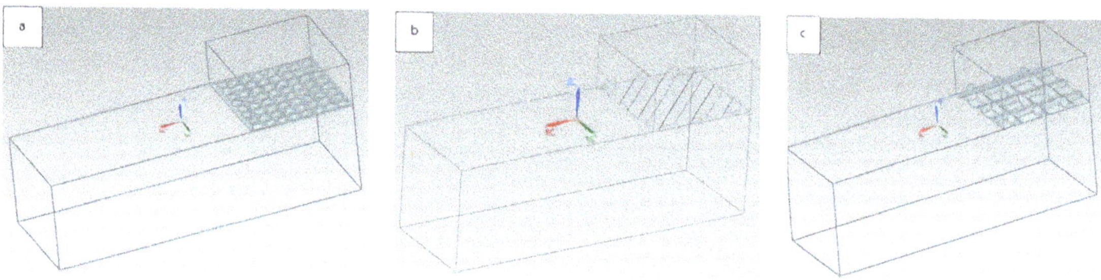

Figure 2. Friction pairs models. (**a**) Upper block with micro-dimples; (**b**) Upper block with micro-grooves; (**c**) Upper block with reticular grooves.

According to the calculation of the Hertz contact theory, the selective sizes of micro-dimples are a diameter of 0.8 mm, the dimple spacing is 1.2 mm, and the area occupancy of dimples is 34.8%. In order to obtain the same area occupancy of textures, the width of the single micro-groove is set at 0.75 mm. The inclination angle is 45°, and the spacing is 2.3 mm. The width of the reticular groove is 0.75 mm, the angle of the reticular groove is 90°, and the spacing is 3.5 mm. The upper and lower sliding blocks used in this experiment are 42CrMo alloy steel. The material properties are shown in Table 1.

Table 1. Material properties of 42CrMo.

Hardness (HV)	Density (kg/cm^3)	Melting Point (C°)	Heat Transfer Coefficient (W/m·K)	Poisson's Ratio	Elastic Modulus (GPa)
410	7.85	1530	10	0.3	210

For the model of friction pairs firstly, the model of the friction pairs was meshed. The mesh units of the upper block specimen and the lower block specimen were set as 0.25 mm and 0.4 mm, respectively. Secondly, initial boundary conditions were set. The positions of all mesh nodes on the bottom of the lower sliding block remained unchanged, and displacement constraints were imposed in X and Y directions. Then load conditions were applied for all mesh units on the upper surface of the upper sliding blocks, and a load of 3 MPa was applied on the surface. All the grid nodes on the lower surface of the upper sliding blocks moved 20 mm along the Y direction. Finally, the contact equivalent stress of the surfaces with different textures was obtained.

2.2. Simulation Results

According to the load conditions of the sliding friction pairs of the above three different types of micro-textured surfaces, the equivalent contact stresses are obtained, as shown in Figure 3. It can be seen from Figure 3 that stress concentration zones are focused on the left and right sides of the three types of micro-textured surfaces. In addition, the equivalent stresses of the left sides are larger than that of the right sides, which indicates that the left sides of the textured surfaces are more worn than the right sides along the sliding direction. The maximum equivalent stress of the three micro-textured surfaces is 10.275 MPa, 17.093 MPa, and 12.885 MPa, respectively. The stress gradient on the surface of the micro-groove textures is the largest compared with other textures. The simulation results show that the wear resistance of the micro-dimples is better than that of the micro-grooves and the reticular grooves.

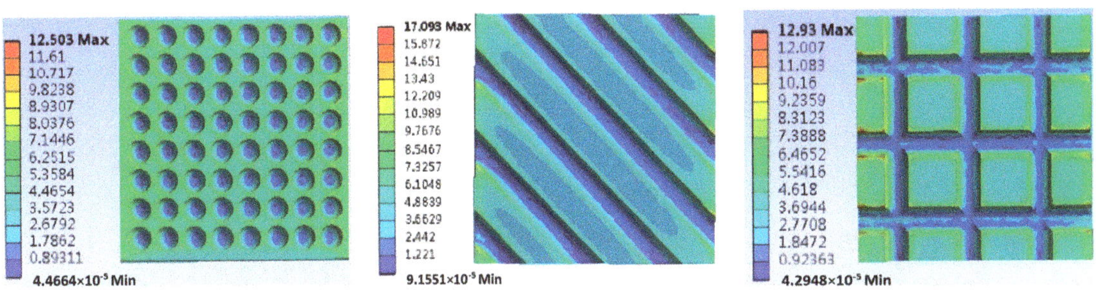

(a) (b) (c)

Figure 3. Equivalent stress cloud diagram of sliding frictional contact with different textured surfaces. (**a**) micro-dimples; (**b**) micro-grooves; (**c**) reticular grooves.

3. Experimental Procedure

3.1. Preparation of Friction Specimen

The dry friction test was conducted by the wear testing machine (MMH-5UMT-2, CETR, San Jose, CA, USA). The mechanism of surface contact friction is shown in Figure 4. The lower surface of the upper block specimen contacted the upper surface of the lower block specimen and slid counterclockwise. The upper specimen is a block (20 mm × 20 mm × 10 mm). The lower specimen is a fixed annular block (inner diameter 340 mm, outer diameter 380 mm) with a surface roughness of Ra12.5 μm.

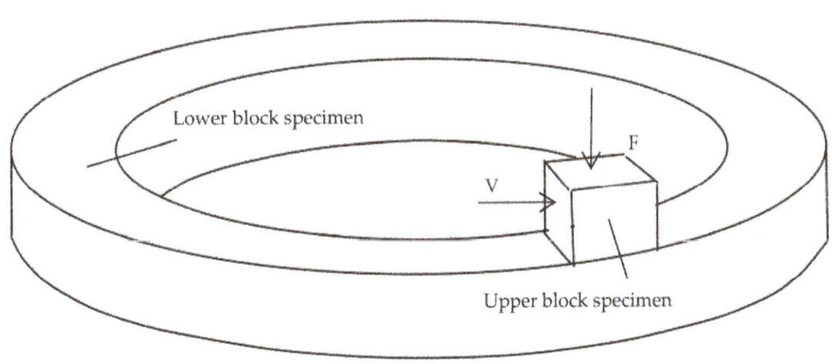

Figure 4. Schematic diagram of the friction motion of ring-block surface contact.

The Nd: YAG laser was used to fabricate the textures at the average laser power of 200 W and a frequency of 5 Hz (DPL20-532, CHUTIAN, Wuhan, China). According to the sizes of the above simulation design, three kinds of micro-structures, which included micro-dimples, micro-grooves, and reticular grooves, were prepared on the friction surface of the upper block specimen. Furthermore, two groups of textured specimens with spacing of 1.5 mm and 1.8 mm were prepared for the micro-dimples. Micro-dimples were fabricated by using the fixed-focusing processing method. The micro-grooves and the reticular grooves were created by utilizing the laser scanning processing method. The work platform was moved during the laser texturing process. Table 2 shows the laser texturing parameters. Due to the existence of molten metal sputtering on the surface of the micro-structure fabricated by laser, the specimen was repeatedly polished with sandpapers (200#–2000#) of different grain sizes after laser texturing. Then, the particles produced from the wear were removed by an ultrasonic cleaning machine (DA-968, DADI, Nanning China). Figure 5 shows the surface morphology of micro-grooves and reticular grooves observed under a three-dimensional microscope with a super depth of field(VHX-2000, Hitachi, Tokyo, Japan). Figure 6 shows the surface morphology of three kinds of micro-dimples with different spacing observed with the scanning electron microscopy (S-3400N, Hitachi, Japan).

Table 2. Laser texturing parameters.

Pulse Duration (ms)	Pulse Energy (J)	Repetition Rate (Hz)	Spot Size (mm)	Focusing Lens Distance (mm)	Scanning Speed (mm/s)
0.5	3.3	60	0.5	150	1

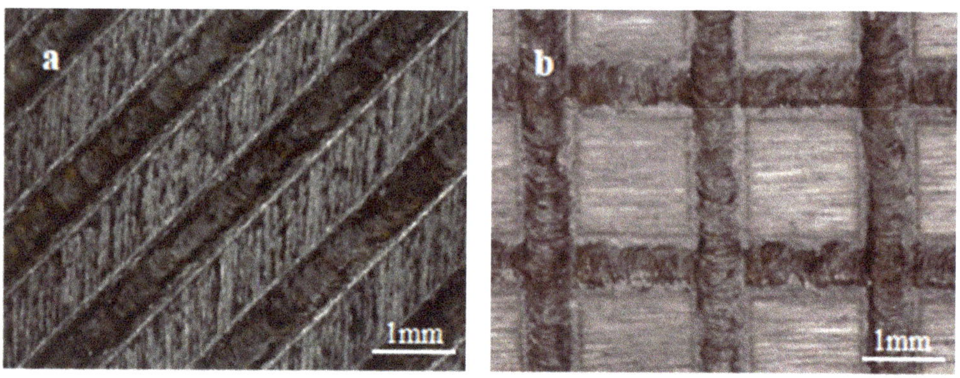

Figure 5. Laser textured surface: (**a**) Micro-grooves with a pitch of 2.3 mm; (**b**) Reticular grooves with a pitch of 3.5 mm.

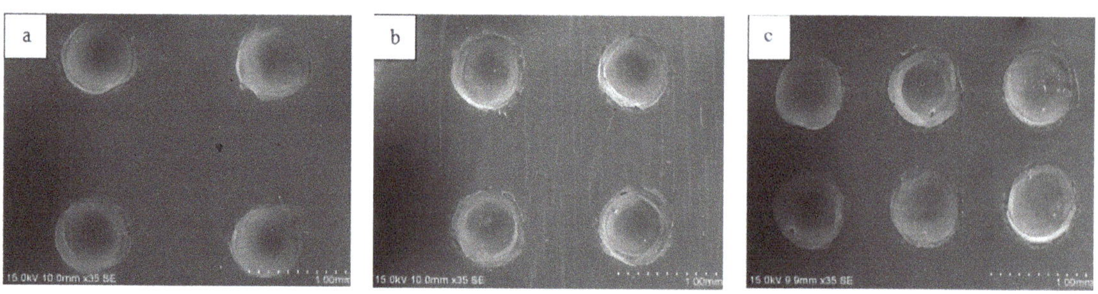

Figure 6. Micro-dimples with different pitch: (**a**) L = 1.8 mm; (**b**) L = 1.5 mm; (**c**) L = 1.2 mm.

3.2. Friction Tests

In the dry friction test, the upper block specimen and the lower block specimen were in dislocation contact. After the friction test, the laser-textured surface of the upper block specimen had an obvious boundary that could distinguish the worn area from the non-worn area. The friction test was designed properly. The initial height of the unworn zone can obtain a uniform reference when the thickness of the worn layer is measured. The load of friction test was set at 50 N, and the rotation speed of the upper block specimen was set at 35 r/min. Each test was repeated 3 times, and each time was 15 min.

For the friction test results, the wear layer thickness can be obtained by observing the three-dimensional morphology of the laser-textured surface of the upper block specimen with an ultra-depth of field three-dimensional microscope. High precision electronic balance was used to weigh the upper block specimen before and after the friction test, and the wear amount could be obtained. The effect of different micro-textures on wear resistance was explored by observing the morphology of the laser-textured surface.

4. Results and Analysis

4.1. Wear Resistance Comparison of Different Laser-Textured Surfaces

Under the condition of the same textured area occupancy, the wear resistance of three types of laser micro-textured surfaces such as micro-dimples, micro-grooves, and reticular grooves, were compared. Figure 7 is the surface topography of three different laser-textured specimens after the friction test.

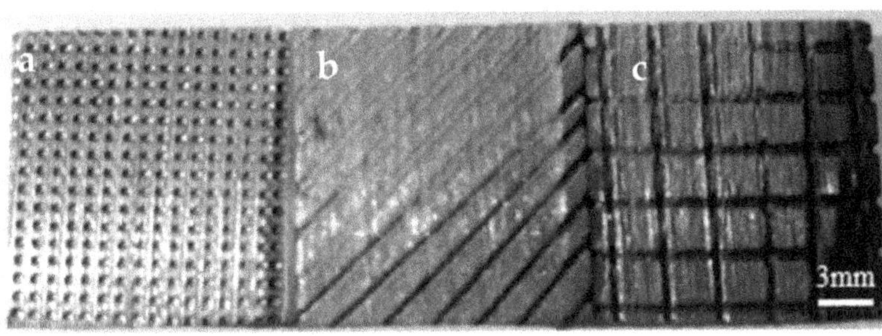

Figure 7. Morphology of different textured surfaces after friction tests. (**a**) Specimen with micro-dimples; (**b**) Specimen with micro-grooves; (**c**) Specimen with reticular grooves.

It can be seen from Figure 7 that the wear of the specimens with the micro-grooves and the reticular grooves are severer than the specimen with micro-dimples. As can be seen from the simulation results above, the equivalent stresses of the micro-grooves and the reticular are also larger than that of the micro-dimples. The experimental results are consistent with the simulation results.

In addition, three types of textured surfaces show the same wear characteristics. It can be seen the wear of the front end is more worn than the rear end in the direction of friction sliding. As can be seen from the simulation results above, the equivalent stress of the front end is larger than that of the rear end. Thus, the wear mechanism is the same for the different micro-textures. And the wear mechanism is abrasive wear. Under the same condition of area occupancy, the micro-dimples improve the tribological performance significantly in comparison with the other textures.

Figure 8 is the three-dimensional surface topography of the untextured specimen surface and three kinds of laser micro-textured surfaces. It can be seen from Figure 8 that the different kinds of specimen surfaces have clear boundaries between the worn areas and the unworn areas. In Figure 8b–d, the worn areas and unworn areas of specimens all have micro-textures. The height differences between the unworn zone and the worn zone in the figure are used to evaluate the thickness of each wearing layer.

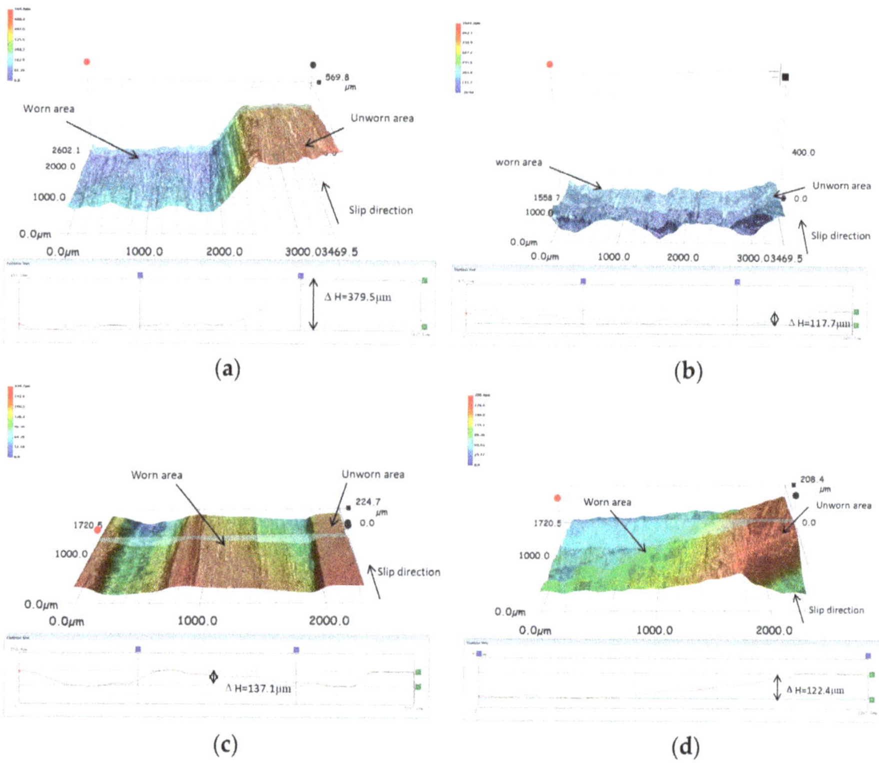

Figure 8. 3D morphology of upper work-pieces surface after the friction test: (**a**) untextured surface; (**b**) micro-dimples surface; (**c**) micro-grooves surface; (**d**) reticular grooves surface.

It can be seen from Figure 8 that the thickness of the wearing layer on the surface of the untextured specimen is the largest, which is 379.5 μm; The wearing layer thickness of three laser-textured surfaces are 117.7 μm, 137.1 μm, and 122.4 μm respectively. Although the bearing areas of the textured surfaces are reduced in comparison with the untextured surface, it can be seen in Figure 8 that the wear of three textured surfaces is less than that of the untextured surface. It shows that the laser micro-textured surface can effectively improve the wear resistance of 42CrMo steel. Under the condition of the same textured area occupancy, micro-dimples are the most reasonable micro-textures for improving the friction properties of 42CrMo steel. The simulation results show that the stress on the surface of the micro-dimple textures is smaller than other surface textures. The wearing layer thickness of the micro-dimple textures is 117.7 μm, which is the lowest in comparison with others. Thus, the wearing layer thicknesses of the friction experiments are consistent with the stress gradients of the simulation.

4.2. Wear Resistance Comparison of Laser Textured Surfaces with Different Spacing of Micro-Dimples

The effect of dimple spacing on the wear resistance of textured surfaces is compared and analyzed under the condition of a certain pit diameter. The distances of the three groups are 1.2 mm, 1.5 mm, and 1.8 mm respectively. The average value of the mass difference of the block specimen before and after three friction tests is taken as the friction and wear amount. Figure 9 shows the wear weight and wear reduction rate with untextured and textured surfaces. Compared with the untextured surface, the wear of the textured surface with 1.2 mm spacing is reduced by more than 80%.

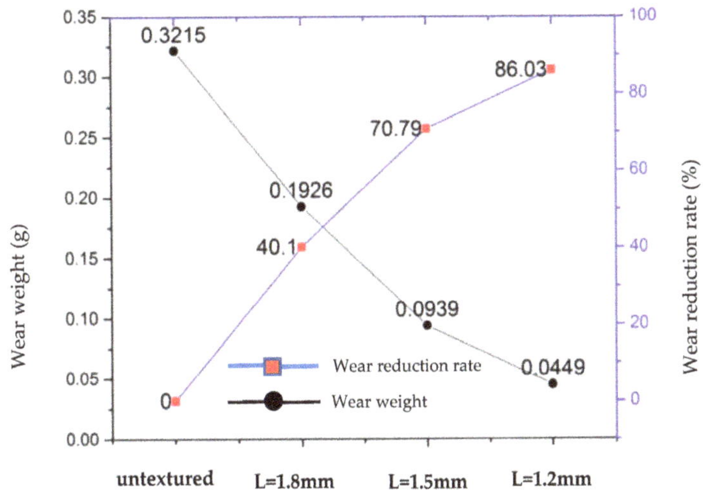

Figure 9. Comparison of wear weight for upper blocks surface friction.

4.3. Wear Mechanism Analysis of Laser-Textured Surface with Micro-Dimples

Figure 10 shows the textured specimen with micro-dimples (the diameter of the dimple is 0.8 mm, the spacing is 1.8 mm) after the friction test for 5 min. After the friction test, the specimen was cleaned by ultrasonic cleaning machinery, and there were obvious particles in the dimples, which indicates that a large number of abrasive particles were produced in the process of friction tests. These abrasive particles play an important role in the wear of the specimen. The micro-dimples have the ability to trap debris, which decreases the effect of abrasive wear and improves the tribological properties at the interface.

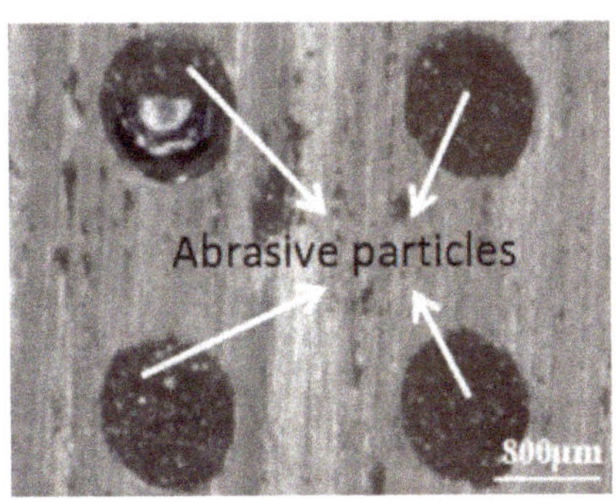

Figure 10. Surface morphology of the textured block specimen with micro-dimples after friction experiment.

Figures 11–13 show the morphology of the specimen after the friction test when the dimple spacings are 1.8 mm, 1.5 mm, and 1.2 mm, respectively. By comparison, it can be found that the surface wear of the three laser-textured specimens with different spacing has similar characteristics. In the vertical direction, which is the friction direction of the specimen, obvious continuous furrows appear in the untextured area among adjacent rows of dimples. In the horizontal direction, there is no furrow among adjacent rows of dimples. It indicates that abrasive wear occurred on the surface of the specimens during the dry friction test, and obvious furrows were formed in the untextured region. While in the textured region, the continuous formation of furrows is prevented by the dimples distributed in the array.

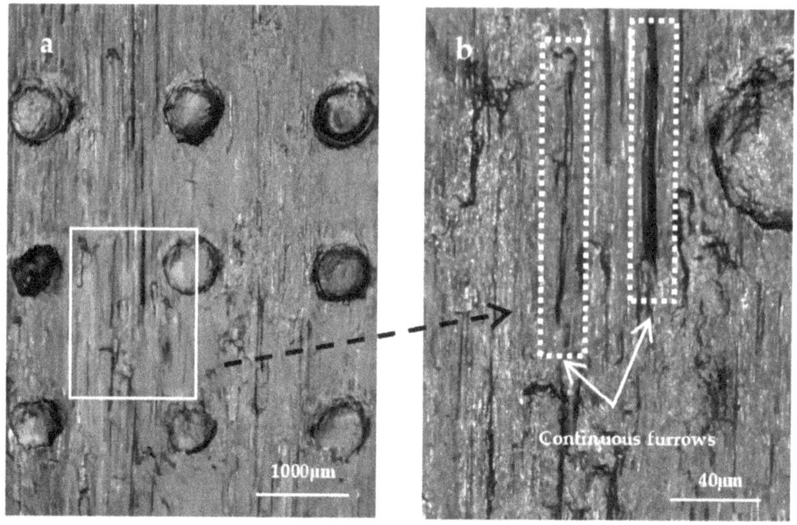

Figure 11. Wear morphology of micro-dimples surface with a spacing of 1.8 mm. (**a**) the overall wear morphology; (**b**) the local wear morphology.

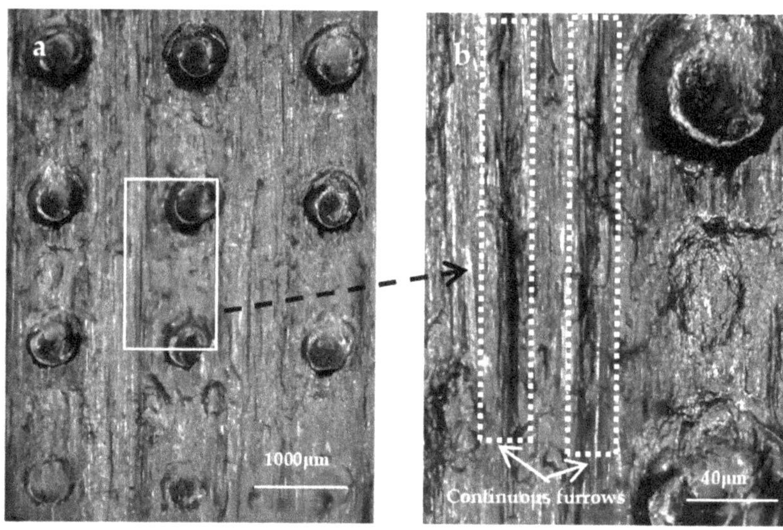

Figure 12. Wear morphology of micro-dimples surface with a spacing of 1.5 mm. (**a**) the overall wear morphology; (**b**) the local wear morphology.

By further comparing the furrow length and depth of the three surfaces, it can be found that the furrow length and depth of the surface with the largest texture area occupancy are smaller than those of the other surfaces, which indicates that with the decrease of the dimple spacing. The abrasive wear on the surface of the specimen is reduced, which is consistent with the test results of the front wear. Based on the above analysis, it can be concluded that the wear resistance mechanism of laser textured surface is that the dimples prevent the continuous formation of furrows, and the effect of micro-dimples increases with the increase of the density of dimples.

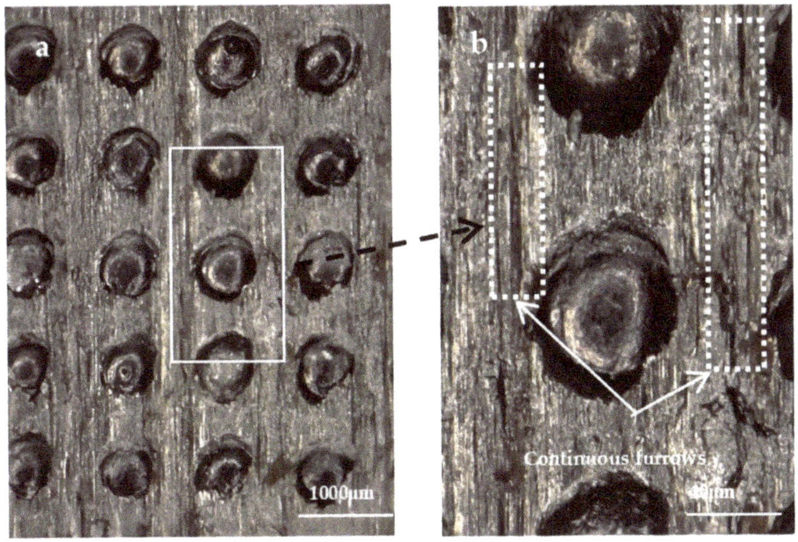

Figure 13. Wear morphology of micro-dimples surface with a spacing of 1.2 mm. (**a**) the overall wear morphology; (**b**) the local wear morphology.

5. Conclusions

In this paper, the wear resistance of 42CrMo steel surface with laser micro-texture has been studied. On the basis of the results from the research, the following conclusion can be drawn:

(1) Under the condition of surface contact dry friction, the front end of the textured surface is more worn than the back end along the sliding direction. The wear resistance of laser-textured surfaces with micro-dimples is better than that of micro-grooves and reticular grooves.
(2) Compared with the untextured surface, the wear loss of the textured surface with 1.2 mm spacing and 34.8% area occupancy is reduced by more than 80%.
(3) The main reason for the improvement of the wear resistance of the laser-textured surface is that the micro-dimples can store the abrasive particles and effectively prevent the continuous formation of furrows.

Author Contributions: Conceptualization, H.Z.; methodology, H.Z. and X.P.; investigation, H.Z. and X.P.; resources, X.J.; data curation, X.P.; writing—original draft preparation, X.P.; writing—review and editing, H.Z.; project administration, H.Z.; funding acquisition, H.Z. All authors have read and agreed to the published version of the manuscript.

Funding: This research was supported by the National Natural Science Foundation of China under Grant No. 51205212, Natural Science Fund of Jiangsu Province of China under Grant No. BK2012233, Nantong City Applied Basic Research Project (JC22022074) and Instruments and Equipment Open Fund of Nantong University (KFJN2235).

Institutional Review Board Statement: Not applicable.

Informed Consent Statement: Not applicable.

Data Availability Statement: Not applicable.

Conflicts of Interest: The authors declare no conflict of interest.

References

1. Schille, J.; Schneider, L.; Mauersberger, S.; Szokup, S.; Hohn, S.; Potschke, J.; Reir, F.; Leidich, E.; Lischner, U. High-Rate Laser Surface Texturing for Advanced Tribological Functionality. *Lubricants* **2020**, *8*, 33. [CrossRef]
2. Agrawal, N.; Sharma, S.C. Micro-grooved hybrid spherical thrust bearing with Non-Newtonian lubricant behaviour. *Int. J. Mech. Sci.* **2023**, *240*, 107940.
3. Tatsuya, S.; Toshiyuki, E. Development of a cutting tool with a nano/micro-textured surface—Improvement of anti-adhesive effect by considering the texture patterns. *Precis. Eng.* **2009**, *33*, 425–429.
4. Tatsuya, S.; Toshiyuki, E. Improving anti-adhesive properties of cutting tool surfaces by nano-/micro-textures. *Precis. Eng.* **2012**, *36*, 229–237.
5. Ryk, G.; Etsion, I. Testing piston rings with partial laser surface texturing for friction reduction. *Wear* **2006**, *261*, 792–796. [CrossRef]
6. Ezhilmaran, V.; Vasa, N.J.; Vijayaraghavan, L. Investigation on generation of laser assisted dimples on piston ring surface and influence of dimple parameters on friction. *Surf. Coat. Technol.* **2018**, *335*, 314–326.
7. Yu, X.Q.; He, S.; Cai, R.L. Frictional characteristics of mechanical seals with a laser-textured seal face. *J. Mater. Process. Technol.* **2002**, *129*, 463–466.
8. Chaudhury, P.; Samantaray, S. Role of Carbon Nano Tubes in Surface Modification on Electrical Discharge Machining—A Review. *Mater. Today Proc.* **2017**, *4*, 4079–4088.
9. Zhou, W.; Tang, J.; Shao, W. Modelling of surface texture and parameters matching considering the interaction of multiple rotation cycles in ultrasonic assisted grinding. *Int. J. Mech. Sci.* **2020**, *166*, 105246.
10. Aurich, J.C.; Egnmann, J.; Schueler, G.M. Micro grinding tool for manufacture of complex structures in brittle materials. *CIRP Ann.—Manuf. Technol.* **2009**, *58*, 311–314.
11. Mao, B.; Siddaiah, A.; Liao, Y.L.; Pradeep, L. Laser surface texturing and related techniques for enhancing tribological performance of engineering materials: A review. *J. Manuf. Process.* **2020**, *53*, 153–173.
12. Wang, M.Z.; Kang, W.; Yue, Z.Q.; Zhu, L.N.; She, D.S.; Wang, C.B. Effects of combined treatment of plasma nitriding and laser surface texturing on vacuum tribological behavior of titanium alloy. *Mater. Res. Express* **2019**, *6*, 066511.
13. Gachot, C.; Rosenkranz, A.; Hsu, S.M.; Costa, H.L. A critical assessment of surface texturing for friction and wear improvement. *Wear* **2017**, *372–373*, 21–41.
14. Duan, R.; Wang, G.; Xing, Y.Q. Investigation of novel multiscale textures for the enhancement of the cutting performance of Al_2O_3/TiC ceramic cutting tools. *Ceram. Int.* **2022**, *48*, 3554–3563. [CrossRef]
15. Xing, Y.Q.; Deng, J.X.; Wang, X.S.; Meng, R. Effect of laser surface textures combined with multi-solid lubricant coatings on the tribological properties of Al_2O_3/TiC ceramic. *Wear* **2015**, *342–343*, 1–12. [CrossRef]
16. Chan, C.W.; Carson, L.; Smith, G.C.; Morelli, A.; Lee, S. Enhancing the antibacterial performance of orthopaedic implant materials by fibre laser surface engineering. *Appl. Surf. Sci.* **2017**, *404*, 67–81. [CrossRef]
17. Wang, T.; Huang, W.F.; Liu, X.F.; Li, Y.J.; Wang, Y.M. Experimental study of two-phase mechanical face Seals with lasersurface texturing. *Tribol. Int.* **2014**, *72*, 90–97. [CrossRef]
18. Etsion, I.; Sher, E. Improving fuel efficiency with laser surface textured piston rings. *Tribol. Int.* **2009**, *42*, 542–547. [CrossRef]
19. Puoza, J.C.; Zhang, T.Y.; Uba, F.; Yakubu, K.; Awudu, I. Optimization of post-treatment honing parameters of laser textured enginecylinder bore and engine performance. *Int. J. Adv. Manuf. Technol.* **2023**, *in press*.
20. Kang, K. Impact of die wear and punch surface textures on aluminum can wall. *Wear* **2009**, *266*, 1044–1049. [CrossRef]
21. Tatsuhiko, A.; Tadahiko, I.; Kenji, W. Femtosecond laser micro-/nano-texturing of stainless steels for surface property control. *Micromachines* **2019**, *10*, 512.
22. Schnell, G.; Studemend, H.; Thomas, R.; Seits, H. Experimental investigations on the friction behavior of partiallyfemtosecond laser-textured journal bearing shells. *Tribol. Int.* **2023**, *188*, 108764. [CrossRef]
23. Zhan, J.; Yang, M.J. Investigation on Dimples Distribution Angle in Laser Texturing of Cylinder–Piston Ring System. *Tribol. Trans.* **2012**, *55*, 693–697. [CrossRef]
24. Liang, X.Y.; Wang, X.H.; Zhang, Z.J. Simulation and experiment investigation on friction reduction by partial laser surface texturing on piston ring. *Tribol. Trans.* **2019**, *11*, 109–113.
25. Hazzan, K.E.; Pacella, M.; See, T.L. Laser processing of hard and ultra-hard materials for micro-machining and surface engineering applications. *Micromachines* **2021**, *12*, 895. [CrossRef] [PubMed]
26. Zheng, K.R.; Yang, F.Z.; Zhang, N. Study on the cutting performance of micro-textured tools on cutting Ti-6Al-4V Titanium alloy. *Micromachines* **2020**, *11*, 137. [CrossRef]
27. Mourier, L.; Mazuyer, D.; Ninove, F.P.; Lubrecht, A.A. Lubrication mechanisms with laser-surface-textured surfaces in elastohydrodynamic regime. *Eng. Tribol.* **2010**, *224*, 697–711. [CrossRef]
28. Liu, S.C.; Sai, Q.Y.; Wang, S.W. Effects of laser surface texturing and lubrication on the vibrational and tribological performance of sliding contact. *Lubricants* **2022**, *10*, 10. [CrossRef]
29. Wang, Z.Q.; Sun, L.T.; Han, B. Study on the thermohydrodynamic friction characteristics of surface-textured valve plate of axial piston pumps. *Micromachines* **2022**, *13*, 1891. [CrossRef]
30. Gaikwad, A.; Vázquez-Martínez, J.M.; Salguero, J.; Iglesias, P. Tribological Properties of Ti6Al4V Titanium Textured Surfaces Created by Laser: Effect of Dimple Density. *Lubricants* **2022**, *10*, 138.

31. Li, C.; Yang, Y.; Yang, L.J.; Zhen, S. Biomimetic Anti-Adhesive Surface Microstructures on Electrosurgical Blade Fabricated by Long-Pulse Laser Inspired by Pangolin Scales. *Micromachines* **2019**, *10*, 816. [CrossRef] [PubMed]
32. Caraguay, S.J.; Pereira, T.S.; Cunha, A.; Pereira, M.; Xavier, F.A. The effect of laser surface textures on the adhesion strength and corrosion protection of organic coatings—Experimental assessment using the pull-off test and the shaft load blister test. *Prog. Org. Coat.* **2023**, *180*, 107558. [CrossRef]

lubricants

MDPI

Article

Tribocorrosion Behavior of γ'-Fe_4N Nitride Layer Formed on Mild Steel by Plasma Nitriding in Chloride-Containing Solution

Yong Sun * and Richard Bailey

School of Engineering and Sustainable Development, Faculty of Computing, Engineering and Media, De Montfort University, Leicester LE1 9BH, UK; richard.bailey@dmu.ac.uk
* Correspondence: ysun01@dmu.ac.uk

Abstract: Nitriding has long been used to engineer the surfaces of engineering steels to improve their surface and subsurface properties. The role of the surface compound layer (γ'-Fe_4N and/or ε-$Fe_{2\text{-}3}N$) in improving the tribological and corrosion-resistant properties of nitrided steels has been established. However, there have been very few studies on the response of the compound layer to tribocorrosion in corrosive environments. In this work, the tribocorrosion behavior of a 5 μm thick γ'-Fe_4N nitride layer produced on mild steel (MS) by plasma nitriding has been studied in a NaCl-containing solution under various electrochemical conditions. The results show that at a cathodic potential of −700 mV (saturated calomel electrode, SCE), where mechanical wear is predominant, the total material removal (TMR) from the γ'-Fe_4N layer is 37% smaller than that from the untreated MS, and at open circuit potential, TMR from the layer is 34% smaller than that from the untreated MS, while at an anodic potential of −200 (SCE), the γ'-Fe_4N layer can reduce TMR from mild steel by 87%. The beneficial effect of the γ'-Fe_4N nitride layer in improving the tribocorrosion behavior of mild steel is derived from its high hardness and good corrosion resistance in the test solution and its ability to resist both mechanical wear and corrosion and to reduce wear–corrosion synergism.

Keywords: nitriding; iron nitride; wear; corrosion; tribocorrosion; mild steel

Citation: Sun, Y.; Bailey, R. Tribocorrosion Behavior of γ'-Fe_4N Nitride Layer Formed on Mild Steel by Plasma Nitriding in Chloride-Containing Solution. *Lubricants* **2023**, *11*, 281. https://doi.org/10.3390/lubricants11070281

Received: 25 May 2023
Revised: 15 June 2023
Accepted: 28 June 2023
Published: 29 June 2023

1. Introduction

Nitriding is a common thermochemical treatment technique used to engineer the surfaces of engineering steels [1–5] and many other metallic materials such as titanium alloys [6,7], aluminum alloys [8,9], nickel-based alloys [10,11] and Co-Cr biomedical alloys [12,13]. In particular, nitriding has long been used in industry to case-harden many steel components such as gears and drive-shafts to improve their performance and durability in real applications [1,2]. This is due to the formation of an iron nitride compound layer on the surface and a nitrogen diffusion zone at the subsurface, which can enhance the wear resistance, fatigue resistance and in many cases, corrosion resistance of steel components [1,14].

During nitriding iron and steels, depending on the nitrogen potential or activity in the treatment media and the steel composition, a compound layer of a few microns thick composed of a single γ'-Fe_4N phase or a mixed γ'-Fe_4N and ε-$Fe_{2\text{-}3}N$ phase is formed at the surface [2–4]. This is followed by a relatively thick nitrogen diffusion zone at the subsurface. If the steel contains nitride-forming elements such as Cr and V, fine precipitates of nitrides of the alloying elements, such as CrN and VN, will form in the diffusion zone to induce precipitation hardening which contributes to the hardening effect of the diffusion zone [3,15]. In pure iron and plain carbon steels such as mild steels, γ'-Fe_4N needles or thin plates form in the diffusion zone, which provides a marginal hardening effect [15,16]. In such steels, the surface-hardening effect is mainly imparted by the iron nitride compound layer at the surface.

The role of the surface compound layer (γ'-Fe_4N and/or ε-$Fe_{2\text{-}3}N$) and the nitrogen diffusion zone in improving the tribological, fatigue and load-bearing capacity properties of nitrided steels has been established [17–20]. The iron nitride compound layer at the surface provides anti-galling properties and good wear resistance, while the nitrogen diffusion zone provides load-bearing capacity and enhances the fatigue strength of steel components, due to the hardening effect and compressive residual stresses [17,19]. In general, a single γ'-Fe_4N phase compound layer of a few microns thick is preferred, because a thick compound layer or a mixed phase compound layer can increase the embrittlement of the surface layer which tends to spall off during service [1–3]. There have also been reports that the iron nitride compound layer can improve the corrosion resistance of steels, such that nitrided steels can be used in more harsh environmental conditions [14,21,22]. Indeed, in many engineering applications, such as in marine and off-shore applications, nitrided components are used in corrosive environments under mechanical contact sliding conditions. Under such conditions, the components are subjected to combined corrosion and mechanical wear actions, i.e., tribocorrosion [23–27]. So far, the studies on the tribocorrosion behavior of nitrided steels have been focused on stainless steels [28,29]. Although there have been a few studies on the tribocorrosion behavior of nitrided low- alloy steels [30,31], the response of the γ'-Fe_4N compound layer to tribocorrosion has not been fully understood. Due to the importance of such a compound layer in determining the performance of nitrided steels, it is necessary to investigate the tribocorrosion behavior of the γ'-Fe_4N layer under various electrochemical conditions.

In the present work, a 5 µm thick γ'-Fe_4N iron nitride layer was produced on mild steel (MS) by plasma nitriding. The tribocorrosion behavior of this γ'-Fe_4N layer was investigated in 1.0% NaCl solution under combined reciprocating sliding wear and electrochemical corrosion conditions. No such systematic study on the response of the technologically important γ'-Fe_4N layer to tribocorrosion has been reported previously. This work provides reference values for researchers and engineers to further explore the application potentials of iron nitride layers to combat material degradation due to tribocorrosion.

2. Materials and Methods

2.1. Substrate Material and Plasma Nitriding

The substrate material was AISI 1020 mild steel with the following chemical composition (in wt%): 0.203 C, 0.463 Mn, ≤0.040 P, ≤0.050 S, and balance Fe. Low-carbon steel was used to ensure the formation of a pure γ'-Fe_4N layer on the surface by controlled plasma nitriding. The steel was in a normalized state with a ferrite + pearlite structure (Figure 1a) with an average ferrite grain size of 26 µm. The normalizing treatment was conducted in a muffle furnace at 910 °C for 20 min, followed by air cooling. Specimens of 25 × 15 × 4 mm dimensions were machined from a steel plate. The surface to be nitrided (25 × 15 mm) was manually ground using SiC grinding papers down to the P1200 grade to achieve a surface finish of 0.2 µm (R_a).

Plasma nitriding was carried out at 550 °C in a plasma atmosphere containing 10% N_2 and 90% H_2, at a treatment pressure of 3 mbar for 5 h. Before nitriding, the surface was cleaned in running water and then in ethanol ultrasonically for 10 min. After plasma nitriding, the specimens were cooled inside the furnace under vacuum (0.1 mbar) down to room temperature.

Figure 1 shows the cross-sectional morphology of the nitrided specimen. Table 1 summarizes the structural features of the nitrided (PN) and un-nitrided (normalized) specimens. It can be seen that plasma nitriding produced a thin "white layer" about 5 µm thick at the surface (Figure 1b) and a relatively thick diffusion zone about 300 µm thick beneath (Figure 1a). X-ray diffraction analysis confirmed that the "white layer" at the surface was composed of γ'-Fe_4N phase, as can be seen from Figure 2, which shows that only γ' and α-Fe were detected, where the α-Fe peaks came from the substrate. In the nitrogen diffusion zone, γ'-Fe_4N needles were formed, which is typical of nitriding pure iron and low- carbon steels [15,16]. The surface hardness of the PN and normalized specimens were also measured under various indentation loads and the results are listed in Table 1. Clearly, the nitrided surface exhibited a higher hardness than the un-nitrided surface. At the small indentation load of 25 g, the surface hardness of the PN specimen was 760 HV, which agrees with the reported hardness value of γ'-Fe_4N [32]. With increasing indentation load, the surface hardness of the PN specimen declined quickly, due to the increasing substrate effect. This also suggests that the precipitation of γ'-Fe_4N needles in the diffusion zone had a limited hardening effect. Microhardness measurements were also made in the diffusion zone below the γ'-Fe_4N layer. The hardness of the diffusion zone ranged from 245 to 255 $HV_{0.1}$, similar to that measured from the normalized substrate. Thus, the surface- hardening effect was mainly derived from the formation of the γ'-Fe_4N layer.

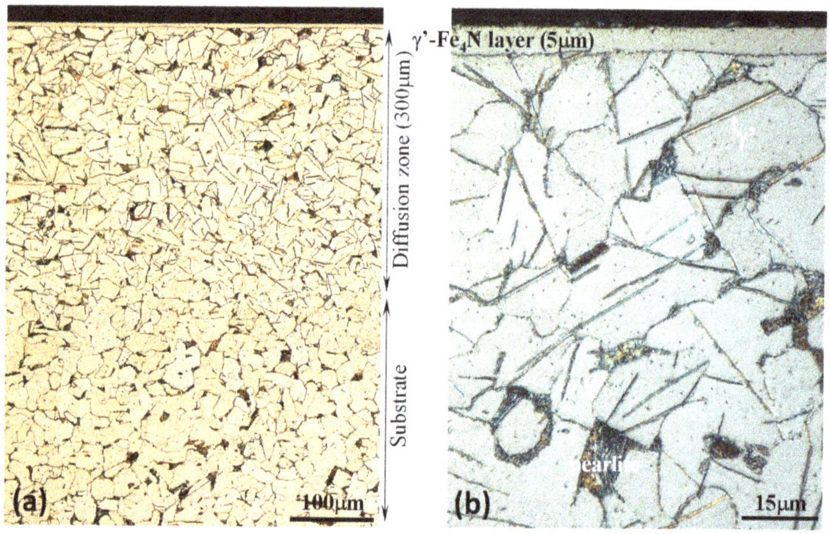

Figure 1. Microscopic images of the nitrided mild steel specimen showing (**a**) the overall view of the cross section from the surface to the substrate and (**b**) details of the 5 μm thick Fe_4N compound layer at the surface and the underneath diffusion zone with needle-like precipitates.

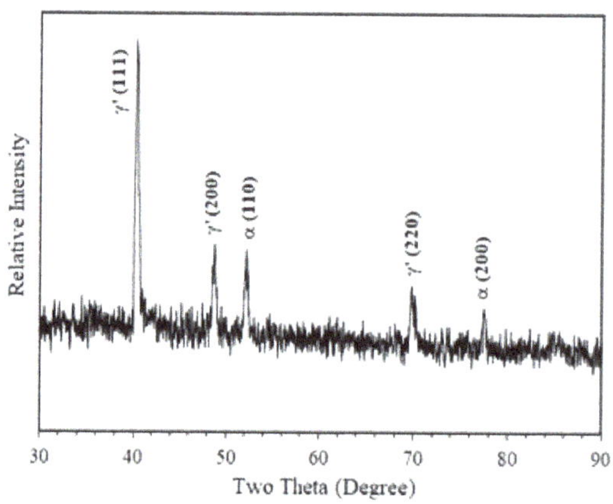

Figure 2. X-ray diffraction pattern generated from the nitrided specimen, showing that the compound layer was composed of γ'-Fe_4N phase (Cu K_α radiation).

Table 1. Summary of structure and surface hardness of the test specimens.

Specimen	Structure		Surface Hardness			
	Surface Layer	Diffusion Zone	$HV_{0.025}$	$HV_{0.05}$	$HV_{0.1}$	$HV_{0.2}$
PN MS	γ'-Fe_4N 5 μm thick	γ'-Fe_4N needles in α + P matrix	760 ± 17	540 ± 14	370 ± 9	292 ± 7
Raw MS	α + P	α + P	266 ± 15	258 ± 13	248 ± 9	245 ± 7

Note: α—ferrite; P—pearlite.

2.2. Corrosion Tests

All corrosion tests were conducted at room temperature in 1.0 wt% NaCl solution by dissolving analytical grade NaCl in double-distilled water. This solution was used because it contains a chloride content similar to that in human body fluids. An ACM Gill AC potentiostat (ACM Gill AC, ACM Ltd., Cumbria, UK) was used to measure the potentials and currents during the tests employing a 3-electrode configuration, i.e., the test specimen as the working electrode, a saturated calomel electrode (SCE) as the reference electrode and a platinum mesh as the auxiliary electrode.

Before corrosion tests, the nitrided and un-nitrided surfaces were slightly polished using 1 μm diamond paste to remove surface contaminants resulting from nitriding and to further smoothen the surface. This was followed by ultrasonic cleaning in ethanol for 10 min. Then the specimen was masked by using an insulating lacquer to leave a test window of 10 mm × 10 mm which would be exposed to the solution.

Corrosion tests were conducted both potentiodynamically to measure the anodic polarization curves of the specimens and potentiostatically to measure the current evolution as a function of time at constant potentials. The potentiodynamic test was conducted at a scan rate of 1 mV/s, by sweeping potential from −200 mV (vs. open circuit potential (OCP)) to 800 mV (vs. OCP). The potentiostatic tests were conducted at a cathodic potential of −700 mV(SCE), at OCP and at an anodic potential of −200 mV(SCE). During all the tests, the current and potential values were recorded continuously at a sampling rate of 1 Hz.

2.3. Tribocorrosion Tests

Tribocorrosion tests were carried out under the same electrochemical conditions and following the same surface preparation procedures as those used in the corrosion tests described in Section 2.2. A laboratory-scale reciprocating wear test machine was used to perform the tests [33]. The tribometer allowed for the reciprocating movement of a flat specimen against a stationary slider (ball) at controlled frequency, stroke amplitude and contact load, and the continuous measurement of friction force by the incorporated load cell. The incorporation of an electrochemical potentiostat with the tribometer allowed for the measurements of potential and current during the sliding wear process. Before the test, the specimen was masked by using an insulating lacquer to leave a test window of 15 mm × 3 mm which would be exposed to the solution. An electrically conducting lead was connected to the specimen to ensure conductivity of the specimen for electrochemical measurements. In order to ensure that electrochemical reactions only occur in the test window, all other fixtures, including the specimen holder, the test cell and the slider holder were made of nylon.

During tribocorrosion testing, the specimen was immersed in the 1.0 wt% NaCl solution. An SCE and a platinum wire were also inserted in the solution as the reference and auxiliary electrode, respectively. An alumina (Al_2O_3) ball of 8 mm diameter (R_a = 0.05 μm) was used as the stationary slider, which made sliding contact with the reciprocating specimen, generating a sliding track on the specimen. All tests were implemented under a contact load of 4 N for a total sliding duration of 7200 s at a reciprocating frequency of 1 Hz and an amplitude of 6 mm. Before and after sliding, the specimen was stabilized in the solution for 300 s under respective conditions, such that potentials and/or currents could be measured before, during and after sliding. The test parameters were selected based on our experience and preliminary tests to ensure that the properties of the γ'-Fe_4N layer were measured.

All corrosion and tribocorrosion tests were duplicated, which was considered sufficient because the results were reproducible. The results from the two tests are presented where applicable. Table 2 summarizes the test conditions under both sliding (tribocorrosion) and no-sliding (corrosion) conditions.

Table 2. Summary of corrosion and tribocorrosion conditions.

	Corrosion	Tribocorrosion
Potentiodynamic	−200 mV to 800 mV, 1 mV/s No sliding	−200 mV to 800 mV, 1 mV/s Sliding at 4 N and 1 Hz
Potentiostatic	−700 mV(SCE) no sliding OCP no sliding −200 mV(SCE) no sliding	−700 mV(SCE) sliding at 4 N and 1 Hz OCP sliding at 4 N and 1 Hz −200 mV(SCE) sliding at 4 N and 1 Hz

2.4. Specimen Characterization

After corrosion and tribocorrosion tests, the corroded surfaces and sliding tracks were examined microscopically under optical microscope (Nikon LV150N, Nikon, Tokyo, Japan) and scanning electron microscope (SEM) (Carl Zeiss EVO LS 15, Carl Zeiss, Jena, Germany). Using the extended depth-of-focus feature of the Nikon microscope, 3D images and surface profiles of the sliding tracks could be captured. The SEM was equipped with EDX facilities for elemental composition analysis. A profilometer (Surtonic Intra Touch, Taylor-Hobson, Leicester, UK) was used to measure the surface profiles across each sliding track at three locations to measure the cross-sectional area of the track and thus obtain the total material removal (TMR) from each specimen due to tribocorrosion.

3. Results

3.1. Potentiodynamic Tests

Figure 3a shows the anodic polarization curves measured without sliding and during sliding for the un-nitrided (normalized) and plasma-nitrided (PN) specimens. Under the condition of corrosion (no-sliding), the normalized specimen showed active anodic dissolution behavior: the current density increased quickly at anodic potentials above the corrosion potential and then increased slowly due to transpassivity. Microscopic examination revealed that the normalized specimen suffered from severe corrosion which etched the grain structures and led to grain detachment (see Figure 4a). Under the same no-sliding condition, the PN specimen showed a higher corrosion potential and much lower current densities in the anodic region. The corroded area only showed some discoloration as compared to the un-corroded area (see Figure 4b). Thus, in the test solution, the normalized specimen was in the active state in the anodic region, while the γ'-Fe_4N layer formed on the surface by nitriding was effective in reducing the dissolution rate of mild steel. Under the condition of tribocorrosion (sliding), the corrosion potentials of both specimens were shifted cathodically and the current densities in the anodic region were increased (Figure 3a). This demonstrates that the sliding mechanical action activated both the normalized surface and PN surface such that corrosion in the sliding tracks was accelerated, which is a common phenomenon observed in tribocorrosion [23–27].

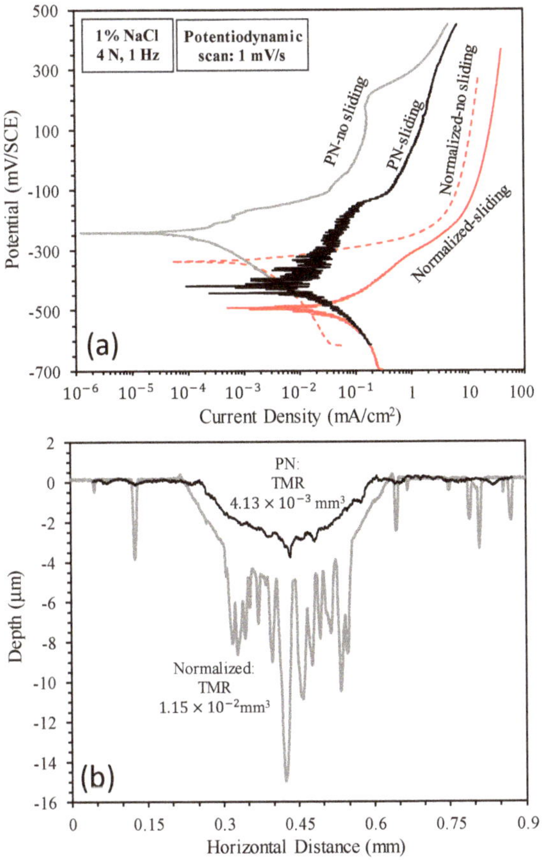

Figure 3. (**a**) Anodic polarization curves measured without sliding and during sliding and (**b**) surface profiles measured across the sliding tracks produced during anodic polarization.

Table 3 lists the corrosion potentials, corrosion current densities and corrosion rates derived from the polarization curves shown in Figure 3a. It can be seen that sliding led to an increase in corrosion current density by an order of magnitude for the normalized specimen and by nearly two orders of magnitude for the PN specimen. As compared to the normalized specimen, the PN specimen showed higher corrosion potentials and lower corrosion current densities, demonstrating that the γ'-Fe_4N layer had the ability to reduce corrosion under both sliding and no-sliding conditions.

Table 3. Corrosion potential (E_{corr}), corrosion current density (i_{corr}) and corrosion rate derived from the polarization curves in Figure 3a.

Specimen	E_{corr} (mV/SCE)		i_{corr} (mA/cm^2)	
	No Sliding	**Sliding**	**No Sliding**	**Sliding**
Raw MS	−335	−490	4.28×10^{-3}	4.95×10^{-2}
PN MS	−241	−418	2.05×10^{-4}	1.91×10^{-2}

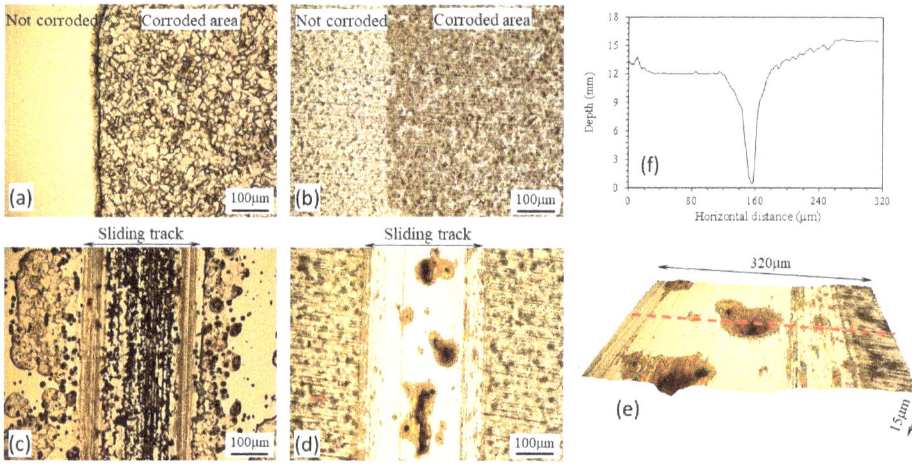

Figure 4. Microscopic images showing the test surfaces after potentiodynamic polarization tests without sliding and during sliding: (**a**) Normalized MS without sliding; (**b**) PN MS without sliding; (**c**) Normalized MS with sliding; (**d**) PN MS with sliding; (**e**) 3D view of corrosion pits in (**d**); (**f**) surface profile measured across a pit shown by the dashed line in (**e**).

Figure 3b shows the surface profiles measured across the sliding tracks after anodic polarization tests during sliding. It can be seen that the sliding track on the normalized specimen was much deeper and rougher than that on the PN specimen. Material removal from outside the sliding track of the normalized specimen was also evident, obviously due to corrosion occurring outside the sliding track, as can be seen from Figure 4c. The deep valleys in the sliding track shown in Figure 3b for the normalized specimen were the results of accelerated corrosion. On the other hand, the sliding track on the PN specimen exhibited a shiny and polished appearance (Figure 4d). The TMR from the PN specimen was 64% smaller than that from the normalized specimen (Figure 3b). The measured surface profile shown in Figure 3b also revealed that the sliding track on the PN specimen was smaller than 4 μm, thus tribocorrosion occurred mainly within the γ'-Fe_4N layer. However, some pits of several tens of microns in size were observed in the sliding track (Figure 4d,e). The penetration depth of the pits was as deep as 15 μm (Figure 4f), suggesting that once the γ'-Fe_4N layer was broken down locally, accelerated corrosion happened in the form of pitting corrosion during the tribocorrosion process.

The results presented in this section provide a general picture of the corrosion and tribocorrosion behavior of the specimens over a wide potential range. In order to further demonstrate the beneficial effect of the γ'-Fe_4N layer in improving the tribocorrosion behavior of mild steel, detailed tests were conducted potentiostatically at different potentials, as discussed in the following sections.

3.2. Cathodic Potential Tests

Figure 5a shows the recorded current versus time curves at the cathodic potential of −700 mV(SCE) in 1.0% NaCl solution under both sliding and no-sliding conditions. Under no-sliding conditions, negative currents were recorded for both specimens, suggesting that cathodic reactions were dominant, while metal oxidation (corrosion) became less dominant [23]. Sliding led to an increase in cathodic currents from around −0.04 mA to around −0.1 mA for both specimens, suggesting that the sliding action activated the surfaces and accelerated cathodic reactions.

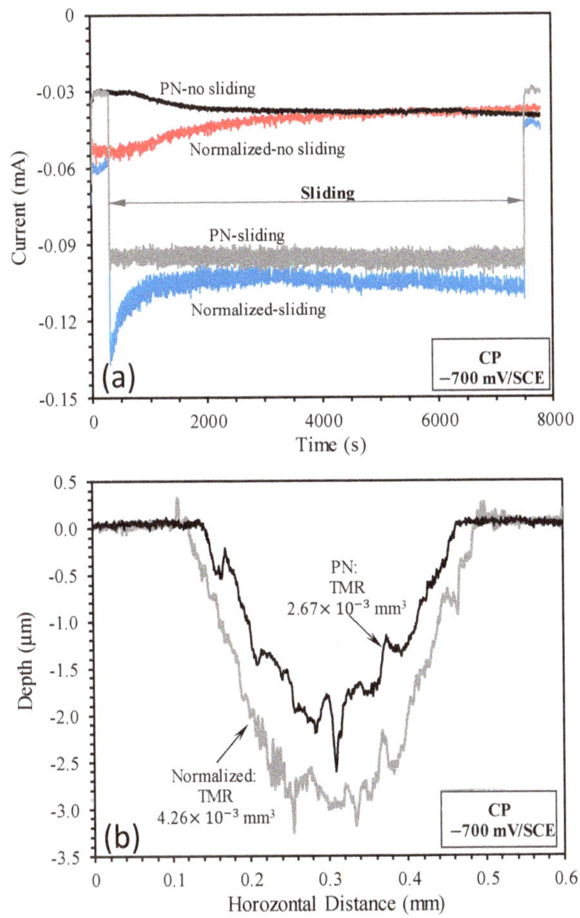

Figure 5. (**a**) Current variation with time measured without sliding and during sliding at the cathodic potential of −700 mV(SCE) and (**b**) surface profiles measured across the sliding tracks produced by sliding at −700 mV(SCE).

Although corrosion is still possible at such a cathodic potential, its contribution to TMR from the sliding track is insignificant [23,33]. Thus, TMR was mainly due to mechanical wear. Figure 5b shows the surface profiles measured across the sliding tracks on the normalized and PN specimens. It can be seen that the sliding track on the normalized specimen was deeper and wider than that on the PN specimen. The wear depth on the PN specimen was smaller than 3 μm, thus wear occurred within the 5 μm thick γ'-Fe_4N layer. The TMR from the PN specimen was about 63% of that from the normalized specimen. Clearly, the γ'-Fe_4N layer had the ability to reduce the mechanical wear of mild steel by about 37%.

Figure 6 shows the SEM images and EDX elemental mappings of the sliding tracks on the normalized and PN specimens. The parallel scratch marks on the sliding tracks are good indications of mechanical abrasion wear arising from the sliding action of the alumina slider. The sliding track on the PN specimen showed a shiny appearance with very few signs of corrosion (Figure 6b). EDX mapping could not find oxygen on the worn surface, instead, nitrogen was found (Figure 6b), further confirming that the γ'-Fe_4N layer was not worn through. On the other hand, the sliding track on the normalized specimen was covered with some dark gray products, which were rich in oxygen (Figure 6a), suggesting that limited corrosion could happen to the normalized specimen at such a cathodic potential.

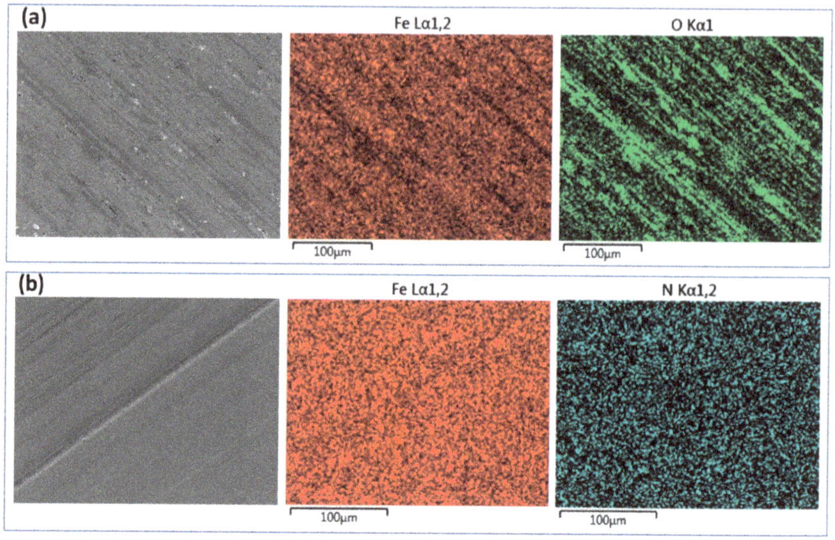

Figure 6. Microscopic images and EDX mapping of the sliding tracks produced at the cathodic potential of −700 mV(SCE) on the (**a**) normalized MS and (**b**) PN MS specimens.

3.3. Open Circuit Potential (OCP) Tests

The OCP versus time curves recorded under open circuit conditions are shown in Figure 7a. Under no-sliding conditions, the OCP of both specimens decreased with increasing time of immersion in the solution. Both specimens were in the active state such that the surfaces became more activated with increasing time of immersion, leading to decreased OCP values. Under sliding conditions, the recorded OCP values also decreased with time and were more negative than those measured under no-sliding conditions, suggesting that sliding led to further activation of the surfaces. In agreement with the potentiodynamic measurements shown in Figure 3a, the OCP values measured for the PN specimen were more anodic than those measured for the normalized specimen under both sliding and no-sliding conditions. Clearly, the normalized surface was in a more active state than the PN surface under open circuit conditions.

The surface profiles measured across the sliding tracks are shown in Figure 7b. The sliding track on the normalized specimen was deeper than that on the PN specimen. Interestingly, material removal also occurred outside the sliding track on the normalized specimen, which was due to corrosion in areas exposed to the solution outside the track, as can be seen from Figure 8a. On the other hand, corrosion outside the sliding track on the PN specimen was very limited (Figure 7b), which was confirmed by microscopic examination (Figure 8c). Inside the sliding track, there were clear signs of corrosion in addition to the parallel abrasion marks (Figure 8b,d). The wear depth of the PN specimen was smaller than 5 μm (Figure 7b), thus wear and corrosion occurred within the γ'-Fe_4N layer. A few pits were also observed inside the sliding track on the normalized specimen, which were several microns in depth (Figure 8b).

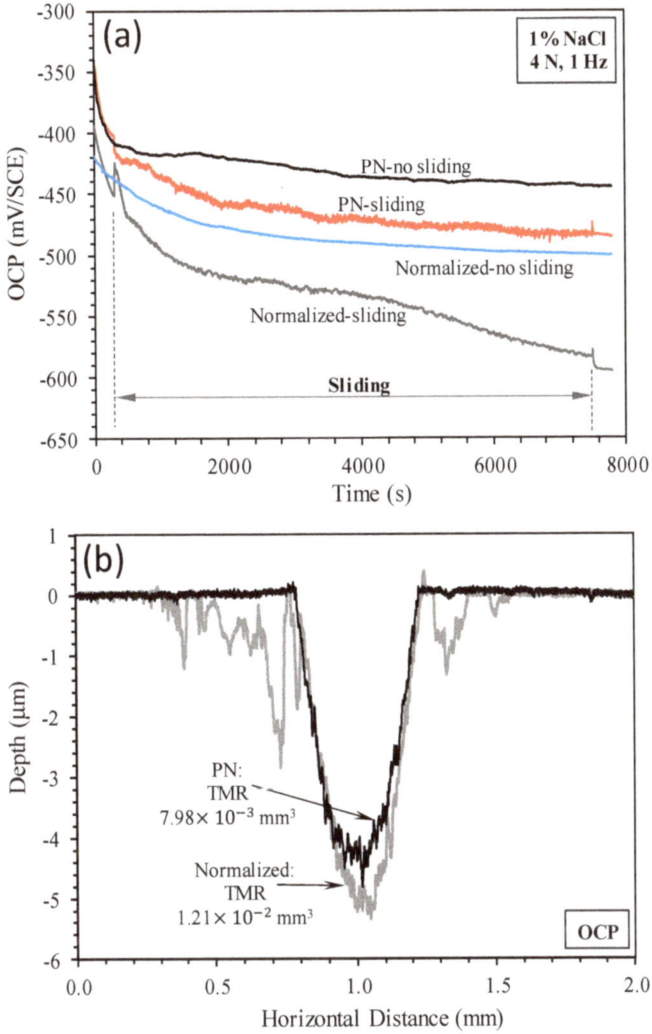

Figure 7. (**a**) Open circuit potential (OCP) variation with time measured without sliding and during sliding at OCP and (**b**) surface profiles measured across the sliding tracks produced by sliding at OCP.

TMRs from the specimens were obviously the combined results of wear (abrasion) and corrosion. TMR from the PN specimen was about 34% smaller than that from the normalized specimen (Figure 7b). When compared with TMRs resulting from sliding at the cathodic potential of −700 mV(SCE) discussed in Section 3.2, TML at OCP was 2.8 times larger for the normalized specimen and 3.6 larger for the PN specimen. This demonstrates the effect of corrosion and wear–corrosion synergism in accelerating TMR from the specimen.

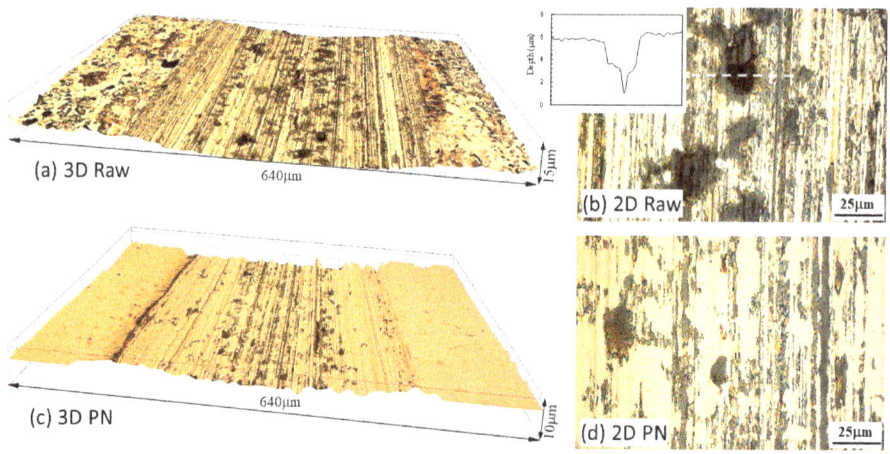

Figure 8. Microscopic images showing 3D (**a**,**c**) and 2D (**b**,**d**) views of the sliding tracks produced at OCP on the normalized MS (**a**,**b**) and PN MS (**c**,**d**) specimens. The inset in (**b**) shows the surface profile measured across the pit indicated by the dashed line.

3.4. Anodic Potential Tests

At the anodic potential of −200 mV(SCE), the normalized and PN specimens behaved quite differently (Figure 9a). Under no-sliding conditions, both the normalized and PN specimens registered continuously increasing the currents with polarization time, confirming that the specimens suffered from anodic dissolution and had no ability to passivate at such an anodic potential. However, the current generated from the PN specimen was more than two orders of magnitude smaller than that generated from the normalized specimen (Figure 9a,b), demonstrating that the γ'-Fe_4N layer had the ability to reduce metal dissolution and improve corrosion resistance.

Interestingly, sliding led to a decrease in currents from the normalized specimen (Figure 9a), which is different from the observations made on passive metals where sliding leads to an abrupt increase in currents due to the depassivation of the surface by the sliding action [24,25,33]. For the active normalized specimen in this study, removal of the corrosion products from the sliding track by the mechanical action obviously helped to reduce metal dissolution. After the termination of sliding, the current from the normalized specimen increased to the level measured under no-sliding conditions. A microscopic examination revealed that corrosion occurred both inside and outside the sliding track of the normalized specimen (Figure 10). Outside the sliding track, the surface became quite rough with peaks and valleys due to grain detachment arising from the corrosion action (Figure 10b). Inside the sliding track, the peaks were flattened due to the wearing action, but the valleys and some corrosion pits could still be observed (Figure 10a).

On the other hand, sliding led to an abrupt increase in currents from the PN specimen, and the current generated during sliding increased slowly for the first 2000 s, and then increased more quickly between 2000 s and 4000 s sliding, and then further quickly after 4000 s sliding (Figure 9b). Microscopic examination (Figure 11) revealed that in the central region of the sliding track on the PN specimen, the γ'-Fe_4N layer was removed, such that pits were formed in this region, which penetrated deep into the substrate (Figures 9c and 11). This explains the quick increase in currents after 2000 s and then 4000 s sliding (Figure 9b).

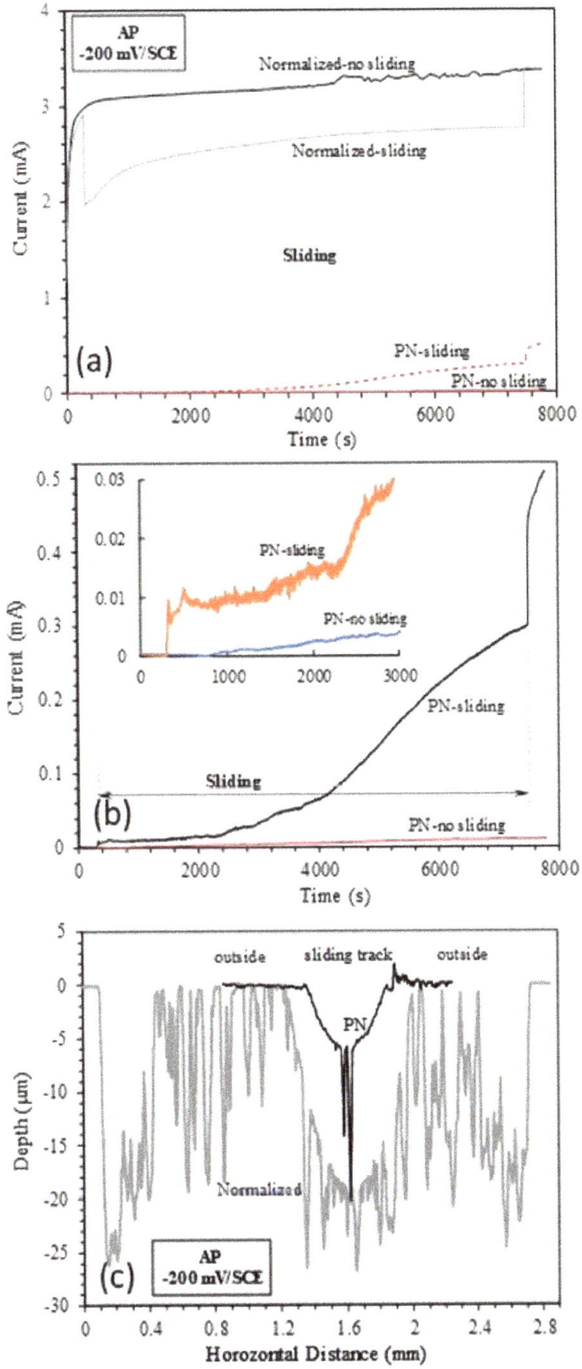

Figure 9. (**a**) Current variation with time measured without sliding and during sliding at the anodic potential of −200 mV(SCE); (**b**) Zoom-in view of current variation of the nitrided specimen during sliding at −200 mV(SCE); (**c**) Surface profiles measured across the sliding tracks produced by sliding at −200 mV(SCE).

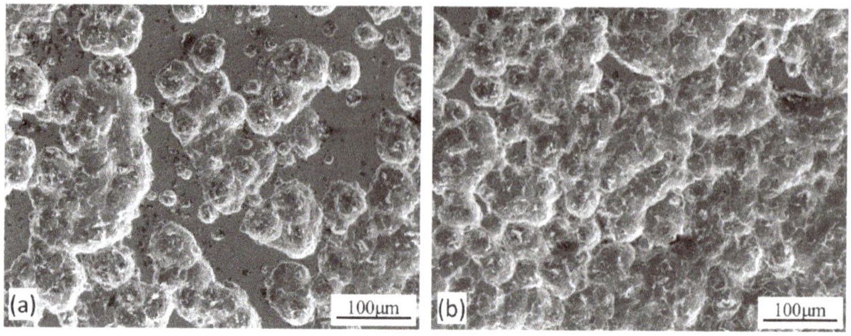

Figure 10. Microscopic images showing (**a**) the worn surface inside the sliding track and (**b**) the corroded surface outside the sliding track, produced at the anodic potential of −200 mV(SCE) on the normalized MS.

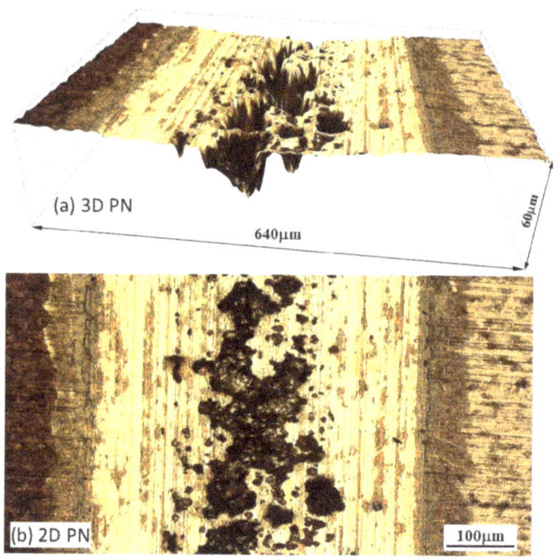

Figure 11. Microscopic images showing (**a**) 3D and (**b**) 2D views of the sliding track, produced at the anodic potential of −200 mV(SCE) on the PN MS.

Surface profile measurements across the sliding tracks showed that TMR from the PN specimen was mainly from the sliding track with limited corrosion loss outside the sliding track (Figure 9c). The wear depth in the central region of the sliding track was more than 5 μm and as deep as 20 μm in the pitting area, suggesting that the γ'-Fe_4N layer was worn through in this region, in line with the microscopic examination (Figure 11). On the other hand, for the normalized specimen, TMR came from both inside and outside the sliding track (Figure 9c). Thus, corrosion played a more significant role in TMR from the normalized specimen.

4. Discussion

4.1. Effect of Potential on Tribocorrosion Behavior

From the results presented in Section 3, the 5 μm γ'-Fe_4N thick layer was not worn through during the tribocorrosion tests. Thus, the measured corrosion and tribocorrosion behavior of the nitride specimen is characteristic of the γ'-Fe_4N layer. It is clear that the γ'-Fe_4N layer produced by nitriding on mild steel has the ability to increase surface hardness, reduce metal dissolution rate and improve tribocorrosion resistance in the NaCl-containing solution. The tribocorrosion behavior of the γ'-Fe_4N layer is affected by the applied potential, as summarized in Figure 12, which shows the variation of the average coefficient of friction (COF) and TMR with potential.

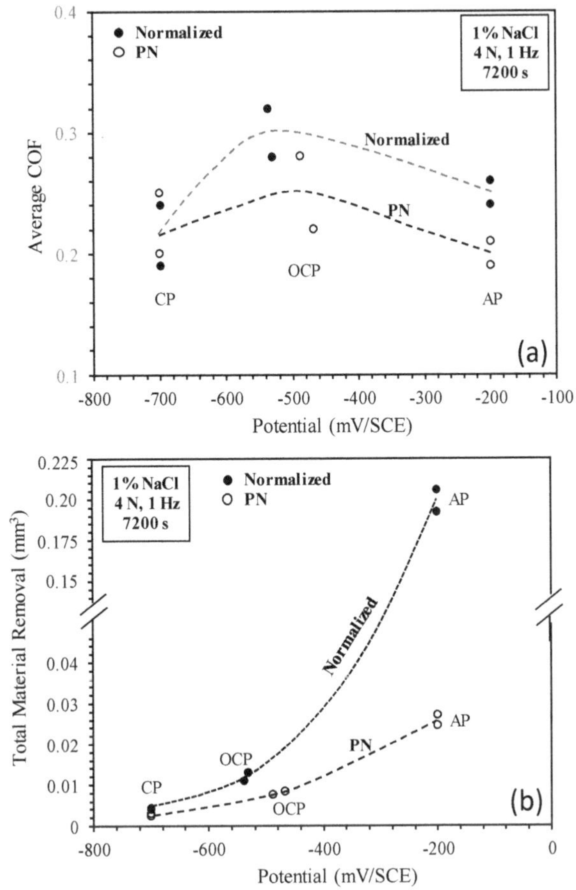

Figure 12. (**a**) Average coefficient of friction (COF) and (**b**) total material removal during sliding at cathodic (CP), open circuit (OCP) and anodic (AP) potentials.

The COF values vary quite largely between two repeated tests under the same conditions (Figure 12a). However, a general trend can be seen that the average COF for both specimens is the highest at OCP and is lower at cathodic and anodic potentials. This behavior can be explained as follows. At the cathodic potential of −700 mV(SCE), mechanical wear is predominant and there is little contribution of corrosion products to friction. At OCP, both mechanical wear and corrosion are involved and there is synergism between wear and corrosion, which could lead to increased friction because of the trapping of corrosion-wear products at the frictional interface. At the anodic potential of −200 mV(SCE), although corrosion is accelerated, the formation of corrosion pits, particularly at the central region of the sliding track on the PN specimen (Figure 11), could reduce the real contact area and thus could lead to reduced friction. As compared to the normalized specimen, the PN specimen shows lower friction at OCP and −200 mV(SCE), thus the γ'-Fe_4N layer is beneficial in reducing friction during tribocorrosion in the NaCl-containing solution.

The variation of TMR with potential is shown in Figure 12b. It can be seen that the TMR of both specimens increases with potential and the TMR of the PN specimen is significantly lower than that of the normalized specimen under all test conditions. As the potential is increased from OCP to −200 mV(SCE), the TMR of the normalized specimen is increased by nearly 17 times, while the TMR of the PN specimen is increased only by 3 times. When comparing the TMR of the two specimens, the TMR of the PN specimen is 37% lower at the cathodic potential, 34% lower at OCP and 87% lower at the anodic potential than the normalized specimen. This demonstrates the beneficial effect of the γ'-Fe_4N layer on improving the resistance to mechanical wear, corrosion and the synergism between wear and corrosion of mild steel, as discussed below.

4.2. Contribution of Mechanical Wear, Corrosion and Wear–Corrosion Synergism

Material removal during tribocorrosion is the result of mechanical wear, corrosion and synergism between wear and corrosion [26,27,34,35]. At the cathodic potential, corrosion is limited and mechanical wear is dominant. The results (Figures 5b and 12b) show that the TMR of the PN specimen is 37% smaller than that of the raw specimen. This can be attributed to the high hardness of the γ'-Fe_4N layer (Table 1), which provides good resistance to mechanical abrasive wear. At OCP and the anodic potential, both mechanical wear and corrosion are involved and there is synergism between wear and corrosion. According to the synergistic approach for tribocorrosion analysis [23,26], the TMR (T) is the sum of material removal due to mechanical wear (W_0), corrosion (C_0) and the synergism between wear and corrosion (S):

$$T = W_0 + C_0 + S \quad (1)$$

where T can be measured after the tests (e.g., Figure 12b), W_0 is the TMR during sliding at the cathodic potential (−700 mV (SCE)) and C_0 can be theoretically estimated by using the Faraday's law [23,26,27]:

$$V_{chem} = \frac{ItM}{nF\rho} \quad (2)$$

where I is corrosion current (A) measured under no-sliding conditions, t (s) is the sliding time, M (g/mol) is the atomic mass of the steel, n is the valence of oxidation, which is 3 to account for FeOOH oxide film on steel, which is the main corrosion product of carbon steel [36,37], F is the Faraday's constant (96,458 C/mol) and ρ is the density of steel.

Equation (2) can be used to calculate C_0 if general corrosion is the predominant form of corrosion. In the present work, at the anodic potential of −200 mV(SCE), significant pitting corrosion occurred inside the sliding track (Figures 10 and 11), particularly in the central region of the sliding track on the PN specimen, thus most of the measured currents (Figure 9a,b) are expected to come from the corrosion pits. In such a case, it is unrealistic to estimate C_0 by using Equation (2). However, at OCP, general corrosion was predominant, thus the corrosion parameter data (i_{corr} under no-sliding conditions) shown in Table 3 can be used to estimate C_0 for both specimens and then Equation (1) can be used to calculate S. The results are summarized in Table 4.

Table 4. Summary of T, W_0, C_0, ΔC and ΔW at OCP.

Specimen	T (mm^3)	W_0 (mm^3)	C_0 (mm^3)	S (mm^3)	S/T×100	DC (mm^3)	DW (mm^3)
Normalized	1.21×10^{-2}	4.26×10^{-3}	5.03×10^{-4}	7.28×10^{-3}	60.1	5.31×10^{-3}	1.97×10^{-3}
PN	7.98×10^{-3}	2.67×10^{-3}	2.39×10^{-5}	5.29×10^{-3}	66.3	2.20×10^{-3}	3.09×10^{-3}

From Table 4, it can be seen that at OCP, as compared to the normalized specimen, the γ'-Fe_4N layer can reduce mechanical wear by 37%, corrosion by 95% and wear–corrosion synergism by 27%. The contribution of corrosion to TMR is smaller than 5% for the normalized specimen and smaller than 1% for the γ'-Fe_4N layer. On the other hand, the synergistic effect has a large contribution to TMR for both specimens, accounting for 60.1% of TMR for the normalized specimen and 66.3% of TMR for the PN specimen (refer to the S/T ×100 column in Table 4). Therefore, it can be concluded that the γ'-Fe_4N layer has the ability to provide improved resistance to mechanical wear, corrosion and wear–corrosion synergism. It is also noted that the contribution of the synergistic effect of the γ'-Fe_4N layer to TMR (66.3%) is larger than that of the normalized specimen (60.1%); to explain this, further analysis is necessary, as follows.

During tribocorrosion, the synergism between wear and corrosion arises from wear-induced corrosion (ΔC) and corrosion-induced wear (ΔW) [23,26]. Thus, the term, S, is the sum of ΔC and ΔW:

$$S = \Delta C + \Delta W \quad (3)$$

where, by using Equation (2), ΔC can be calculated from the measured corrosion current during sliding (subtracted by the corrosion current measured under no-sliding conditions). Using the current density data in Table 3, ΔC is calculated for both specimens and then ΔW is obtained from Equation (3). The results are listed in the last two columns of Table 4. It can be seen that ΔC and ΔW have positive values, suggesting that, at OCP, corrosion is accelerated by wear and wear is accelerated by corrosion in both specimens. For the normalized specimen, the contribution of wear-accelerated corrosion (ΔC) to the synergism is more significant than the contribution of corrosion accelerated-wear (ΔW). On the other hand, for the γ'-Fe_4N layer, corrosion accelerated-wear (ΔW) contributes more to the synergism. The ΔW value from the γ'-Fe_4N layer is even larger than that from the normalized specimen. This suggests that the corrosion-wear products trapped at the contact interface can serve as third-body particles to cause accelerated wear of the γ'-Fe_4N layer, as evidenced by the many wide scratch marks and the existence of corrosion-wear products on the sliding surface shown in Figure 8c,d.

Although it is not possible to quantitatively estimate the relative contribution of various TMR components at the anodic potential by using Equations (1) to (3), judging from the measured currents (Figure 9) and sliding track morphology (Figures 10 and 11), it can be deduced that corrosion and wear–corrosion synergism play increasing roles in TMR from both specimens. For the PN specimen, the local breakdown of the γ'-Fe_4N layer can lead to much accelerated corrosion in the form of pitting corrosion, which in turn can lead to accelerated mechanical wear [38]. Thus, like other surface coatings for corrosion and wear protection, there is always an issue concerning the sustainability of the coating system under harsh loading and environmental conditions [39].

5. Conclusions

1. In the NaCl-containing solution, both the normalized mild steel and γ'-Fe_4N layer are in the active state in the anodic region. The γ'-Fe_4N layer has the ability to reduce metal dissolution and improve corrosion resistance of mild steel.
2. At the cathodic potential, where mechanical wear dominates, the γ'-Fe_4N layer can reduce total material removal by 37% due to its higher hardness than that of normalized mild steel.
3. At open circuit potential, where both mechanical wear and corrosion are involved, the γ'-Fe_4N layer has the ability to reduce mechanical wear, corrosion and the synergy between wear and corrosion due to its higher hardness and better resistance to metal dissolution, such that the total material removal is reduced by 34% as compared to that from the normalized mild steel.
4. At the anodic potential, where corrosion plays an increasing role, the γ'-Fe_4N layer can reduce total material removal by 87%. However, local breakdown of the γ'-Fe_4N layer can happen in the sliding track, leading to accelerated pitting corrosion.
5. The γ'-Fe_4N layer has the ability to improve the tribocorrosion behavior of mild steel in the NaCl-containing solution under all test conditions, provided that the layer is not worn through or broken down locally during the tribocorrosion process. However, there is a concern regarding the sustainability of the layer when localized breakdown or wearing-through occurs, which can lead to accelerated pitting and accelerated material removal.

Author Contributions: Methodology, Y.S.; Experimentation, Y.S. and R.B.; Investigation, Y.S. and R.B.; writing—original draft preparation, Y.S.; writing—review and editing, R.B. All authors have read and agreed to the published version of the manuscript.

Funding: This research received no external funding.

Data Availability Statement: The original data of this work is available from the authors upon reasonable request.

Conflicts of Interest: The authors declare no conflict of interest.

References

1. Staines, A.M.; Bell, T. Technological importance of plasma-induced nitrided and carburized layers on steel. *Thin Solid Films* **1981**, *86*, 201–212. [CrossRef]
2. Spies, H.J.; Thien, H.L.; Biermann, H.B. Controlled nitriding. *Met. Sci. Heat Treat.* **2004**, *46*, 272–276. [CrossRef]
3. Lightfoot, B.J.; Jack, D.H. Kinetics of nitriding with and without white-layer formation. In *Heat Treatment '73*; The Metals Society: Metals Park, OH, USA, 1973; Republished in the Source Book on Nitriding; American Society for Metals: Metals Park, OH, USA, 1977; pp. 248–254.
4. Mittemeijer, E.J.; Somers, A.L. (Eds.) *Thermochemical Surface Engineering of Steels*; Elsevier Woodhead Publishing: Sawston, Cambridge, 2015.
5. Menthe, E.; Rie, K.T.; Schultze, J.W.; Simson, S. Structure and properties of plasma-nitrided stainless steel. *Surf. Coat. Technol.* **1995**, *74–75*, 412–416. [CrossRef]
6. Zhang, L.; Shao, M.; Zhang, Z.; Yi, X.; Yan, J.; Zhou, Z.; Fang, D.; He, Y.; Li, Y. Corrosion behaviour of nitrided layer of Ti_6Al_4V titanium alloy by hollow cathodic plasma source nitriding. *Materials* **2023**, *16*, 2961. [CrossRef] [PubMed]
7. Zhecheva, A.; Malinov, S. Titanium alloys after surface gas nitriding. *Surf. Coat. Technol.* **2006**, *201*, 2467–2474. [CrossRef]
8. Buchwalder, A.; Bocker, J.; Hegelmann, E.; Klemm, V. Investigations on the microstructure of an aluminium nitride layer and its interface with the aluminium substrate (Part I). *Coatings* **2022**, *12*, 618. [CrossRef]
9. Visuttipitukul, P.; Aizawa, T.; Kuwahara, H. Advanced plasma nitriding for aluminium and aluminium alloys. *Mater. Trans.* **2003**, *44*, 2695–2700. [CrossRef]
10. Sun, Y. Kinetics of layer growth during plasma nitriding of nickel based alloy Inconel 600. *J. Alloys Compd.* **2003**, *351*, 241–247. [CrossRef]
11. Tao, X.; Kavanagh, J.; Li, X.; Dong, H.; Matthews, A.; Leyland, A. An investigation of precipitation strengthened Inconel 718 superalloy after triode plasma nitriding. *Surf. Coat. Technol.* **2022**, *442*, 128401. [CrossRef]
12. Luo, X.; Li, X.; Sun, Y.; Dong, H. Tribocorrosion behaviour of S-phase surface engineered medical grade Co-Cr alloy. *Wear* **2013**, *302*, 1615–1623. [CrossRef]
13. Purandare, Y.; Shukla, K.; Sugumaran, A.; Ehiasarian, A.; Khan, I.; Hovsepian, P. Improving tribocorrosion resistance of a medical grade CoCrMo alloy by the novel HIPIMS nitriding technique. *J. Sci. Adv. Mater. Devices* **2023**, *8*, 100570. [CrossRef]
14. Spies, H.J. Corrosion behaviour of nitrided, nitrocarburised and carburised steels. In *Thermochemical Surface Engineering of Steels*; Mittemeijer, E.J., Somers, A.L., Eds.; Elsevier Woodhead Publishing: Sawston, Cambridge, 2015.
15. Mittemeijer, E.J. Nitriding of binary and ternary iron-based alloy. In *Thermochemical Surface Engineering of Steels*; Mittemeijer, E.J., Somers, A.L., Eds.; Elsevier Woodhead Publishing: Sawston, Cambridge, 2015.
16. Jasinski, J.J.; Kurpaska, L.; Frazek, T.; Lubas, M.; Sitarz, M. Structural characterisation of fine γ'-Fe_4N nitrides form by active screen plasma nitriding. *Metals* **2020**, *10*, 1656. [CrossRef]
17. Bell, T.; Loh, N.L. The fatigue characteristics of plasma nitrided three Pct Cr-Mo steel. *J. Heat Treat.* **1982**, *2*, 232–237. [CrossRef]
18. Kato, H.; Eyre, T.S.; Ralph, B. Wear mechanism map of nitrided steel. *Acta Metall. Mater.* **1994**, *42*, 1703–1713. [CrossRef]
19. Bell, T.; Sun, Y. Load bearing capacity of plasma nitrided low alloy steel under combined rolling-sliding contact. *Surf. Eng.* **1990**, *6*, 133–139. [CrossRef]
20. Boztepe, E.; Alves, A.C.; Ariza, E.; Rocha, L.A.; Cansever, N.; Toptan, F. A comparative investigation of the corrosion and tribocorrosion behaviour of nitrocarburized, gas nitrided, fluidized-bed nitrided, and plasma nitrided plastic mould steel. *Surf. Coat. Technol.* **2018**, *334*, 116–123. [CrossRef]
21. Ram Mohan Rao, K.; Nouveau, C.; Lakshman, S.; Muralidhar, P.; Trinadh, K. Effect of low and high temperature plasma nitriding on electrochemical corrosion of steel. *Mater. Today Proc.* **2020**, *39*, 1367–1371. [CrossRef]
22. Valdes, J.; Huape, E.; Oseguern, J.; Ruiz, A.; Ibarrra, J.; Bernal, J.L.; Medina, A. Effects of plasma nitriding in corrosion behaviour of an AISI 4140 steel using a seawater medium solution. *Mater. Lett.* **2022**, *316*, 131991. [CrossRef]
23. Watson, S.W.; Friedersdorf, F.J.; Madsen, B.W.; Gramer, S.D. Methods of measuring wear-corrosion synergism. *Wear* **1995**, *181–183*, 476–484. [CrossRef]
24. Ponthiaux, P.; Wenger, F.; Drees, D.; Celis, J.P. Electrochemical techniques for studying tribocorrosion processes. *Wear* **2004**, *256*, 459–468. [CrossRef]
25. Munoz, A.I.; Espallargas, N.; Mischler, S. (Eds.) *Tribocorrosion*; Springer: Cham, Switzerland, 2020.

26. Mischler, S. Triboelectrochemical techniques and interpretation methods in triborrosion: A comparative evaluation. *Tribol. Int.* **2008**, *41*, 573–583. [CrossRef]
27. Landolt, D.; Mischler, S.; Stemp, M. Electrochim. Electrochemical methods in tribocorrosion: A critical appraisal. *Electrochim. Acta* **2001**, *46*, 3913–3929. [CrossRef]
28. Mindivan, F.; Aktas, G.R.; Bayram, A. Influence of plasma nitriding on dry wear, corrosion and tribocorrosion performance of 17-4 precipitation hardening stainless steel. *Mater. Werkst.* **2022**, *53*, 963–978. [CrossRef]
29. Li, X.; Dou, W.; Tian, L.; Dong, H. Combating the tribo-corrosion of LDX2404 lean duplex stainless steel by low temperature plasma nitriding. *Lubricants* **2018**, *6*, 93. [CrossRef]
30. Hacisalihoglu, I.; Yildiz, F.; Celik, A. Tribocorrosion behavior of plasma nitrided Hardox steels in NaCl solution. *Tribol. Int.* **2018**, *120*, 434–445. [CrossRef]
31. Wong-Angel, W.D.; Martinez-Trinidad, J.; Campos-Silva, V.; Hernandez-Hernandez; Silva-Rivera, U.S.; Garcia-Leon, R.A. Wear-corrosion synergy on Din-16MnCr$_5$ steel under nitriding and post-oxidizing treatments. *J. Bio-Tribo-Corros.* **2021**, *7*, 83. [CrossRef]
32. Takahashi, T.; Burghaus, J.; Music, D.; Dronskowski, R.; Schneider, J.M. Elastric properties of γ'-Fe_4N probed by nanoindentation and ab initio calculation. *Acta Mater.* **2012**, *60*, 2054–2060. [CrossRef]
33. Sun, Y.; Bailey, R. Comparison of wear performance of low temperature nitrided and carburized 316L stainless steel under dry sliding and corrosive-wear condition. *J. Mater. Eng. Perform.* **2022**, *32*, 1238–1247. [CrossRef]
34. Casar, J.; Malia, B.; Mazzonello, A.; Karl, A.; Buhagiar, J. Improved tribocorrosion resistance of a CoCrMo implant material by carburising. *Lubricants* **2018**, *6*, 76. [CrossRef]
35. Rasool, G.; Shafei, Y.E.; Stack, M.M. Mapping tribo-corrosion behaviour of Ti-6Al-4V Eli in laboratory simulated hip joint environments. *Lubricants* **2020**, *8*, 69. [CrossRef]
36. Antunes, R.A.; Ichikawa, R.U.; Martinez, L.G.; Costa, I. Characterisation of corrosion products on carbon steel exposed to natural weathering and to accelerated corrosion tests. *Int. J. Corros.* **2014**, *2014*, 419570. [CrossRef]
37. Fonna, S.; Ibrahim, I.B.M.; Gunawarman; Huzni, S.; Lksan, M.; Thalib, S. Investigation of corrosion products formed on the surface of carbon steel exposed in Banda Aceh's atmosphere. *Heliyon* **2021**, *7*, e06608. [CrossRef]
38. Jiang, J.; Stack, M.M.; Neville, A. Modelling the tribo-corrosion interaction in aqueous sliding conditions. *Tribol. Int.* **2002**, *35*, 669–679. [CrossRef]
39. Sun, Y.; Dearnley, P.A.; Bertram, M. Response of duplex Cr(N)/S and Cr(C)/S coatings on 316L stainless steel to tribocorrosion in 0.89% NaCl solution under plastic contact conditions. *J. Biomed. Mater. Res. Part B* **2017**, *105*, 1503–1513. [CrossRef] [PubMed]

www.ingramcontent.com/pod-product-compliance
Lightning Source LLC
Chambersburg PA
CBHW040316270926
51850CB00007BC/1497
9783725809530